Illustrated Dictionary of Parasitology in the Post-genomic Era

https://doi.org/10.21775/9781910190678

Edited by

Hany M. Elsheikha

School of Veterinary Medicine and Science
Faculty of Medicine and Health Sciences
University of Nottingham
Nottingham
UK
hany.elsheikha@nottingham.ac.uk

Edward L. Jarroll

Department of Biological Sciences
Lehman College
City University of New York
New York, NY
USA
edward.jarroll@lehman.cuny.edu

Caister Academic Press

Caister Academic Press
Norfolk, UK

www.caister.com

British Library Cataloguing-in-Publication Data
A catalogue record for this book is available from the British Library

ISBN: 978-1-910190-67-8 (paperback)
ISBN: 978-1-910190-68-5 (ebook)

Cover design adapted from the following figures: balantidiasis, copulatory bursa and midges (courtesy: Parasitic Diseases, 6th edn. Despommier, D.D., D. Griffin, R.W. Gwadz, P.J. Hotez, C. Knirsch. Parasites Without Borders, Pubs. 2017); hard ticks (courtesy: Elsheikha H.M., Khan N.A. Essentials of Veterinary Parasitology, 1st edition, Caister Academic Press. 2011.); erratic parasite (courtesy: Vito Colella, University of Bari, Bari, Italy); and giant cell.

Contents

Current Books of Interest

Bacterial Evasion of the Host Immune System	2017
Next-generation Sequencing and Bioinformatics for Plant Science	2017
The CRISPR/Cas System: Emerging Technology and Application	2017
Brewing Microbiology: Current Research, Omics and Microbial Ecology	2017
Metagenomics: Current Advances and Emerging Concepts	2017
Bacillus: Cellular and Molecular Biology (Third Edition)	2017
Cyanobacteria: Omics and Manipulation	2017
Foot-and-Mouth Disease Virus: Current Research and Emerging Trends	2017
Brain-eating Amoebae: Biology and Pathogenesis of *Naegleria fowleri*	2016
Staphylococcus: Genetics and Physiology	2016
Chloroplasts: Current Research and Future Trends	2016
Microbial Biodegradation: From Omics to Function and Application	2016
Influenza: Current Research	2016
MALDI-TOF Mass Spectrometry in Microbiology	2016
Aspergillus and *Penicillium* in the Post-genomic Era	2016
The Bacteriocins: Current Knowledge and Future Prospects	2016
Omics in Plant Disease Resistance	2016
Acidophiles: Life in Extremely Acidic Environments	2016
Climate Change and Microbial Ecology: Current Research and Future Trends	2016
Biofilms in Bioremediation: Current Research and Emerging Technologies	2016
Microalgae: Current Research and Applications	2016
Gas Plasma Sterilization in Microbiology: Theory, Applications, Pitfalls and New Perspectives	2016
Virus Evolution: Current Research and Future Directions	2016
Arboviruses: Molecular Biology, Evolution and Control	2016
Shigella: Molecular and Cellular Biology	2016
Aquatic Biofilms: Ecology, Water Quality and Wastewater Treatment	2016
Alphaviruses: Current Biology	2016
Thermophilic Microorganisms	2015
Flow Cytometry in Microbiology: Technology and Applications	2015
Probiotics and Prebiotics: Current Research and Future Trends	2015
Epigenetics: Current Research and Emerging Trends	2015
Corynebacterium glutamicum: From Systems Biology to Biotechnological Applications	2015
Advanced Vaccine Research Methods for the Decade of Vaccines	2015
Antifungals: From Genomics to Resistance and the Development of Novel Agents	2015

Preface

The number of infectious diseases in the world today seems to be increasing rather than decreasing as one would hope, especially new and emerging viral diseases that either did not exist previously or were not recognized. Despite this, malaria, a parasitic disease caused by *Plasmodium* spp., remains the number one infectious disease worldwide. Almost 12 years have elapsed since the *Dictionary of Parasitology* by Peter J Gosling was published, more than a decade marked by extensive discoveries and developments related to the field of parasitology. The dramatic shift from a parasite-focused subject to a field in which the emphasis is on the interaction between the parasite and its host organism has led to major breakthroughs in understanding the pathobiology of many parasites at a more in-depth level. More exciting is the discovery of the mechanisms by which some helminthic parasites modulate host immune response and their applications in medicine to cure chronic diseases, which are now being translated into therapies that are not only symptomatic but also potentially disease modifying. These research advances have stimulated a broader interest in parasitology among many researchers in other fields, such as medicine and bioscience. Historically, parasitologists have maintained a unique parasite nomenclature (e.g. *Dictyocaulus*, *Dracunculus*, *Leishmania*, *Angiostrongylus*) and terminology (e.g. hypobiosis, vector, xenodiagnosis) that has often puzzled their colleagues in other fields, causing them to associate parasitology to a black box. In the meantime, like most scientific disciplines, parasitology has become more cross-disciplinary with fast-growing vocabularies, which are new to parasitologists. Therefore, it is necessary to develop a new dictionary to help both parasitologists and non-parasitologists to grasp the contemporary concepts and terms used in modern parasitology and associated scientific areas.

The first edition of the *Illustrated Dictionary of Parasitology in the Post-genomic Era* comes at a time when the topic of 'one health' and its medical, veterinary and environmental impacts are at the forefront of news stories and political discussions. Successful implementation of the 'one health' initiative requires cross-fertilization between different disciplines, which will improve our understanding of the mechanisms underlying host–pathogen dynamic interactions. With over 4500 entries , illustrated with more than 170 images and line drawings, and reflecting ground-breaking advances in parasitology research this dictionary provides, in a single-volume, up-to-date resources for the many terms encountered in contemporary parasitological literature. The dictionary also covers many pertinent terms from related fields of veterinary medicine and life sciences, such as microbiology, genetics, biochemistry, biotechnology, infectious diseases, epidemiology, zoonosis, public health, molecular biology, zoology, pharmaceutical science, environmental science, taxonomy, and population genetics.

This dictionary will serve as a guide for students, academic staff, medical and veterinary professionals, and life scientists, as well as for members of industrial establishments, governmental agencies and research foundations involved in research activities relating to parasitology and associated scientific fields.

Finally, this book could not have been delivered in such a professional way without the help of the professional staff of Caister Academic Press and Prepress Projects Ltd. Their help is deeply appreciated.

Hany M. Elsheikha
School of Veterinary Medicine and Science
Faculty of Medicine and Health Sciences
University of Nottingham
Nottingham
UK

Edward L. Jarroll
Department of Biological Sciences
Lehman College
City University of New York
New York, NY
USA

Photo Credits

Courtesy: Parasitic Diseases, 6th edn. Despommier, D.D., D. Griffin, R.W. Gwadz, P.J. Hotez, C. Knirsch. Parasites Without Borders, Pubs. 2017.

- amoebic liver abscess, babesiosis, balantidiasis, Bancroftian filariasis, bed bugs, Chagas disease, clonorchiasis, coccidia, copulatory bursa, cryptosporidiosis, crypts, cutaneous leishmaniasis, cyclophyllidea, cyclozoonosis, cysticercoid, egg capsule, egg membrane, *Enterobius vermicularis*, erythrocytic asexual-stage schizonts, gametocyte, hepatosplenic schistosomiasis, inner shell, Kato–Katz method, lymphatic filariasis, metacyclic promastigotes, midges, miracidia hatching, nurse cell, *Onchocerca cervicalis*, *Onchocerca ochengi*, operculum, polar plugs, proglottid, promastigote, protoscolex, pulicide, rostellum, salivarian, scolex, soil-transmitted helminths, sporocyst, strongyloidiasis, sucking disk, taeniasis, toxocariasis, triatomid bug, *Trichinella spiralis*, *Trichuris trichura*, trypomastigote, tsetse fly, urinary schistosomiasis, vacuole, xenodiagnosis.

Courtesy: Elsheikha H.M., Khan N.A. Essentials of Veterinary Parasitology, 1st edition, Caister Academic Press. 2011.

- *Anoplocephala perfoliata*, bottle jaw, bursa, cestodes, cyst, *Haematobia irritans*, haematophagous, hard ticks, hexacanth embryo, hydatid cyst, hydatid sand, paramphistome, psoroptic mange, pyriform apparatus, strobilation, thelaziosis, tick pyaemia of sheep, whipworms.

Courtesy: Veterinary Business Development Ltd.

- acetabulum, blowfly, Branchiura, *Ctenocephalides felis*, ciliata, Crustacea, insect pupa, leishmaniasis, macronucleus, myiasis, nemathelminthes, puparium, *Varroa*, zoonosis.

Courtesy: Bristol University tick ID online resource with the kind permission of Professor Richard Wall.

- *Dermacentor reticulatus*, *Ixodes ricinus*, *Rhipicephalus sanguineus*.

Courtesy: Bayer Animal health.

- angiostrongylosis, aelurostrongylosis.

Courtesy: Dr JP Dubey, USDA, Beltsville, Maryland, USA.

- bradyzoite.

Courtesy: Emma Drinkall, University of Nottingham, Nottingham, UK.

- life cycle, *Ligula intestinalis.*

Courtesy: Vito Colella, University of Bari, Bari, Italy.

- erratic parasite.

A B C D E F G H I J K L M N O P Q R S T U V W X Y Z

AA+ superfamily proteins ATPases found in all living organisms where they participate in a variety of activities, including membrane fusion, proteolysis/protease digestion, DNA replication, and act as chaperones.

abattoir/slaughterhouse a facility where animals are killed, and meat is prepared by a butcher for sale.

ABC (antibiotic-based combination) a therapy used for the treatment of severe malaria. It is a standardized (possibly co-formulated) intravenous combination of an antibiotic with antimalarial activity with a traditional antimalarial (e.g. quinine or artesunate).

aberrant host a dead-end host in which the parasite cannot complete its development or continue its life cycle.

aberrant migration the migration of larvae of a parasite into sites not typically found in its life cycle. This commonly happens in accidental hosts.

abiotic used primarily for the non-living parts in the environment, such as chemical and physical factors (e.g. soil, water, air, temperature and sunlight) that affect ecosystems.

abiotic factor a non-living or physical component of an ecosystem, such as climate or habitat.

abiotic niche the set of physical conditions (e.g. temperature, moisture) that permit a positive turnover of the population of an organism. Very generally, it is the set of features of the weather that allow the persistence of populations of living organisms. Geographic features, such as altitude, are not part of the abiotic niche of an organism.

abluminal surface the surface of the endothelium facing the basement membrane, that is, opposite their luminal surface in contact with the circulating blood.

abomasum the fourth/final and glandular compartment of the ruminant's stomach.

abopercular end in eggs having an operculum, the end opposite the operculum is referred to as the 'abopercular end'.

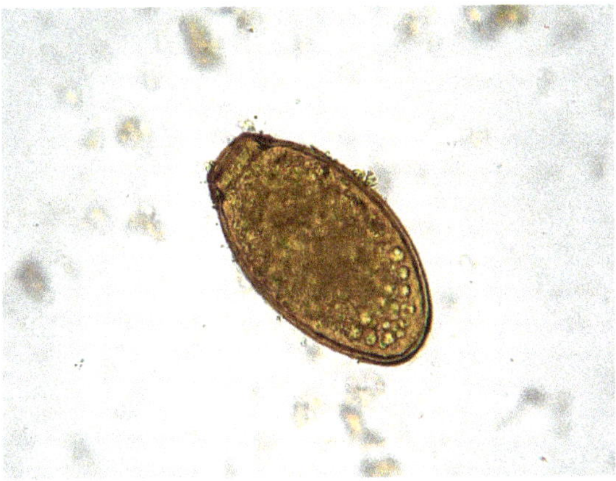

aboral away from or opposite the mouth in those groups of animals that have no clear-cut dorsal or ventral surfaces.

absolute risk the observed or calculated probability of an event in the population under study.

absolute risk difference the difference in the risk for disease or death between an exposed population and an unexposed population.

absolute risk reduction (ARR) the difference in the absolute risk (rates of adverse events) between study and control populations.

Acanthamoeba a protozoan genus that includes many species of free-living amoebae common in soil and water bodies. Some of these amoebae are opportunistic pathogens that can cause *Acanthamoeba* keratitis, disseminated infections (skin and lungs mostly involved), and *Acanthamoeba* granulomatous encephalitis (AGE).

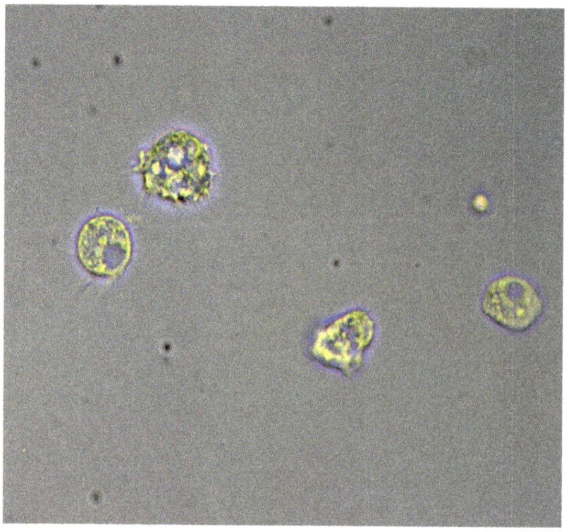

acanthella acanthocephalan larva following the acanthor and prior to the cystacanth.

Acanthocephala a phylum of elongate, mostly cylindrical organisms with an anterior proboscis covered with many hooks, parasitic in the intestines of most classes of vertebrates. Acanthocephalans are not encountered as commonly as parasitic flatworms (trematodes and tapeworms) or nematodes, but are found in many species of fishes, amphibians, birds, and mammals. The most distinguished morphological characteristic to separate acanthocephalans from other parasitic worms is the presence of an anterior, protrusible proboscis that is usually covered with hooks. It is this characteristic that gives the acanthocephalans their common name, the thorny-headed worms.

acanthocephalans members of the phylum Acanthocephala.

acanthocephaliasis infection with worms of the phylum Acanthocephala.

acanthopodia spiky projections present on the pseudopods of *Acanthamoeba* trophozoites.

acanthor acanthocephalan larva with blade-like hooks and develops inside the egg capsules.

Acari a subclass of the arthropod class Chelicerata and is divided into two super orders containing a total of seven orders between them. Superorder Actinotrichida contains the orders Prostigmata, Astigmata and Oribatida, and the superorder Anactinotrichida contains the orders Notostigmata, Holothyrida, Ixodida and Mesostigmata. Acari comprise the mites; the largest group within the arthropod class Arachnida. This number is misleading since it is estimated that only between 5% and 10% of all mite species have been formally described. Mites are distinctive in both their small size (adult body length ranging from 0.1 to 30 mm), ecological diversity and the fact that the first nymphal stage has only six legs instead of the eight present in the remaining nymphal stages and the adults. Many species are serious pests of agricultural crops, either through direct damage or indirectly as vectors of plant pathogens. Other species are parasitic on domestic animals and cause losses in meat, egg, and fibre production. Others, such as the human scabies mite, are direct agents of human disease or, as in the case of chiggers and ticks, vectors of pathogens. Other mites may affect humans by infesting stored food products.

acariasis infestation with arthropod parasites of the order Acarina including the ticks and mites.

acaricide a pesticide that destroys ticks and mites. Common examples includes the organophosphorus compounds, the synthetic pyrethroids and the carbamates.

acarid a tick or a mite of the order Acarina.

acarina arthropods (class Arachnida), including mites and ticks.

acarinosis any disease caused by mites. Also termed acariasis.

accessory cell cell required for, but not actually mediating, a specific immune response. Often used to describe antigen-presenting cells.

accidental or incidental host a host in which the parasite is not commonly found, nevertheless it is suitable for the parasite's development. In some instances (e.g. cysticercosis) the accidental host becomes a 'dead end' because even though the parasite develops through its appropriate stages, it fails to find a portal of exit and is thus blocked from continuing its life cycle.

accidental parasite a parasite that infects an organism other than the usual host.

accidental transmission the direct or indirect transmission of an otherwise vector-dependent parasite arising ultimately as a result of artificial environmental conditions, such as cannibalism among birds kept at high stocking densities. Also parasites transmitted by sharing dirty needles and syringes such as may occur in drug users.

acclimatization way in which an organism adapts to a different or changing environment.

accuracy the percentage of correct results obtained by a test under evaluation compared with the results of a reference or 'gold standard' test.

acetabulum the ventral muscular sucker or holdfast of some diagenetic trematode worms.

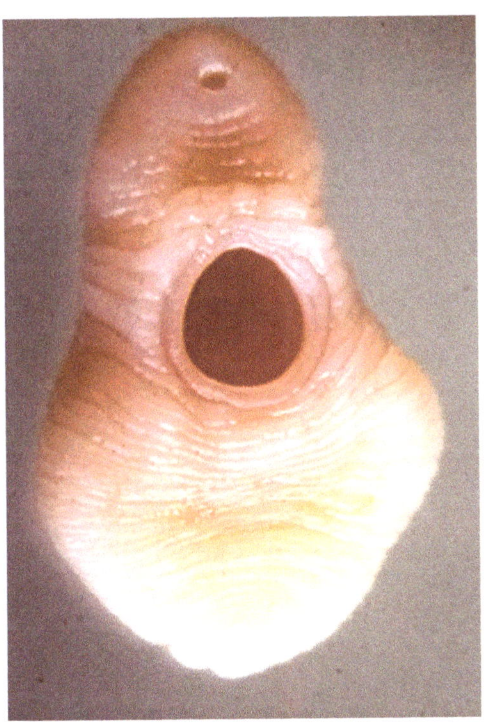

acetyl-CoA carboxylase (ACC) is responsible for the synthesis of malonyl-CoA, the two-carbon donor for fatty acid synthesis.

ACGT an acronym for the four types of bases found in a DNA molecule: adenine (A), cytosine (C), guanine (G), and thymine (T). A DNA molecule consists of two strands wound around each other, with each strand held together by bonds between the bases. Adenine pairs with thymine, and cytosine pairs with guanine. The sequence of bases in a portion of a DNA molecule, called a gene, carries the instructions needed to assemble a protein.

acidocalcisomes acidic calcium stores rich in polyphosphate present in different organisms from bacteria to humans.

acinar refers to liver parenchyma, in close association with hepatocytes.

acoelomate a condition in which a body cavity is lacking, as in the members of the phylum Platyhelminthes, where the organs lie embedded in parenchyma.

acquired cell-mediated immunity an immune state mediated by T and B cells and characterized by the development of activated macrophages.

acquired immune deficiency syndrome (AIDS) a disease caused by the human immunodeficiency virus that causes destruction of key components of the immune system. As a result, infected individuals become very susceptible to infections and cancers.

acquired immune response (acquired immunity) immunity mediated by lymphocytes and characterized by antigen-specificity and memory. The response by the adaptive immune system develops much less rapidly than the innate immune response but culminates in the subsequent activation of parasite-specific effector cells and antibodies and the generation of memory cells (dormant populations of parasite specific lymphocytes which will be activated rapidly following secondary exposure to the parasite). The principal cell types involved in this arm of the immune system are T and B lymphocytes and antibody-secreting plasma cells. Adaptive immunity is however linked to innate immunity by innate antigen-presenting cells, such as dendritic cells and macrophages, which process and present antigens to lymphocytes in conjunction with relevant MHC molecules. Lymphocyte populations consist of two major types, CTL, which express CD8 on their surface, and helper T (Th) lymphocytes, which express CD4 instead of CD8 on their surface.

acquired partial immunity an enhancement of the defence mechanism of the host against infection or clinical disease either as a result of previous infection with the pathogen, by the prenatal or postnatal transfer of protective substances from mother to child, or by the injection of such substances. Immunity from malaria is partial because it does not fully protect against infection and may wane over time in the absence of repeated infections.

acquired protection acquired ability of organisms to prevent or minimize infection (in terms of parasite establishment or development) by a particular set or species of parasite after a first infection.

acrocentric chromosome a chromosome with the centromere near one end. It is a chromosome where the 'p' arm is present but is too small to be observed. *See* **centromere**.

actin polymerization the addition of actin monomers to actin filaments; actin polymerization adjacent to the plasma membrane generates force that moves the plasma membrane forward. Many regulatory molecules can promote either the formation of new actin filaments or the extension of existing ones. If these regulators are associated with the plasma membrane, then actin polymerization occurs in a polarized manner.

actinomycin D an antibiotic commonly used to inhibit mRNA synthesis. It intercalates into DNA and blocks transcription by RNA polymerase.

activated macrophage a macrophage in a state of enhanced metabolic and functional activity.

activation-induced cytosine deaminase (AICDA) an enzyme critical for protecting against malaria and other infectious diseases and has recently been found to play a role in triggering Burkitt's lymphoma. Prolonged malaria infection was implicated in increasing AID production, which may subsequently lead to Burkitt's lymphoma.

activator protein 1 (AP-1) transcription factor composed of subunits belonging c-Fos, c-Jun, ATF, and JDP families; AP-1 regulates cellular responses to stress, including infections.

active case detection (ACD) the detection of malaria infections at community and household levels in population groups that are considered to be at high risk. It can be conducted as fever screening followed by parasitological examination of all febrile patients or as parasitological examination of the target population without prior fever screening.

active cysticercosis an infection with viable cysticerci, the infection can be transmitted to the final host.

active efflux net movement of a substance out of a cell in a direction opposite to that of the prevailing electrochemical gradient; requires coupling to an energy source.

active immunity production of antibodies by the body's own immune system due to the exposure to antigen.

active infection detection (active ID) the detection of malaria (clinical and subclinical) infections at community and household levels in population groups that are considered to be high risk.

active transmission this may take either of two forms: (1) When an animate vector is involved, the vector itself seeks out the host and brings the parasite to it, as in African trypanosomiasis. (2) When an active, aggressive parasite is involved, the parasite itself seeks out the host and enters it, as in schistosomiasis.

activity-based probes (ABPs) chemical probes designed to bind covalently the active site of a specific enzyme or enzyme class. They harbour a reactive moiety and a recognition element for specific labelling of enzymes and a reporter element for detection.

actomyosin basic force-producing or structural unit in cells consisting of myosin motors that bind and pull on actin filaments.

actomyosin contraction myosin motors can interact with actin filaments. Hydrolysis of ATP by myosin moves the actin filament relative to myosin. Many myosins are dimeric and can therefore move two actin filaments relative to one another.

acute any process which has a sudden onset and has a relatively brief duration.

acute disseminated encephalomyelitis (ADEM) an immune-mediated disease of the brain. It usually occurs after a viral infection but may appear in bacterial or parasitic brain infections, after vaccination and even spontaneously. ADEM produces multiple inflammatory lesions in the brain and spinal cord, particularly in the white matter.

acute infection is caused by an extensive expansion of pathogens soon after primary infection. It is usually symptomatic and is often controlled by the development of an efficient immune response specific for the pathogen.

acute lung injury diffuse heterogeneous lung injury characterized by hypoxaemia ($PaO_2/FiO_2 < 300$ mmHg), non-cardiogenic pulmonary oedema, low lung compliance, and widespread capillary leakage.

acute malaria acute presentation of a symptomatic *Plasmodium* infection.

acute phase proteins serum proteins whose levels increase during infection or inflammatory reactions, such as C-reactive protein.

acute schistosomiasis syndrome that occurs during the first six months after exposure to contaminated water; typically occurs upon first exposure.

adaptation an umbrella term describing any biological change (genetic or phenotypic) that increases the fitness of an individual in a given ecological setting.

adaptive evolution an evolutionary process that is directed by positive selection, which makes a population better adapted to live in a particular environment.

adaptive immune responses antigen-specific antibody responses that include the development of memory and clonal selection.

adaptive radiation the evolution of ecological and phenotypic diversity within a rapidly multiplying lineage.

additive genetic effects when the combined effects of alleles at different loci are equal to the sum of their individual effects. *See* **complex trait**.

addressin cell adhesion molecule presents on the luminal surface of blood and lymph vessel endothelium, and recognized by homing molecules which direct leucocytes to tissues with the appropriate 'address'.

adenine (A) is one of four chemical bases in DNA, with the other three being cytosine (C), guanine (G), and thymine (T). Within the DNA molecule, adenine bases located on one strand form chemical bonds with thymine bases on the opposite strand. The sequence of four DNA bases encodes the cell's genetic instructions. A form of adenine called ATP serves as an energy storage molecule and is used to power many chemical reactions within the cell. *See* **bp**, **nucleotide**.

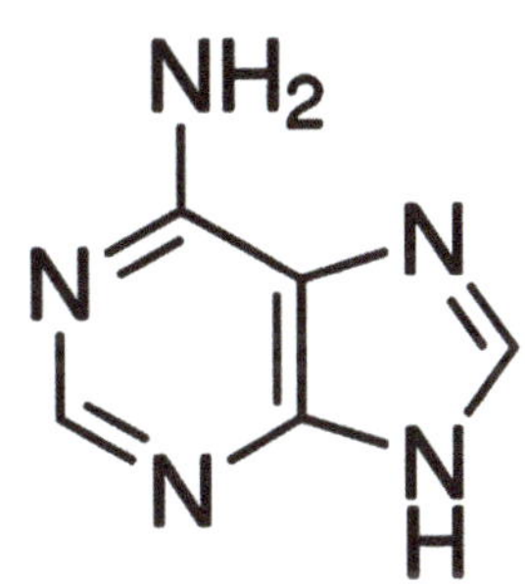

adenopathy an enlargement of lymph nodes in response to some stimulus, such as inflammation, toxicity or infection. *See* **lymphadenopathy**.

adenylate cyclase enzyme converting ATP into cAMP.

adenylate/uridylate-rich element (ARE) a sequence of nucleotides located in the 3′-UTRs of mRNAs that is rich in adenosine and uridine bases. It is generally consists of several AUUUA pentamers, overlapping or within a short distance from each other; regulates the stability of host mRNA.

adequacy comparison of the performance or impact of the programme against previously established targets.

adherence the ability of pathogens to adhere (stick) to host surfaces.

adhesins merozoite surface proteins involved in host cell interactions and induction of invasion. Adhesins include the micronemal erythrocyte binding antigens (EBA) family, EBA175, EBA140, EBA181, and the rhoptry-localized reticulocyte-binding homologues (Rh) family, Rh1, Rh2a, Rh2b, Rh4, and Rh5.

adhesive (or cellophane) tape test the gold standard diagnostic test for detecting pinworm (*Enterobius vermicularis*) infection. The tape is applied to a glass slide with the adhesive side out and then pressed against the anus and perianal area of the patient causing pinworm eggs to stick to the tape surface. The tape is then pressed against the slide with the adhesive side down and used as a cover slip for microscopic examination.

adhesome the combined molecular composition of focal adhesions.

adjustment a summarizing procedure in which the effects of differences in composition of the populations being compared have been minimized by statistical methods. *See* **confounding variable**.

adjuvant immunological or pharmacological substances (such as bacterial toxins) that alter the response induced by either a drug or vaccine; often used to enhance immune responses induced by a vaccine, particularly recombinant protein based vaccines.

adoptive immunity the development of immunity as a result of the transfer of cells from an immunized animal to an unimmunized recipient.

adsorption accumulation of molecules or cells on the surface of a substance.

adult cestode a long, segmented, flatworm located in the digestive tract (e.g. *Taenia saginata* in humans). The body of the worm is composed of numerous independent segments (proglottids), and distal segments (typically gravid) detach from the worm and are excreted with faeces. The head (scolex) is attached to the intestinal mucosa by means of suckers.

adult stage the final stage in the life cycle in which a parasite has fully developed, grown to full size, and became sexually mature and reproductively active.

adult stem cells undifferentiated cells in a tissue. These cells can grow into any of the types of specialized cells in that tissue.

adulticides chemotherapeutic preparations used to eliminate both immature (L3) and adult heartworms. Different treatment protocols are used depending on disease severity. Care and support should be provided to the animal when treated with an adulticide.

Aedes albopictus a highly anthropogenic mosquito species; best known for transmitting dengue and chikungunya viruses but has also been found naturally infected with West Nile virus, eastern equine encephalitis and Japanese encephalitis.

aelurostrongylosis the disease caused by nematodes of the genus *Aelurostrongylus*. Cat aelurostrongylosis produced by *A. abstrusus* is characterized by varying respiratory and general clinical signs: mild to intense chronic cough, sneezing, wheezing, mucopurulent nasal discharge, dyspnoea, tachypnoea, tachycardia, open-mouthed abdominal breathing, lethargy, depression, loss of body weight, and even death. Specific treatment relies on the administration of benzimidazoles or macrocyclic lactones.

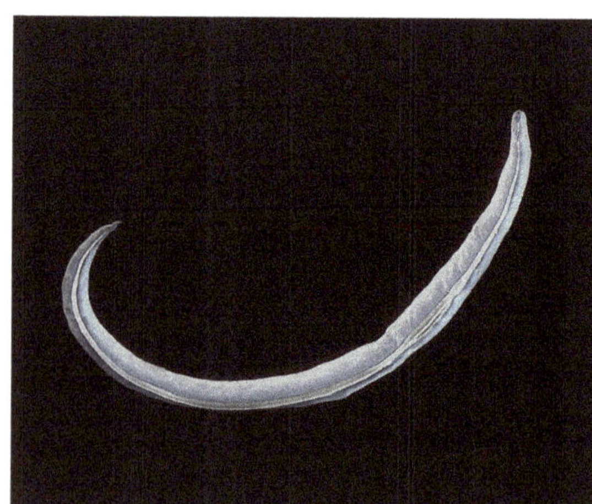

aequorin fluorescent protein from the jellyfish *Aquora victoria* used to detect calcium *in vivo*.

aerobic requiring oxygen.

aerobic glycolysis predominant fermentation of glucose to lactate typically even under oxygen pressures considered to be aerobic. Fractions of glycolytic intermediates that are not fermented are redirected and are seemingly sufficient to sustain biosynthetic pathways such as the pentose phosphate pathway, shikimate pathway, and lipid biosynthesis.

aerosols airborne solid or liquid particles (typically no larger than a few micrometres) that can remain in the atmosphere for hours to days. Aerosols impact climate by scattering or absorbing radiation, initiating cloud formation, or altering the optical properties of clouds.

aerotaxis the movement of an animal in response to oxygen concentration.

aestivate to remain dormant by inactivity and lowered metabolic rate, as when habitats dry out for freshwater snails.

aetiology the study of the cause or origin of a disease.

afebrile relating to the absence of fever.

affinity the strength of binding (affinity constant) between a receptor (e.g. one antigen-binding site on an antibody) and a ligand (e.g. epitope on an antigen).

affinity chromatography a method to isolate antigens or antibodies based upon antigen–antibody binding.

affinity maturation the memory population is generated from those B cells whose receptors bind to antigen on follicular dendritic cells with sufficient avidity to compete with existing circulating antibody. These B cells present antigen to T cells that induce B-cell proliferation and expansion in germinal centres. The mechanism ensures that B cells with high-affinity receptors will contribute to subsequent antibody responses.

affordable medicines facility an innovative financing mechanism designed to expand access to the most effective treatment for malaria, artemisinin-based combination therapies (ACTs), hosted and managed by the Global Fund. It aims to enable countries to increase the provision of affordable ACTs through the public, private, and non-governmental organization sectors. To achieve this aim, the Global Fund has negotiated lower ACT prices with manufacturers and will pay a large proportion of this (a 'co-payment') directly to manufacturers on behalf of buyers across the public, private-for-profit, and not-for-profit sectors.

African animal trypanosomosis (AAT) a parasitic disease that causes serious economic losses in the livestock production system. Diverse clinical signs are observed from anaemia, loss of body condition and emaciation to death when untreated. In sub-Saharan Africa, the disease is caused by *Trypanosoma congolense*, *Trypanosoma vivax* and *Trypanosoma brucei*. Its distribution overlaps with the presence of the tsetse fly except for *T. vivax* and *Trypanosoma evansi* that can be transmitted mechanically by other haematophagous biting flies. AAT restricts agricultural production, limits the availability of food and contributes to poverty across rural sub-Saharan Africa.

African trypanosomiosis (also referred to as Old World trypanosomiasis) a lethal human disease that is prevalent in sub-Saharan Africa. Also known as 'sleeping sickness', is caused by microscopic parasites of the species *Trypanosoma brucei*. Two specific subspecies of *Trypanosoma brucei*, known as *Trypanosoma brucei* gambiense and *Trypanosoma brucei* rhodesiense, cause disease in humans. It is transmitted by the tsetse fly (*Glossina* species), Sleeping sickness is curable with medication, but is fatal if left untreated. A third subspecies, *Trypanosoma brucei brucei*, and the related trypanosomes *Trypanosoma congolense* and *Trypanosoma vivax*, infect only non-primates, causing wasting disease, which limits economic development in endemic areas.

agammaglobulinaemia agammaglobulinaemia was used in earlier years before the development of methods sufficiently sensitive to detect relatively small quantities of gamma globulin in the blood (hypogammaglobulinaemia).

age related toxicity organophosphates are not given to very young animals because of the narrow margin of safety.

agent-based model a class of computational models for simulating the actions and interactions of autonomous agents with a view to assessing their effects on the system as a whole. For example, an agent-based model identifies active entities (i.e. the agents, including pathogen, T cells, APCs, and antibodies), defines their behaviour (e.g. recognition and presentation) through simple rules, places them in a certain environment (e.g. host), establishes connections, and runs a simulation. The global behaviour then emerges as a result of interactions of many individual entities.

agglomerates of haemozoin form of haemozoin, composed of several crystal aggregates. It is possible to observe this after its deposition in organs.

agglutinating antibodies any immunoglobulin isotype that cause parasites to clump together when bound to an antigen on the parasite surface. They are potentially important as part of the immune response needed to control parasite numbers.

agglutination the aggregation of particulate antigen by antibodies. Agglutination applies to red blood cells as well as to bacteria and inert particles covered with antigen. It is different from precipitation.

AICD activation-induced cell death apoptosis triggered following T-cell activation, which induces the expression of ligands for death receptors.

AIDS a collection of symptoms known as acquired immunodeficiency syndrome. It is caused by infection with the human immunodeficiency virus (HIV), which leads to loss of immune cells and leaves individuals susceptible to other infections and the development of certain types of cancers.

airborne transmission transmission by air of infectious agents from respiratory secretions.

airway submucosal glands mucus-secreting glands found in the connective tissue that provide fluid for hydrating the surface of the airway epithelial cells and enabling ciliary function.

airway surface liquid fluid interface between the air and the cells in the lungs that confers protection from infection and facilitates removal of foreign particles.

ala (-ae) flattened, wing-like expansions of the cuticle of certain nematodes (cervical, caudal or lateral).

albendazole a benzimidazole anthelminic with high efficiency in sheep and cattle against all intestinal nematodes, except *Trichuris* spp., and intestinal tapeworms and lungworms and, with increased dose rates, will kill adult liver flukes. It is approved for use against mature liver flukes (e.g. *F. hepatica)* in beef and non-lactating cattle. Because albendazole is a teratogen, it cannot be used in pregnant cattle during the first 45 days of gestation or in female dairy cattle of breeding age.

algae photosynthetic, aquatic, eukaryotic organisms that contain chlorophyll but lack terrestrial plant structures (e.g. roots, stems, and leaves). Algae can exist in many sizes ranging from single cells to giant kelps several feet long.

algorithm a set of instructions that tells a computer how to solve a problem or execute a task. A computer program typically consists of several algorithms.

alignment-based approach a computational approach that uses whole-genome multiple alignments to find conserved RREs.

alignment-free approach a computational approach that looks for conserved RREs from unaligned sequences by comparing only the functional regions of the 3′-UTRs of orthologous genes between related organisms.

all-cause mortality rate the number of deaths per population per unit time from any cause. In emergencies, this is usually per 10 000 persons per day and can be age stratified.

Allee effect a positive effect of population density on fitness of the members of that population. A component effect refers to any measure of individual fitness.

Allee threshold a critical density below which a population declines to extinction in response to an environmental variable, and above which it can increase.

allele one of two or more versions of a gene. An individual inherits two alleles for each gene, one from each parent. If the two alleles are the same, the individual is homozygous for that gene. If the alleles are different, the individual is heterozygous. Though the term allele was originally used to describe variation among genes, it now also refers to variation among non-coding DNA sequences. *See* **gene expression, locus**.

allele-specific polymerase chain reaction a polymerase chain reaction in which amplification depends upon the exact match of a primer to the DNA sequence of only one of two or more alleles of interest, thus resulting in amplification of products that are specific to particular alleles.

allelic relating to one of a series of two or more alternative forms of a gene that occupy the same position or locus on a specific chromosome.

allelic association statistical comparison of marker allele frequencies between a disease group and a control group, used to suggest genes and regions that might influence disease risk.

allelic exclusion the phenomenon whereby, following successful rearrangement of one allele of an antigen receptor gene, rearrangement of the other parental allele is suppressed.

allelic heterogeneity reflects different allelic mutations at the same locus associated with multiple disorders.

allergen an antigenic substance that induces an allergy (mould, grasses, certain food, antibiotics, etc.).

allergic reaction response by the immune system against a foreign substance. It involves a specific antibody called Immunoglobulin E (IgE).

allergy **1** abnormal sensitivity of the body to a substance known as an allergen. Contact with the allergen causes symptoms such as skin rashes, watery eyes, rhinitis, etc. **2** IgE-mediated hypersensitivity, e.g. asthma, eczema, hay fever and food allergy. Characterized by the release of pharmacological agents as a result of mast cell and basophil degranulation, which is usually mediated by antibodies of the IgE class. Also called Immediate (Type I) hypersensitivity.

allogamy fertilization between gametes derived from genetically different individuals.

allogeneic intraspecies genetic variations, i.e. a genetic dissimilarity within the same species.

allograft **1** a tissue/organ graft between two genetically non-identical members of a given species. **2** the transplant of an organ or tissue from one individual to another of the same species with a different genotype.

allopatric having non-overlapping geographic ranges. Cf. sympatric.

allopatric populations conspecific populations occurring in geographically distinct locations.

allopatric speciation the origin of two or more species resulting from divergent evolution of populations that are geographically isolated from one another.

allopatry two or more species (or populations of the same species) are in allopatry when they occur in separate, not overlapping geographic areas (the opposite term is sympatry). One species may be the vector of a pathogen under laboratory conditions. However, if they occur in allopatry there is not the possibility of contact, and therefore the transmission of the pathogen by the vector is not possible. Cf. sympatry.

allopurinol a drug usually administered in combination with meglumine antimonite for the first month of treatment of leishmaniasis and thereafter alone as maintenance for the dog's life. The mechanism of action for leishmaniasis is thought to be due to its incorporation into the parasite purine salvage pathway. This leads to the formation of a toxic analogue of adenosine triphosphate, which is incorporated into ribonucleic acid. Adverse effects may include erythema, hypersensitivity and predisposition to xanthine calculi.

allosteric regulation regulation of protein activity by remote-site covalent protein modifications (e.g. phosphorylation or acetylation) or non-covalent interactions with effector ligands, which change the functional site by the propagation of subtle conformational changes.

allotype an allelic variant of an antigen which, because it is not present in all individuals, may be immunogenic in members of the same species which have a different version of the allele.

allozyme a variant form of an enzyme that is encoded by a different allele of the same genetic locus. Allozyme analysis, using appropriate staining and starch-gel electrophoresis, was used to infer genetic variation before direct DNA sequencing became widely used. *See* **multilocus enzyme electrophoresis**.

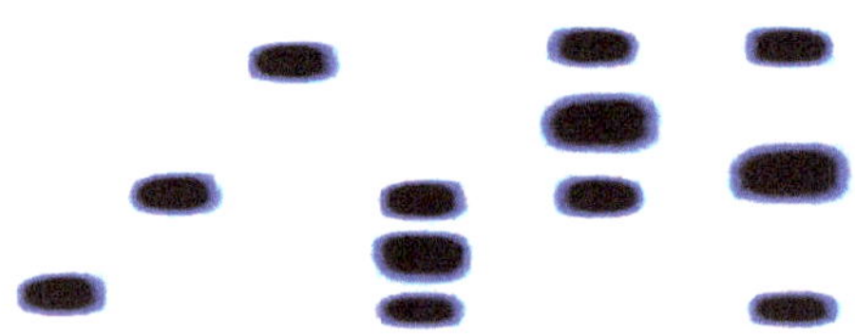

alpha-1-acid glycoprotein (AGP) one of the major acute phase proteins in humans, rats, mice and other species. Its serum concentration increases in response to systemic tissue injury, inflammation or infection. The biological function of AGP remains unknown; however, a number of activities of possible physiological significance, such as various immunomodulating effects, have been described. AGP also has the ability to bind and to carry numerous basic and neutral lipophilic drugs from endogenous (steroid hormones) and exogenous origin; one to seven binding sites have been described. It can also bind acidic drugs such as phenobarbital.

alternate grazing the technique of periodically changing the species of livestock that grazes on a given pasture. This practice reduces parasite loads on pasture since different animals are affected by different parasites.

alternate host one which alternates with another in the life cycle of a parasite. For example, snails and humans are alternate hosts of schistosomes.

alternation of generations this describes the life cycle of a species in which two or more different forms occur in successive generations, one sexual, the other asexual.

alternative complement pathway a series of enzyme reactions triggered by interactions on activating surfaces leading to activation of the complement system.

alternative pathway (of complement activation) activation pathway that involves complement components C3, Factor B, Factor D and Properdin which, in the presence of a stabilizing activator surface such as microbial polysaccharide, generates the alternative pathway C3 convertase C 3 bBb.

alternative splicing **1** different ways of combining a gene's exons to make variants of the complete protein. **2** during pre-mRNA processing, particular exons of a gene may be included or excluded from the final mRNA. This process allows for the generation of multiple proteins with potentially multiple functions from a single gene in the genome.

alternatively activated macrophages (AAMs) are activated by Th2 cytokines such as IL-4 and IL-13 to induce the production of proline, an amino acid essential for the synthesis of collagen. AAMs are thought to be involved in promoting collagen deposition and the onset of fibrosis.

altruism when one individual provides a benefit to another while gaining no benefit itself (or even while suffering a detriment).

alveolar oedema leakage of plasma fluid in the interstitial tissue and alveoli of the lungs.

alveolar–capillary membrane separation between air and blood in the lungs, formed by type 1 pneumocytes of the alveolar wall, the endothelial cells of the capillaries, and the interstitium, with a basement membrane between the two cell types.

alveolus the air sac in the lung where gaseous exchange occurs.

AMA1 the apical membrane antigen 1 of *P. falciparum*, a micronemal protein that is important for the host-cell invasion process of the parasite.

amastigote aflagellated intracellular developmental form in the life cycle of *Leishmania* species and *Trypanosoma cruzi* protozoan parasites that resides intracellularly within vertebrate host cells.

amastigote antigen 2 amastigote stage specific protein of *Leishmania* that plays a role in its survival within the mammalian host. It is induced by stress, and following heat shock it complexes with endoplasmic reticulum binding protein, BiP. On the contrary, the A2 gene in *L. major* is a non-expressed pseudogene. A2 may also play a role in protecting *L. donovani* from stress associated with infection involving visceral organs, including fever typically associated with visceral leishmaniasis.

amastin small surface glycoproteins in trypanosomatids, implicated in disease progression.

ameiotic recombination reciprocal (cross-over) or non-reciprocal (gene conversion) exchange of genetic material in the absence of meiosis in asexual lineage.

American trypanosomiasis *See* **Chagas disease**.

Ames test widely employed *Salmonella*-based cell assay used in medicinal chemistry projects to determine whether a given compound is mutagenic.

ametabolous type of metamorphosis in insects in which there is no external change as they proceed through a series of moults to the adult; ametabola refers to the taxonomic group.

amiloride-sensitive epithelial sodium channel (ENaC) membrane-bound channel permeable for Na^+ and localized on the apical membrane of epithelial cells. It regulates fluid efflux from the alveoli in the lungs.

amino acids the building blocks of proteins in living things. There are 20 known amino acids found in living organisms. Proteins consist of one or more chains of amino acids called polypeptides. The sequence of the amino acid chain causes the polypeptide to fold into a shape that is biologically active. The sequence of amino acids in a protein determines its function. This sequence of amino acids is determined by genetic code (i.e. the sequence of bases found in the gene coding for that protein).

aminoquinoline a family of drugs belonging to aminoquinoline with an amine at the 8-position of quinoline. This family contains three members, primaquine, tafenoquine, and pamaquine, and is used in the treatment of malaria. They may be used to eradicate malaria hypnozoites from the liver and have been used for malaria prophylaxis. The 8-aminoquinoline drugs must not be given to patients with G6PD deficiency because they can cause fatal haemolysis in these patients.

amitotic division nuclear division by constriction without the formation of chromosomes yielding two equal nuclei.

amitraz a formamidine that is selective towards acarians (ticks and mites) and has been used since the late 1960s to control ticks. It interferes with the octopaminergic system of arthropods, similar to the adrenergic system in mammals, by binding to the octopamine receptors, which leads to the stimulation of monoamine oxidases (adenylate cyclase activity) and the G protein. It thus induces cAMP and cGMP synthesis, which have various intracellular actions. At a sublethal dose, ticks already attached will fall off their host, and those that infest a host do not attach nor feed. At sublethal doses, reproduction is impaired and most of the eggs do not hatch.

amitraz toxicity amitraz is used in dogs in the form of collars to prevent tick infestation, in the form of lotions to treat canine demodicosis, and in the form of spot-on solutions in combination with other insecticides/acaricides (metaflumizone or fipronil) to treat and prevent flea and tick infestations. Interaction between amitraz and the adrenergic system could occur in carnivores. Intoxication can be caused by ingesting, chewing, or sucking amitraz-impregnated collars or by licking spot-ons or solutions. Intoxication produces an α2-agonist effect which induces hyperglycaemia, a neurosedative effect (lethargy and ataxia), and more rarely cardiovascular disorders (bradycardia, hypotension).

amniocentesis a procedure in which a small amount of amniotic fluid (the fluid surrounding a baby in the womb) is drawn out using a needle and syringe. Because amniotic fluid contains cells from the baby, it can be used to test for chromosomal and genetic disorders, as well as certain birth defects.

amniote a subclass of vertebrate species characterized by the presence of an amnion, which is a protective envelope that encloses the embryo. They include reptiles, birds and mammals, but not amphibians.

amodiaquine (AQ) a common antimalarial drug.

amoeba a single-celled mass of protoplasm which changes shape by extending cytoplasmic processes called pseudopodia by which it moves and absorbs nourishment. The majority of amoebae are free-living in soil and water.

amoebic liver abscess a frequent extraintestinal manifestation of *Entamoeba histolytica* infection. The infection enters the portal venous system from the colon. Prompt recognition and appropriate treatment of amoebic liver abscess lead to improved morbidity and mortality.

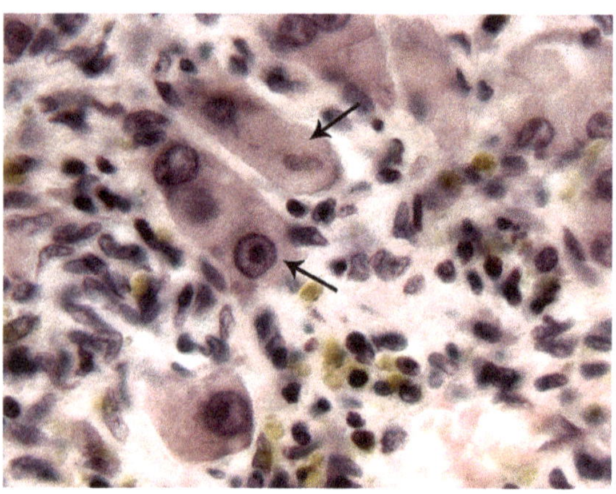

amoeboid cells of no fixed shape where movement involves protrusion of cytoplasm of the cell to form pseudopodia are referred to as amoeboid.

amorph/hypermorph/hypomorph/neomorphmutant alleles exhibiting no activity, increased activity or expression, reduced activity or expression, and some novel activity, respectively.

amphidial canals two channels that are contiguous with the external environment and encase the amphidial (sensory) nerves and accessory cells as they track posteriorly from the head.

amphids a pair of anterior chemosensory structures in the nematode head that are open to the environment enabling the detection of chemical cues.

amphiphilic capable of dissolving either in water or in an organic solvent.

amphipods an order of crustaceans, characterized by a shrimped-like body flattened laterally, and the presence of three pairs of uropods (tail-limbs).

amplification an increase in the number of copies of a specific DNA fragment; can be *in vivo* or *in vitro*. *See* **polymerase chain reaction**.

amplification bias uneven representation of regions of the DNA template in amplified DNA.

amplified fragment length polymorphism (AFLP) selective amplification of genomic restriction fragments (obtained by restriction fragment length polymorphism) by PCR using randomly selected primers. It is used for detection of genetic polymorphism. *See* **random amplified polymorphic DNA**.

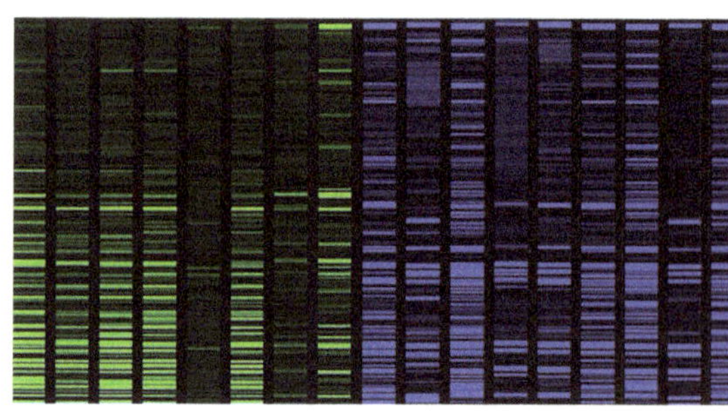

amplifying host a host in which the level of pathogen can become high enough that a vector (e.g. a tick) that feeds on it will probably become infectious.

amprolium an analogue of vitamin B_1, thiamine. It acts on the first-generation coccidial schizonts to prevent development to the merozoite stage. It has some activity against the sexual stages and the sporulating oocysts of coccidia. Prolonged use can lead to the development of thiamine deficiency and the development of CNS dysfunction.

anabolic reactions the synthesis of complex molecules in living organisms.

anabolism the energy-dependent formation of building blocks and macromolecules in a cell.

anaemia a condition caused by a reduced number of red blood cells or haemoglobin secondary to blood loss from the vascular space. This results in reduced oxygen-carrying capacity of the blood, and reduced aerobic activity in body cells.

anaerobic lacking or not requiring oxygen.

anagenesis evolution within a lineage, usually expressed by morphological change, but not associated with a splitting event.

analogous enzymes two enzymes in different species that share the same catalytic function but have unrelated amino acid sequences.

anamnestic response the enhanced immune response that occurs against an antigen as a result of previous host exposure to a related antigen.

anaphylatoxic (proinflammatory) protein fragments generated during the classical, lectin or alternative (C5a) pathways of complement activation.

anaphylatoxin some peptides from the complement system (e.g. C3a, C4a or C5a) capable of triggering mast cell degranulation and smooth muscle contraction.

anaphylaxis an often fatal hypersensitivity reaction, triggered by IgE or anaphylatoxin-mediated mast cell degranulation, leading to anaphylactic shock due to vasodilatation and smooth muscle contraction.

Anaplasma marginale a tick-borne intraerythrocytic pathogen that causes bovine anaplasmosis, a disease affecting cattle in temperate, subtropical and tropical regions of the world. Besides cattle other domestic and wild ruminants such as water buffalo (*Bubalus bubalis*) can become persistently infected and carriers of *A. marginale*.

anaplasmosis a tick-borne disease caused by rickettsiae of the genus *Anaplasma*.

anaplerosis the process of replenishment of depleted metabolic cycle or pathway intermediates. Most commonly referring to the TCA cycle, this concept is also used to describe glycolysis and glutaminolysis generated substrates for macromolecular biosynthesis or anabolism.

anapolysis the process in which terminal, gravid proglottids are not shed in certain tapeworms.

anasarca stage acute phase of bovine besnoitiosis, characterized by fever, weakness, swelling of the superficial lymph nodes, and oedemas that occur as a consequence of replication of tachyzoites in blood vessels.

ancestry-informative markers sets of polymorphisms for a particular DNA sequence that appear in substantially different frequencies between populations from different geographical regions of the world. These markers can be used to estimate the geographical origins of the ancestors of an individual typically by continent of origin.

androgen a male sex hormone (e.g. testosterone).

anergy potentially reversible specific immunological tolerance in which the lymphocyte becomes functionally non-responsive.

aneuploid a defective cell that contains only part of the full complement of either haploid (one copy) or diploid (two copies) genes. When defective for essential genes, aneuploid cells die.

aneuploidy aneuploid cells or nuclei have different numbers of different chromosomes, usually as a result of non-disjunction. A well-known example is Down's syndrome (trisomy 21) in humans in which affected individuals have three copies of chromosome 21, but two copies of all other chromosomes.

angiogenesis complex physiological process involving the organization of proliferating endothelial cells into new blood vessels from pre-existing vessels. It is also known as 'neovascularization'. This is a normal process in growth and development (as well as in wound healing). However, this is also a fundamental step in the transition of tumours from being dormant to becoming malignant. It is also a fundamental step in various inflammatory processes.

angiostrongyliosis canine angiostrongyliosis is a snail-borne infection caused by metastrongylid nematode known as *Angiostrongylus vasorum*. Adult worms live in the heart and pulmonary arteries of domestic dogs and wild canids causing severe respiratory and circulatory distress ending often with death. *A. vasorum* has an indirect life cycle, which requires an intermediate gastropod (i.e. slugs and snails) and/or paratenic (e.g. amphibians, lizards, mice and rats) hosts for the development of first-stage larvae (L1) into the infective third-stage larvae (L3). Dogs become infected by ingesting the intermediate or paratenic hosts containing L3. This infective stage penetrates the gastrointestinal wall and young nematodes follow a lymphatic-hepatic-cardiopulmonary migratory route till reach the right ventricle of the heart and pulmonary arteries, where sexual reproduction and oviposition take place.

animal model a non-human species used in medical research because it can mimic aspects of a disease found in humans. Animal models are used to obtain information about a disease and its prevention, diagnosis, and treatment. By using animals, researchers can carry out experiments that would be impractical or ethically prohibited with humans.

anisakiasis a parasitic disease caused by anisakid nematodes (worms) that can invade the stomach wall or intestine of humans. The transmission of this disease occurs when infective larvae are ingested from fish or squid that humans eat raw or undercooked. In some cases, this infection is treated by removal of the larvae via endoscopy or surgery.

anisogamete morphologically different male and female gametes.

ankyrin-repeat domain a common protein–protein interaction platform assembled with multiple repeats of the ankyrin repeat, which is a 33 amino acid sequence that adopts a helix–loop–helix structure.

annotation extraction, definition and interpretation of features on the genome sequence derived by integrating computational tools and biological knowledge. *See* **bioinformatics**.

annual parasite index the number of confirmed parasite cases during 1 year/total population × 1000.

annualized estimate (burden) of morbidity and mortality the mean burden of malaria epidemics per year, expressed as total episodes and/or deaths or in terms of attack rate or cumulative mortality rate per PAR. Case fatality rate the proportion of episodes resulting in death.

anopheline mosquito mosquitoes are classified within the Culicidae family, which is, based on morphology, divided into two subfamilies, the Anophelinae and the Culicinae.

Anoplocephala perfoliata a common intestinal tapeworm (Cyclophyllidea, Anoplocephalidae) that affects horses worldwide. Mounting evidence suggests that *A. perfoliata* is responsible for several intestinal disorders, and its accurate detection is crucial for both clinical and epidemiological purposes.

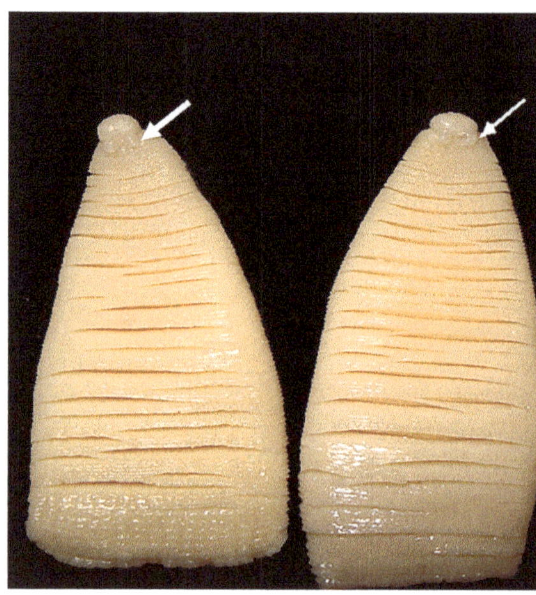

anoplocephalid cestodes belonging to the anoplocephalid family, including *Moniezia* spp. in ruminants and *Anoplocephala* spp. in horses and donkeys (*Anoplocephala perfoliata*, *Anoplocephala magna*, *Paranoplocephala mammillana*). Most tapeworms are found in the small intestine of their host as adults or, with *Thysanosoma* spp., have access to the intestine. All tapeworms use an intermediate host; and both vertebrate and invertebrates frequently serve in this capacity. *A. perfoliata* is the most prevalent and most frequently associated with clinical disease. All the three *Anoplocephala* species are similar in that the scolex is devoid of rostellum, hooks, or hooklets, and suckers are unarmed. Transmission is by ingestion of infected free-living pasture (oribatid) mites. The mechanical damage caused by the suckers on the scolex of the tapeworms at the site of attachment and parasite antigens play a role in the pathogenic process. Lesions caused by the tapeworm attachment may alter the pattern of intestinal motility and inflammatory changes at the site of parasite attachment and changes in bowl wall diameter at the ileocecal junction may be a potential cause of intussusception. Ulceration at the site of attachment may be seen on necropsy. Tapeworms compete with the host for nutrients, hinder normal gut motility, and excrete some toxic wastes into the host's GI tract.

anorexia lack of appetite.

anseriform birds an order of birds that contains swans, geese, ducks and screamers.

anterior station protozoan development in the anterior part of an insect vector, for example, Salivaria of the genus *Trypanosoma*; transmission in this case occurs by biting.

anthelmintic a chemical agent destructive to worms. They are classified as antinematicidal, antitrematicidal, and anticesticidal. Also termed dewormer.

anthelmintic resistance (AR) arises when a greater frequency of individuals in a parasite population no longer responds to the normal clinical dose of treatment. This impacts agricultural incomes and poses a clear threat to animal health and welfare. A critical aspect of resistance is that it is a heritable trait. Nematodes have a range of different strategies to achieve a state of reduced susceptibility towards a given anthelmintic drug. These include modification of drug target (e.g. binding site), so that the drug is no longer recognized by the target and is thus ineffective, increasing target site numbers (e.g. neuronal receptors) or amplification of target genes to overcome the drug action, increasing drug efflux (e.g. through transmembrane pumps) and modification in metabolism that expels the drug, for instance, up-regulation of P-glycoprotein exporter in the helminth.

anthropocentrism a view that regards humans as the central element of the universe. Proponents believe that we should only protect and replenish the environment so that it serves human purposes such as producing food and drugs, and that the fate of animals and plants are not morally significant except in terms of sustaining human wellbeing.

anthropogenic the term refers to the human impact on the environment and includes impacts on the environment and biodiversity. In a wider sense, the term refers to everything that is originated from human actions.

anthroponoses human diseases that are transmissible to animals.

anthroponotic an infectious disease caused by microorganisms that normally exist in a transmission cycle involving only humans, for example, tuberculosis and malaria.

anthropophilic from Greek anthropos (human) and philein (to like); usually said of parasites that prefer human beings to animals. For example mosquitoes exhibiting a preference for blood feeding on humans.

anthropophilic behaviour mosquito preference for breeding sites created by human activities, using human dwellings as resting places and humans as a source of bloodmeal.

anthropozoonosis a disease of either animals or humans that may be transmitted from one species to the other.

anthropozoophilic attracted to both human beings and animals, for example certain species of mosquitoes.

antiadhesion antibodies prevent parasite adhesion to receptors when bound to the surface of infected erythrocytes.

antibiosis an association between two populations of organisms that is detrimental to one of them, or between one organism and an antibiotic produced by another.

antibiotic gradients the gradual increases or decreases in antibiotic concentration that are observed between two spatially segregated sites (for example, two tissues in the body).

antibiotics compounds having antibacterial activity, including naturally occurring antibacterial compounds and their synthetic derivatives and novel synthetic antibacterial drugs.

antibody-dependent cellular cytotoxicity (ADCC) a cytotoxic reaction in which an antibody-coated target cell is directly killed by an Fc receptor-bearing leucocyte, e.g. NK cell, macrophage or neutrophil.

antibody detection (AbD) indirect methods for parasitic infection diagnosis using immunoassays for the detection of anti-*Schistosoma* antibodies.

anticestode drugs drugs that kill tapeworms are often called cestocides. The killed worms may then be digested by the host animal, and thus may not be evident in the faeces. Segments, which are fragile, may not be evident, but eggs can still be shed until gone. Control of intermediate hosts (e.g. fleas for *Dipylidium*, rodents for *Taenia* and *Echinococcus*, and mites for *Anoplocephala* and *Moniezia*) should also be considered.

anticoagulant substance that prevents blood from clotting.

anticodon a sequence of three bases in a molecule of transfer RNA (tRNA) that binds to a complementary codon in messenger RNA (mRNA). Each anticodon designates a specific amino acid to be added to a growing polypeptide.

antigen a substance (foreign or self) that can stimulate a detectable immune response (specific antibodies of T cell response) when introduced into the animal's body. For example, pollen grains, dust, parasite, bacteria and viruses are recognized by the body as being foreign and the body responds by producing specific antibodies (immunoglobulins) to the antigen.

antigen depot reservoir of pathogen-derived antigen or antigen fragments that is slowly released and could help sustain host immune responses against that pathogen.

antigen presentation the process, by which naive T cells are primed by the presentation of antigen by an antigen-presenting cell via the MHC/peptide–T cell receptor interaction.

antigen-presenting cells (APC) cells that can present the antigen to lymphocytes, through their cell surface class II MHC and stimulate the lymphocytes to provide a specific immune response. Monocytes, macrophages, dendritic cells, B cells and endothelial cells are APCs.

antigen processing large molecules are broken down (processed) within macrophages into peptides and presented within the groove of MHC spell out molecules.

antigen receptor the specific antigen-binding receptor on T or B lymphocytes; these receptors are transcribed and translated from rearrangements of V genes.

antigen binding site the part of an immunoglobulin molecule that binds antigen specifically.

antigenic determinant epitopes on the surface of an antigen that stimulates a specific immune response and against which that response is directed.

antigenic mimicry acquisition of or production of host antigens by a parasite so that it is not recognized as non-self, such as in *Schistosoma*.

antiglobulin antibody directed against an immunoglobulin. Usually obtained by injecting immunoglobulin into an animal of another species.

antimalarial antibody any antibody that recognizes *Plasmodium* antigens.

antimalarial resistance the ability of a malarial parasite strain to survive and multiply despite the administration and adsorption of a drug given in doses equal to or higher than those usually recommended but within tolerance of the subject.

antimicrobial peptides (AMPs) short peptides produced by the host as part of the innate immune response to kill pathogens.

antimicrobial resistance the ability of parasites or other microbes to mutate (change form) and become insensitive to (resistant to) the effects of a chemotherapeutic agent that was previously effective. Drug resistance occurs when parasites change in some way that reduces or eliminates the effectiveness of drugs, chemicals, or other agents designed to cure or prevent infections. *See* **anthelmintic resistance**.

antimonials sodium stibogluconate and meglumine antimonate are used in the treatment of leishmaniasis.

antimycin a potent inhibitor of the respiratory chain at the level of cytochrome b–c1.

antinuclear antibodies antibodies directed against nuclear constituents (usually nucleoproteins) found in the blood of patients with some disorders, mainly lupus erythematosus.

antiparasitic drugs usually classed as endoparasiticides, ectoparasiticides or endectocides.

antiparasitic drugs for heartworm chemotherapeutic management of heartworm infection involve three aspects: (i) removal of adult heartworms using an adulticide, (ii) interruption of the cycle using a microfilaricide to eliminate the source of adult heartworm, and (iii) prevention of infection using a larvicide.

antipredator responses behaviours exhibited by prey animals to avoid or escape potential predators.

antiprotozoal drug resistance drug resistant protozoa have been documented and, there is sound evidence that resistance does occur in a number of protozoal species, e.g. *Plasmodium* sp., *Giardia* sp., *Eimeria* sp.

antipyretics any drugs or herbs that are used to reduce fever.

antisense the non-coding DNA strand of a gene. A cell uses antisense DNA strand as a template for producing messenger RNA (mRNA) that directs the synthesis of a protein. Antisense can also refer to a method for silencing genes. To silence a target gene, a second gene is introduced that produces mRNA complementary to that produced from the target gene. These two mRNAs can interact to form a double-stranded structure that cannot be used to direct protein synthesis. *See* **transcription**.

antisense RNAs single-stranded RNAs that are complementary to an mRNA.

antisense transcription transcription of an (antisense) RNA from the opposite DNA strand as the template encoding a sense mRNA transcript.

antiserum (plural **antisera**) serum containing antibodies against a specific antigen.

antitoxins protective antibodies that neutralize soluble toxins.

apical complex a complex set of organelles found at the anterior end of the apicomplexan protozoans. The apical complex has a role in the penetration and invasion of host cells and has been used as the best defining features for the Apicomplexa, from which its name is derived. It encompasses polar rings, subpellicular microtubules, conoid, rhoptries, and micronemes.

apical membrane antigen-1 (AMA-1) a type 1 transmembrane protein unique to apicomplexans, represents one of the best-studied invasion proteins in *Plasmodium* plays an essential role in host cell invasion by apicomplexan parasites.

apical polar ring (APR) pivotal component of apical complex of all apicomplexa; a microtubule-organizing centre.

apicomplexan parasites belong to apicomplexa (synonym Sporozoa), an obligate single-cell parasitic group of eukaryotes. Many are pathogenic to humans like *Plasmodium*, the causative agent of malaria, and *Toxoplasma*, which induces severe manifestations in immunocompromised individuals. They are obligate intracellular parasites that invade host cells by developing into specialized stages called zoites. These unicellular organisms have a system of organelles at their apical end that produces substances to help the organism penetrate host cells and attain an intracellular location. Some apicomplexan species have the capacity to produce tissue cysts.

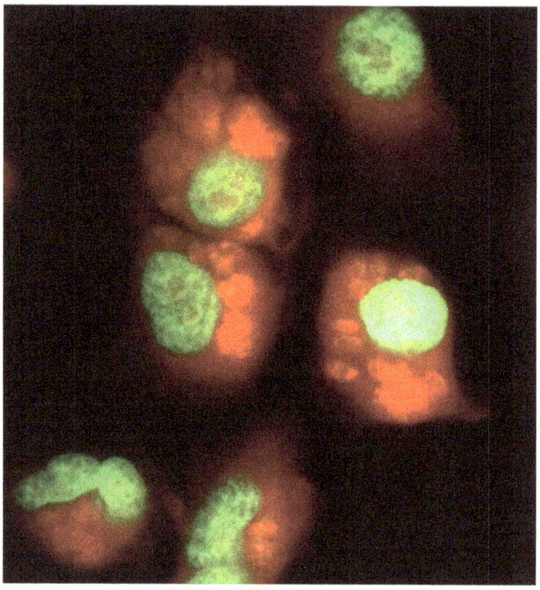

apicoplast a non-photosynthetic plastid organelle with its own circular genome, unique to apicomplexan parasites. It is surrounded by multiple membranes and probably arose from the endosymbiosis of a red alga.

apolipoprotein a lipid-binding protein found associated with lipoprotein particles in the bloodstream. An example is apolipoprotein A1 (ApoA1), a cofactor of lecithin cholesterolacyltransferase, which recruits free cholesterol for esterification to allow its addition to the hydrophobic lipid core of the lipoprotein particle.

apolysis the process in which terminal, gravid proglottids are detached and shed by certain tapeworms.

apomorphy a derived characteristic of a taxon; for example, any feature novel to a taxon.

apoptosis a form of programmed cell death, characterized by endonuclease digestion of DNA.

apoptosome a ring-like platform formed through the assembly of multiple monomers of APAF-1 in humans, CED-4 in *Caenorhabditis elegans*, or ARK in *Drosophila*. This platform enables activation of caspases. Differences in platform symmetry and the requirement for cytochrome *c* in apoptosome assembly occur in mammals, flies, and worms.

apoptotic protease-activating factor 1 (APAF-1) a large cytoplasmic adaptor protein that assembles into the apoptosome to enable caspase activation.

aptamers DNA, RNA or peptide molecules that bind to specific ligands. Aptamers that are composed of RNA are called RNA aptamers.

aptazymes complex RNA devices that are composed of a ligand-sensing domain and a ribozyme domain that controls gene expression.

aquaculture the farming of aquatic organisms such as fish, mollusks and aquatic plants.

Arachnida (Chelicerata) includes terrestrial forms, such as mites, ticks, spiders, and scorpions. Unlike insects, arachnids have no wings or antennae, and adults have four pairs of legs, compared with three pairs for insects. Arachnid adults have eight legs and two body parts (head and abdomen). The arachnid life cycle has four stages, namely egg, larva, nymph, and adult. The larval arachnid (mites and ticks) is unusual because it is only six-legged. Of medical importance are those serving as vectors for microbial diseases (mites and ticks) or as venomous animals that bite (spiders) or sting (scorpions).

arbovirus a group of viruses that are transmitted by arthropod vectors, such as a mosquito, tick, sand fly, etc.

Archaea single-celled prokaryotic microbes that are structurally and metabolically similar to bacteria but share some features of their molecular biology with eukaryotes. The Archaea are a distinct branch of life from the Bacteria and Eukarya.

archigregarines gregarine parasites of the intestines of marine invertebrates exhibiting three types of multiplication (merogony, gametogony and sporogony); considered to be among the earliest diverging apicomplexan parasites.

area under the receiver–operating curve (AUC) the receiver–operating curve for a predictor (for example, a genetic test) plots the proportion of cases correctly identified by the test against the proportion of controls that are incorrectly classified as cases. The AUC indicates the probability that a factor (for example, a genetic risk score) will predict a higher risk of disease in a randomly selected case than in a control.

Arf-like protein 1 (ARL1) Golgi protein involved in trafficking and expressed in bloodstream form *T. brucei*; also encoded in *L. major* and *T. cruzi* genomes.

argonaute protein family includes the Ago and Piwi clades. *C. elegans* contain specific group 3 argonaute proteins, termed WAGO proteins, as well as the Ago and Piwi clade. Parasitic flatworms evolved a group 4 Ago clade.

***Argulus* (fish louse)** a crustacean belonging to the subclass Branchiura and is found worldwide in both freshwater and marine environments. *Argulus* is easily identified by its oval appearance due to the 'U-shaped' carapace of the head with a pair of compound eyes, two circular suckers and a proboscis used for attachment and feeding, respectively. The thorax has four carapace-fused segments with a pair of swimming biramous appendages attached to each segment. A small bi-lobed abdomen containing the testes in males is present caudally.

arithmetic mean the sum of a collection of numbers divided by the number in the collection, often simply called the mean or average.

ARM/HEAT repeats repeat motifs initially described from the *Drosophila* Armadillo protein and four cytoplasmic proteins in mammals and yeast, respectively, which often act in protein–protein interactions.

aromatic diamidines imidocarb dipropinate has activity against *Babesia*, *Hepatozoon*, *Cytauxzoon* sp., and *Ehrlichia* sp. The drug interferes with DNA formation and parasite proliferation. Imidocarb has a low therapeutic index. Parenteral administration causes vomiting. Other side effects are pain on injection and signs of cholinergic stimulation (hypersalivation, tachycardia, vomiting, restlessness, and diarrhoea). These latter signs may be due to inherent anticholinesterase activity and can be controlled with atropine. Imidocarb should not be used concurrently with other cholinesterase inhibitors. Imidocarb is also associated with dose-dependent increases in liver enzyme activity and hepatic damage. Diminazene diaceturate is available for parenteral administration for treatment of trypanosomiasis, babesiosis, and cytauzoonosis.

ARP2/3 actin nucleation complex actin-related protein 2 (ARP2) and ARP3 form part of a complex that nucleates the formation of new actin filaments. This complex preferentially initiates new filaments from the side of existing filaments, leading to a branched actin network. ARP2/3 function is particularly associated with the formation of planar actin-driven protrusions known as lamellipodia.

arrayed library individual primary recombinant clones (hosted in phage, cosmid, YAC, or other vector) that are placed in two-dimensional arrays in microtitre dishes. Each primary clone can be identified by the identity of the plate and the clone location (row and column) on that plate. Arrayed libraries of clones can be used for many applications, including screening for a specific gene or genomic region of interest. *See* **genomic library**, **gene chip technology**, **library**.

arrested development *See* **hypobiosis**.

artemether (ATM) a methyl ether derivative of artemisinin and currently a preferred treatment for malaria, particularly in combination with lumefantrine.

artemisinin combination therapy (ACT) the combination of artemisinin or one of its derivatives (e.g. artesunate, arthemeter, etc.) with an antimalarial or antimalarials of a different class (e.g. mefloquine, lumefantrine, amodiaquine, piperaquine, and pyronaridine).

artemisinin monotherapy an antimalarial medicine that has a single active compound, where this active compound is artemisinin or one of its derivatives.

arterial system large system of blood vessels that carry oxygenated blood from the heart to the organs.

arthropod an invertebrate animal of the large phylum Arthropoda; for example, insects, ticks and mites, spiders, and crustaceans. Phylum Arthropoda is the largest group of organisms in the kingdom Animalia. Arthropods are complex, multicellular organisms that may be involved directly in causing an invasive or a superficial (infestation) disease process or indirectly as intermediate hosts and vectors of many infectious agents, including protozoan and metazoan parasites.

artificial chromosome (AC) heterologous DNA molecule capable of replicating in a given host, bacterium, yeast, or mammalian cell, that contains all functional elements required for auto-replicating and segregating normally upon cell division.

asaccharolytic a microorganism unable to metabolize carbohydrates and therefore must use other carbon sources (e.g. peptides) for energy.

ascarid infection nematode worms from the ascarid family: *Ascaris suum* in pigs, *Ascaris lumbricoides* in humans, *Parascaris equorum* in horses and *Toxocara vitulorum* in buffalo and cattle. In *A. suum*, adult male and females live in the small intestine; an unembryonated egg is passed in the faeces; a larva develops within the egg to the infective L2 stage; pig ingests embryonated egg; the larva is digested free of the egg; the larva penetrates the intestinal wall and migrates through the liver, is carried in blood to lungs, breaks into alveoli, ascends respiratory tree is swallowed; develops to an adult in the host's intestine. Transmission is by ingestion of highly resistant and long-lived larvated eggs. *P. equorum* is an unusual cause of coughing in foals, weanlings and yearlings, and is associated with the migration of larvae through the lungs, resulting in mild eosinophilic pneumonia. Diagnosis is based on findings ascarid eggs in faecal samples using faecal flotation to demonstrate characteristic ascarid-type eggs with thick mamillated wall.

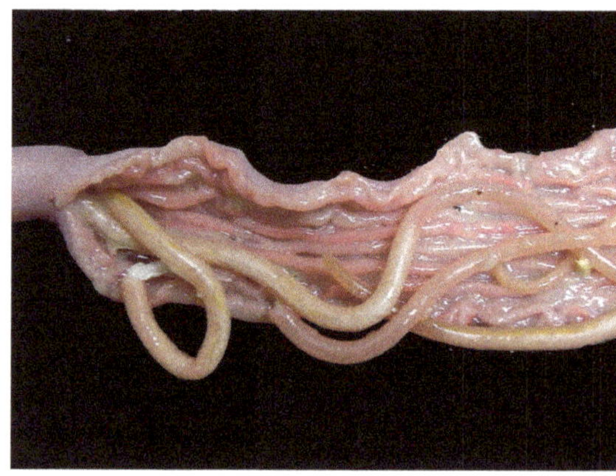

ascarosides water-soluble, small-molecule pheromones that mediate developmental diapause, aggregation, and mating attraction in nematodes. These pheromones are derivatives of the dideoxy-L-sugar ascarylose, which are linked to fatty acid-derived side chains of varying lengths and are often decorated further with building blocks derived from amino acids, folate, and other primary metabolites.

ascertainment bias deviation in the result of a test due to a bias in the sampling process (i.e. due to non-random sampling).

asexual blood-stage trophozoite an asexual form of the intra-erythrocytic *Plasmodium* parasite that is undergoing cell growth and nuclear division, in preparation for parasite differentiation into a mature schizont that produces individual progeny (known as merozoites). These merozoites burst from the infected cell, ready to initiate new rounds of intracellular development.

asexual parasite prevalence the proportion of a population that harbours asexual blood-stage parasites, most commonly determined by microscopy.

asexual reproduction reproduction without the formation and fusion of gametes, usually male and female. Formation of new individuals by any non-sexual means, e.g. binary fission, budding, schizogony, etc. Reproduction that does not involve gametes or fertilization. There is a single parent, and all offspring are genetically identical. This commonly occurs in prokaryotes, fungi, protozoa and some plants.

aspartate-glutamate carrier transporter that exchanges aspartate for glutamate located at the mitochondrial outer membrane.

asRNAs (antisense ribonucleic acids) antisense RNAs are non-coding RNAs that have the potential to bp with complementary mRNA molecules. Notably, this group of RNAs may include sRNAs and many RNAs that are produced by pervasive transcription.

assemblage a monophyletic group of, for example, *Giardia intestinalis* isolates. There are currently three assemblages with genome sequences: the humans isolates WB (assemblage A) and GS (assemblage B), and the pig isolate P15 (assemblage E).

assembly putting sequenced fragments of DNA into their correct chromosomal positions.

assignment (test) a method of assigning individuals to the populations from which they were most likely to have originated (regardless of where they dispersed to or were sampled).

association statistical dependence between two or more events, characteristics, or other variables. An association may be fortuitous or may be produced by various other circumstances; the presence of an association does not necessarily imply a causal relationship.

association or linkage disequilibrium (LD) mapping the basic principle behind LD mapping is that alleles at loci nearest a particular disease-influencing locus will show strong statistical associations with the disease, whereas alleles at distant loci will not. The intuition behind this principle is that specific alleles at loci that immediately neighbour the disease locus found on the (ancestral) chromosome upon which the disease mutation arose will tend to remain on that chromosome as it is transmitted from generation to generation. However, recombination can assort alleles at more distant loci. Thus, older ancestral mutations will have greater opportunity to be disassociated by recombination than will more recent mutations.

assortative mating non-random choice of a mate, either directly through the evaluation of some attribute, or indirectly through habitat choice (for taxa that mate within their habitat).

astrocytes are the most abundant cells in the human central nervous system, the only organ where they are found. They are glial cells and are also known as astroglia. These represent a major component of the blood–brain barrier and have foot processes that are in direct contact with the endothelial cells.

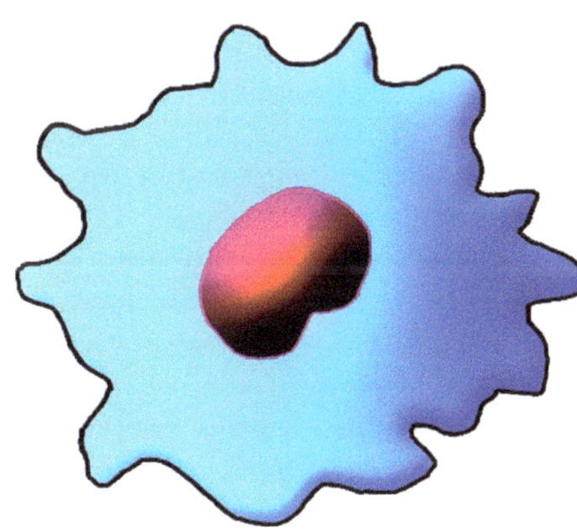

astrogliosis abnormal proliferation of astrocytes leading to the formation of a glial scar.

asymptomatic infection infection that does not display any clinical symptoms, but may still be capable of transmitting disease.

asymptomatic malaria infections malaria infections that lack typical clinical symptoms but are detectable by microscopy, rapid diagnostic test or molecular methods.

asymptomatic microfilaremics individuals who do not complain of any symptoms can be attributed to filarial infection. However, their blood contains circulating microfilariae, often in high numbers. Although they have no overt signs or symptoms, sensitive imaging techniques (e.g. ultrasound) reveal that they do indeed manifest some pathology that can be attributed to ongoing filarial infection.

atmospheric pressure chemical ionization (APCI) technique used in mass spectrometry to produce ions at atmospheric pressure. APCI is adapted to the ionization of non-polar (e.g. triacylglycerol) to very polar molecules, with a molecular weight ranging from around 50 to 1000 kDa.

atomic force microscopy (AFM) a high-resolution type of scanning probe microscopy that can be used to assess the surface topography of biological molecules. This technique that uses a mechanical probe to characterize and magnify surface features with atomic detail.

atopic allergy IgE-mediated hypersensitivity, i.e. asthma, eczema, hay fever and food allergy.

atopy a term used by allergists to describe IgE-mediated anaphylactic responses in humans, usually genetically determined.

ATP (adenosine triphosphate) a multifunctional nucleotide responsible for cellular energy transfer and storage.

attenuated rendered less virulent.

atypical cutaneous leishmaniasis non-ulcerated benign skin lesion caused by *Leishmania chagasi*.

autoantibody antibody to self antigens.

autocatalytic set collection of molecules that mutually cooperate in the sense that none of them can replicate without all the others, such that the reactions that form the components of the set are catalysed by other components of the set.

autochthonous originating in the same area in which it is found.

autocorrelation correlation of an observation with itself.

autocrine a term used to describe the activation of cellular receptors on the same cell that produces the ligand.

autocrine signalling a form of cell signalling in which molecules secreted by the cell are recognized by the cell's own receptors to bring about changes in differentiation or behaviour.

autocrine versus paracrine local mediators can induce their biochemical effects on the producing cell (autocrine) or on surrounding cells (paracrine); classical hormones, which are transported via the bloodstream to exert their respective effects on certain organs, are considered endocrine.

autogamy fertilization between gametes derived from the same individual.

autograft a tissue graft from one area to another on a single individual.

autoimmune disease a disease caused by an immune reaction against an individual's own tissues.

autoimmunity (autoallergy) immune reactions of a host against its own self components.

autoinfection a process in which the progeny of a parasite reinfect the host without passing out of it, for example, *Taenia solium*. It is a reinfection by a parasite derived from within the host and which is not exposed to the outside environment.

autologous from the same individual.

autophagic body the inner-membrane-bound structure released into the vacuole lumen after the outer membrane of the autophagosome fuses with vacuolar membrane. The autophagic bodies only appear in yeast and plants due to the large size of the vacuole, and are not found in mammalian lysosomes.

autophagic cell death one main type of programmed cell death that leads to a total destruction of the cell, also termed type II cell death.

autophagosome double membrane surrounded organelles that form in the endoplasmic reticulum (ER) in response to cellular stress (e.g. nutrient depletion). Their main role is to target substrates to the lysosomes for degradation.

autophagy a process by which cells digest parts of their own cytosolic material. This allows the recycling of molecules under conditions of nutritional limitation and remodelling of intracellular structure for cell differentiation. It includes sequestering of long-lived proteins and organelles in a double-membraned vesicle (the autophagosome), which is delivered to the lysosome for protein degradation. Important regulators of autophagy are the target of rapamycin (TOR) kinase and autophagy-specific gene 8 (Atg8, also called microtubule-associated protein 1 light chain 3, LC-3).

autoradiography a technique that uses X-ray film to visualize radioactively labelled molecules or fragments of molecules; used in analysing length and number of DNA fragments after they are separated by gel electrophoresis.

autosomal dominance is a pattern of inheritance characteristics of some genetic diseases. 'Autosomal' means that the gene in question is located on one of the numbered, or non-sex, chromosomes. 'Dominant' means that a single copy of the disease-associated mutation is enough to cause the disease. This is in contrast to a recessive disorder, where two copies of the mutation are needed to cause the disease. Huntington's disease is a common example of an autosomal dominant genetic disorder.

autosomal dominant autosomal' refers to a non-sex chromosome. Autosomal dominance is when one particular form of a gene (one allele) dominates over other alleles, and is always expressed when present in an individual, whether they are homozygous or heterozygous for that allele.

autosome a non-sex chromosome not involved in sex determination. Humans have 22 pairs of autosomes and one pair of sex chromosomes (normally XX for female XY for male). Autosomes are numbered roughly in relation to their sizes. That is, Chromosome 1 has approximately 2800 genes, while chromosome 22 has approximately 750 genes.

autotroph an organism that biochemically synthesizes its own organic materials from inorganic compounds using light or chemical energy.

average the mean value, which is the total amount divided by the number of contributors.

avermectins/milbemycins produce GABA-mimetic effects by binding to glutamate-gated chloride channels. The main nerve ganglia affected are the pharyngeal ganglia in nematodes, which leads to cessation of feeding and death. Avermectins/milbemycins have a sublethal effect on female nematodes and arthropods, with cessation of egg laying. After spot-on application, avermectins/milbemycins are absorbed transcutaneously, circulate in plasma, and are stored in fat tissues. Elimination occurs gradually, mainly in faeces.

avidin a glycoprotein found in egg white who has a very high affinity and binds to biotin (a water-soluble vitamin).

avidin–biotin peroxidase complex technique a method useful for the localization of peptide hormones or other antigens in formalin-fixed tissues.

avidity (functional affinity) the binding strength between two molecules [e.g. antibody (Ab) and antigen (Ag)] taking into account the valency of the interaction. Thus the avidity will always be equal to or greater than the intrinsic affinity. *See* **affinity**.

avirulence describes the inability of a pathogen to reproduce or cause a disease in a host.

avirulence gene (AVRs; Avr gene) genes that encode a pathogen effector that is recognized by a host resistance nucleotide-binding domain and leucine-rich repeat-containing protein (NLR).

axoneme the intracellular portion of a flagellum.

axostyle a longitudinal rod-like or tube-like supporting structure running the length of a flagella in some flagellates that gives rigidity to the body of members of the flagellated protozoan in the order Trichomonadida (*Trichomonas* spp.); probably serves as a part of the cytoskeleton.

A **B** C D E F G H I J K L
M N O P Q R S T U V W
X Y Z

B cells (also called B lymphocytes) play a major role in the humoral (antibody)-mediated immune responses as an activated B cell (plasma cell) responsible for production of antibodies. Also, it can act as antigen-presenting cells.

B lymphocyte-induced maturation protein 1 (BLIMP-1) is required for the differentiation of CD8 cytotoxic effector T cells, but suppresses primary memory CD8 T cell development. BLIMP-1 also acts as a transcriptional repressor of T cell proliferation and as a transcription factor responsible for plasma B cell differentiation.

B-1/B-2 cells the two major subpopulations of B lymphocytes. B-1 cells bear high levels of surface IgM, lower levels of surface IgD, are $CD43^+$, $CD23^+$ and most express the cell surface antigen CD5; they are self-renewing, and frequently secrete high levels of antibody which binds to a range of antigens ('polyspecificity') with a relatively low affinity. The majority of B cells, however, are B-2 which express low levels of surface IgM, higher levels of surface IgD, do not express CD5, and are $CD43^+$, $CD23^+$; they are directly generated from precursors in the bone marrow, and secrete highly specific antibody.

Babesia bovis a tick-borne apicomplexan intraerythrocytic parasite that causes bovine babesiosis, a severe disease that has considerable economic impact on cattle health and production worldwide, especially in tropical and subtropical regions. Animals that have recovered from clinical babesiosis remain persistently infected, with no clinical signs, and become carriers. The major vectors of *B. bovis* are *Rhipicephalus* (*Boophilus*) spp. ticks, *R.* (*Boophilus*) *microplus* and *R.* (*Boophilu*) *annulatus*. The ticks can acquire *B. bovis* from infected cattle during acute and persistent infection and transmit the infectious pathogen into other hosts.

babesiosis a tick-borne disease caused by protozoa of the genus *Babesia* that infect red blood cells and are spread by certain ticks. Babesiosis is preventable, if simple steps are taken to reduce exposure to ticks. *Babesia microti* is transmitted by the bite of infected *Ixodes scapularis* ticks – typically, by the nymph stage of the tick, which is about the size of a poppy seed.

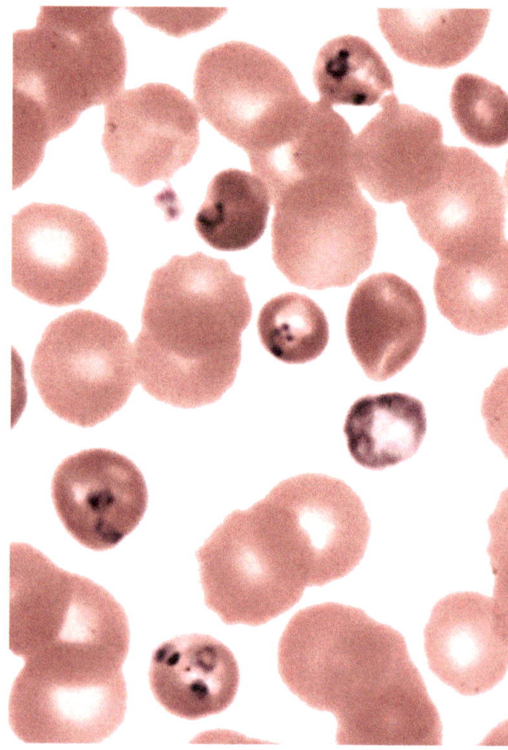

backcross a cross between a eukaryote that is heterozygous for alleles obtained from two parental strains and a second eukaryote from one of those parental strains. Also used to describe the breeding protocol of an outcross followed by a backcross.

bacteria a large group of single-celled, prokaryotic organisms that do not have organelles enclosed in membranes. Most of their DNA is located in one circular chromosome and the remainder in small circular plasmids. They have a cell wall composed of protein and complex carbohydrate (polysaccharide) over a plasma membrane.

bacterial and archaeal species a monophyletic group of organisms with a high degree of coherence in their genetic and phenotypic traits, which differentiate it from its close relatives.

bacterial and archaeal tree of life the phylogenetic tree of all known bacteria and archaea based on the 16S ribosomal RNA gene.

bacterial artificial chromosome (BAC) a circular episomal DNA molecule present in bacteria, which may harbour up to 300 kb of DNA. A BAC also contains functional elements for self-replication and segregation in bacteria. *See* **cloning vector**.

bacteriocin antimicrobial proteins that are produced by one type of bacterium and toxic to other bacteria.

bacteriophage a virus that infects bacteria. It can be lysogenic, remaining stable in the genome of the host bacterium until triggered (e.g. by stress) to replicate or lytic, in which case they immediately copy themselves within the host and kill the host bacterium. *See* **phage**.

bacteriophage recombineering molecular biology methodologies that involve DNA engineering through homologous recombination, mediated by bacteriophage proteins, rather than traditional systems reliant on cutting and religating strands with restriction enzymes and DNA ligase. These methods allow very high efficiency in assembling and modifying DNA constructs.

baculovirus a type of virus that specifically infects insect cells.

baculovirus inhibitor of apoptosis (IAP) repeat (BIR) a zinc-binding domain of approximately 70 amino acids that was first identified through sequence homology to IAP.

Baermann test the copro-microscopic technique used for the isolation of hydrophilic nematode larvae from different biological samples (e.g. faeces and lung tissues). It is the gold standard test for detecting metastrongyloid, lungworm, larvae in the faeces of an infected animal.

bait the hybridization probe designed to capture effectively the coordinates of the target region to be sequenced. The bait design differs by manufacturer and method. Some methods use baits that tile the target region, whereas others use baits that do not overlap and differ in distance between the baits.

Balamuthia mandrillaris is a free-living amoeba (a single-celled living organism) naturally found in the environment. *Balamuthia* can cause a very rare but serious infection of the brain and spinal cord called granulomatous amoebic encephalitis (GAE).

balancing selection a form of natural selection that maintains genetic polymorphisms (or multiple alleles) within a population. Unlike positive selection, balancing selection increases levels of nucleotide variation at linked sites. A variety of mechanisms can result in elevated levels of polymorphism including over-dominance (heterozygote advantage), frequency-dependent selection, and temporally or spatially variable selection.

balantidiasis intestinal parasitic infection caused by the protozoan *Balantidium coli*. This parasite can be transmitted to humans through the faecal–oral route by contaminated food and water. *B. coli* infection is mostly asymptomatic, but people with other serious illnesses can experience persistent diarrhoea, abdominal pain, and sometimes a perforated colon. When travelling to endemic tropical countries, *B. coli* infection can be prevented by following good hygiene practices. Wash all fruits and vegetables with clean water when preparing or eating them, even if they have a removable skin. Often associated with pigs and guinea pigs.

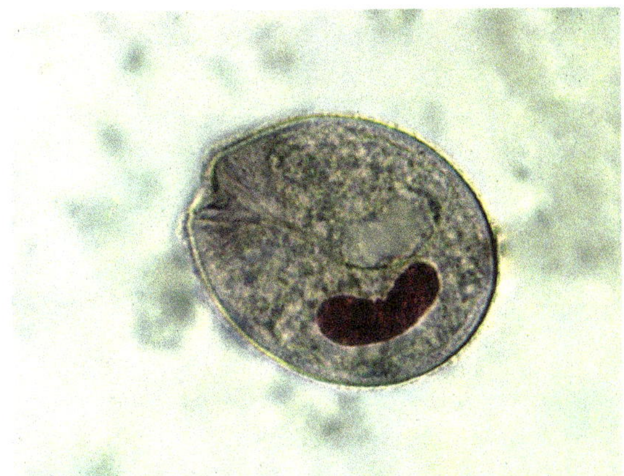

BAM diagram a Venn diagram in which the interactions among the Biotic niche, the Abiotic niche, and the Mobility are expressed. This diagram has profound implications for distributional ecology, but it summarizes the zones where populations of an organism can have positive turnover, where the relationships with organisms allow the survival of the focus species, and that area reachable from areas where permanent populations exist. Vectors, reservoir hosts, pathogens, and humans are a transmission system that responds to the BAM diagram; the disease is transmitted only where the BAM conditions of all component species are fulfilled.

Bancroftian filariasis lymphatic filariasis caused by *Wuchereria bancrofti*. A major cause of human elephantiasis.

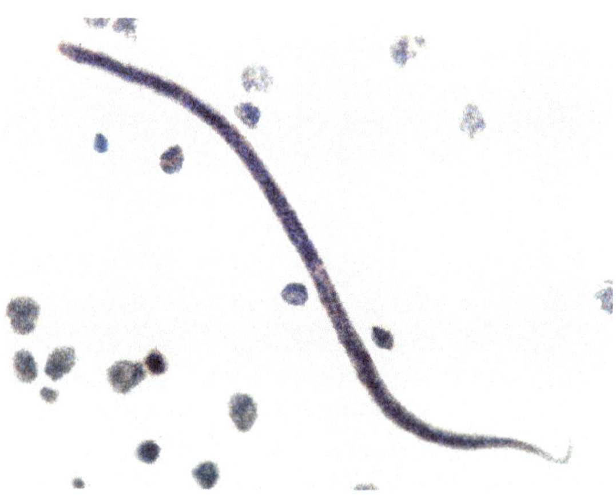

BAR proteins (Bin1/Amphiphysin/Rvs167 proteins) a family of proteins characterized by a long and slightly curved shape that can interact with membranes.

barcode short nucleotide sequences (usually three to ten nucleotides) that are added to the primers (the forward and/or reverse primer) prior to sequencing, either through a ligation step or by directly incorporating the nucleotides to the 5′ extremity of the primer. Each barcode used in the same sequencing run typically differ from each other by two or more nucleotides. Barcodes are used in the multiplexing methodology to pool and sequence several samples in a single sequencing run.

barcode index number (BIN) an operational taxonomic unit assigned to a group of similar DNA barcode sequences through the sequence clustering method built into BOLD and used as a proxy for species-level identification.

barcoded primers unique DNA sequences that are incorporated into adaptor sequences for tagging of different samples before sample pooling and shotgun sequencing.

barcoding taxonomic method that uses DNA sequence of a particular gene to identify an organism to a particular species. The goal is not classification but simply to identify an unknown sample in terms of a known classification.

basal clade a relative term in phylogenetic analyses referring to the earliest clade to branch in a larger clade.

basal granule in ciliates, the granule-like body from which each cilium arises. Comparable to the blepharoplast from which a flagellum arises.

basal lamina (BL) layer of extracellular matrix secreted by epithelial cells on the basal side of the epithelium, serving as support for the epithelial cells. Two major components of the BL are the proteins laminin and collagen.

basal radiation the most ancient divergence event in a phylogeny.

base one of four types of simple molecules or nucleotides (adenine, thymine, cytosine and guanine) that are the subunits (building blocks) of DNA and RNA. *See* **base sequence**, **nucleotide**.

base J or beta-D-glucopyranosyloxymethyluracil a new base identified first in 1993 in the nuclear DNA of *Trypanosoma brucei*. It is the first hypermodified base found in eukaryotic DNA. It is present in all kinetoplastid flagellates analysed and some unicellular flagellates closely related to trypanosomatids, but it has not been found in other protozoa or in metazoa. Base J is formed by hydroxylation of thymidine and subsequent glycosylation by glycosyltransferase enzyme.

base quality log (base 10; $\log_{10}$) transformed measure of the probability that a given base call is incorrect multiplied by 10.

base sequence the order of the chemical units (nucleotide bases) adenine, thymine, cytosine and guanine in a DNA molecule bases that forms the genetic code. The sequence of the bases will determine what protein is produced.

base sequence analysis a method used for determining the base sequence.

basic leucine zipper transcription factor (BATF) a transcription factor that directly regulates differentiation and function of IL-17 and IL-21-producing T cells (T helper 17 and T follicular helper cells) as well as antibody class switch in B cells.

Basic Local Alignment Search Tool (BLAST) a fast algorithm that compares a DNA or protein sequence to a library or database of sequences and identifies those which resemble the query sequence above a specified threshold.

basic reproduction/reproductive number (R0) the standard measure of parasite transmissibility. R0 refers to the expected number of infections resulting from the introduction of a single infected individual into a susceptible population. An increase in R0 corresponds to an increase in disease prevalence.

basigin (Ok blood group antigen, CD147, EMMPRIN, M6) an erythrocyte receptor that interacts with the rhoptry adhesin Rh5.

basophil a granular leucocyte., whose granules contain heparin, histamine and other vasoactive amines. Within tissues, these cells are known as mast cells.

Bateson–Dobzhansky–Muller incompatibility hybrid incompatibilities observed when crossing two close species or divergent strains of a species with one another. The incompatibility is caused by the cosegregation of non-parental allele combinations and results in a dysfunctional genetic interaction (negative epistasis).

Batf-3 basic leucine zipper transcription factor, ATF-Like 3 (Batf-3) is a transcription factor protein encoded by the *batf3* gene that is required for the formation of $CD8^+$ and dermal $CD103^+$ DC subsets.

Batty approach a method for testing efficacy of antimalarial drugs in mouse models of malaria, developed by Kevin Batty and colleagues, wherein drug is administered after the number of malarial parasites in the mouse has climbed to a high level.

Bauplan (building plan) the shared characteristics of members of a systematic group (taxon). The concept is frequently used in phylogenetics and evolution.

Bauplan constraint the evolutionary constraints imposed by the shared characteristics of a particular taxon.

Bax/Bak proapoptotic proteins that can permeabilize the outer mitochondrial membrane following their activation and oligomerization. The presence of both Bax and Bak is essential for apoptosis signalling to occur in mammals.

Bayes' theorem a probability theory by Thomas Bayes to calculate conditional probabilities based on prior distributions of parameters in a model and the observed experimental data.

Bayesian based on Bayes' theorem, a statistical inference in which evidence or observations are used to update or newly infer the probability that a hypothesis might be true.

Bayesian analysis a method of calculating the statistical probability that the probability of one treatment is superior based on the observed data and prior beliefs.

Bayesian clustering analysis a method to infer population structures without previous knowledge about the population. Various Bayesian approaches using neutral molecular markers have been proposed, which include using Markov chain Monte Carlo (MCMC) methods to infer population structures. A commonly used programme, STRUCTURE, can infer the assignment of individuals to populations or the admixture proportions of individuals for a given number of populations (K).

Bayesian phylogenetic method a phylogenetic reconstruction method based on Bayesian statistics. This method estimates the phylogeny or set of phylogenies that is more likely to be explained by the data (aligned sequences).

Baylisascaris an infection caused by a roundworm found in raccoons. This roundworm can infect people as well as other animals, including dogs. Human infections are rare, but can be severe if the parasites invade the eye (ocular larva migrans), organs (visceral larva migrans) or the brain (neural larva migrans).

B-cell lymphoma-2 (Bcl-2) the first-identified pro-survival protein of the intrinsic cell-death pathway.

BCG (bacille Calmette–Guérin) an attenuated strain of *Mycobacterium tuberculosis* var. bovis.

Bcl-2 (B cell lymphoma 2) family a family of apoptosis regulator proteins.

Bcl-2 homology (BH) domains regions of sequence homology that define all Bcl-2 proteins. There are four BH domains, known as BH1, BH2, BH3 and BH4, which are each characterized by unique sequence motifs.

Beavis effect systematic overestimation of true quantitative trait locus effects.

bed bugs are small blood-sucking insects that live in cracks and crevices in and around beds. There are two primary species of bed bugs that infest humans, *Cimex lectularius* (distributed worldwide) and *Cimex hemipterus* (distributed in the tropics and subtropics). They crawl out at night and bite exposed skin to feed on blood. Inflammatory responses can be severe, resulting from allergic reactions to the bugs' salivary proteins. The best way to prevent bed bugs is regular inspection for signs of an infestation.

behaviour the branch of science that deals with the way a species reacts or responds to a variety of stimuli.

behavioural epigenetics an emerging multidisciplinary field of research that aims to understand how epigenetic processes transform transient environmental cues into persistent molecular patterns of gene expression in order to modulate animal behaviour.

Bence-Jones proteins (BJ proteins) immunoglobulin light chain dimers, normally produced by plasma cells. BJ proteins are sufficiently small to be excreted by the kidney. It is a characteristic protein found in the urine of most patients with multiple myeloma.

benign tertian malaria malaria caused by *Plasmodium vivax*.

benthic and limnetic morphs morphologically distinguishable, reproductively isolated forms of the same species that are morphologically and behaviourally adapted for life in the benthic zone or the limnetic zones of water bodies.

benthos organisms that live in or on the bottom of a body of water.

BH3-mimetic a drug that mimics the activity of a BH3-only protein by antagonizing pro-survival proteins to trigger apoptosis. Currently BH3-mimetics are in clinical trials for the treatment of cancer.

BH3-only protein a proapoptotic protein that is related to other Bcl-2 proteins only by the presence of the BH3 domain, and which initiates the apoptotic cascade.

bias or systematic error deviation of results or inferences from the truth, or processes leading to such deviation. *See* **referral bias**, **selection bias**.

bicistronic mRNA mRNA containing two independent coding sequences separated by an internal ribosome entry site allowing ribosomes to translate the second cistron or by a peptide sequence specifically cleaved by an endogenous protease.

bidirectional best hits two genes from different organisms that are each other's closest matches in reciprocal sequence similarity searches and thus might be supposed to be functional orthologues.

bifurcated divided into two branches.

bimodal age distribution a frequency distribution, in this case of infection, that shows peaks at two different ages.

binary fission reproduction by division of an individual into two individuals each a duplicate of the other. Division by which two equal and similar individuals are formed, e.g. amoeba (*Entamoeba histolytica*). Division may be transverse, e.g.

ciliates (*Balantidium coli*) or longitudinal, e.g. flagellates (*Trypanosoma brucei gambiense*). It usually occurs in the free-living state but may occur in the encysted state where several divisions may occur within a single cyst, e.g. *E. histolytica* produces four daughter amoebae in the cyst stage. Division may be either amitotic or mitotic.

binary phenotypes disease traits that have two major states on the observed scale: diseased or healthy. They may nonetheless be complex traits in which transition to the disease state is influenced by continuous variation on an underlying liability scale for disease that is controlled by many genetic loci and environmental effects.

binning a computational technique frequently used to reduce noise by grouping sequencing reads mapped to contiguous genomic segments.

binocular pertaining to both eyes.

binocular microscope a microscope with two eyepieces, permitting use of both eyes simultaneously.

binomial classification a system of nomenclature describes each living organism by two names: genus and species. These names are either underlined or italicized and the genus is capitalized while the species is lower case.

binuclear (binucleate) a cell having two nuclei.

bioactive plants or nutraceuticals plants containing secondary metabolites that are considered for their beneficial effects on health rather than for their direct nutritional value for animals.

bioactives secondary metabolites with any kind of biological activity.

biochar biomass-derived black carbon.

biocide any chemical agent that can kill a living organism. For instance, pesticides kill insects.

biocontainment a process aimed at keeping biological organisms within a limited space or area. For example, if an outbreak of foot and mouth disease is found on one farm, a biocontainment process would aim at stopping the disease from spreading to other farms.

biocrystallization process of forming crystals from organic macromolecules carried out by living organisms.

bioCyc collection of pathway/genome databases containing information for individual organisms. Enzyme annotations (generated by PathwayTools software) are sometimes manually refined by curators of the database.

bioethics the study of the ethical and moral implications of applications of biomedical research and biotechnology.

biofilm a layered aggregate of microorganisms. These adherent cells are frequently embedded within a self-produced extracellular matrix that is generally composed of DNA, proteins and polysaccharides.

biogeochemical model a type of ecosystem model used to represent biologically mediated transformations and flows of carbon and other materials within an environment.

biogeochemistry study of how interactions among biological and geochemical processes influence the global cycling of such essential elements as carbon, nitrogen, phosphorus, and sulfur.

biogeographical model a type of ecosystem model used to determine how populations in a particular region change over long time scales.

bioindicator quantifiable response in biochemical, physiological, or ecological parameters that has been related to effects at organism, population, community, or ecosystem levels of biological organization.

bioinformatics a subdiscipline of biology and computer science concerned with the acquisition, storage, analysis, and dissemination of biological data, most often DNA and amino acid sequences. Bioinformatics uses computer programs for a variety of applications, including determining gene and protein functions, establishing evolutionary relationships, and predicting the three-dimensional shapes of proteins. *See* **informatics**.

biolistics the use of small particles, e.g. colloidal gold, as a vehicle for carrying agents (e.g. drugs, nucleic acid) into a cell. Following coating with the desired agent(s), the particles are fired into the dermis of the recipient using a helium-powered gun.

biological chassis an autonomous genetic and/or biochemical scaffold that functions as a dynamic platform for implanting forward-designed biological devices.

biological control the control of a population of one organism by another organism. Generally the controlling organism is a predator or disease-causing organism of the species being controlled.

biological fitness the average reproductive success of a genotype in a particular environment relative to other genotypes.

biological ontology biological ontologies describe concepts and relationships that can exist for a biological entity in a computer-readable fashion.

biological pump collection of biological ocean processes that regulate the uptake, storage, transformation, and release of carbon.

biological vector a living obligate host in which the parasite undergoes morphological change and/or multiplication. There are three classes of biological vectors: (1) *propagative* in which the parasite multiplies but does not undergo morphological change, (2) *cyclopropagative* in which the parasite undergoes cyclic, morphological changes and also multiplies,(3) *cyclodevelopmental* in which the parasite undergoes cyclic morphological change, but does not multiply.

bioluminescence light generation by living organisms.

biomarker a biological marker (DNA, RNA, protein, protein modification or metabolite) that reflects a biological state. It is a characteristic that is objectively measured and evaluated as an indicator of normal biological or pathogenic processes or pharmacological responses to a therapeutic intervention.

biomarker discovery the phase of research in which candidate biomarkers are identified, often with the help of '-omics' technologies, such as genomics, proteomics or metabolomics, or genetics. These approaches allow a non-targeted discovery of biomarkers correlated to certain biological processes or states.

biomarker qualification accumulating evidence about the utility and limitations of a biomarker for use in a specific context.

biomarker verification the correlation of the biomarker candidate with biological processes or states using a more targeted technology [e.g. reverse transcriptase (RT)-PCR assays or protein assays].

biomass the total quantity or weight of organisms in a given area or volume. The measurement of biomass production is important when studying metabolic reactions that are required for growth.

biomass (cellulosic) plant stalks, trunks, stems, and leaves.

biometric (estimation of gene number) an approach to estimating the minimum number of genes underlying a trait difference that relies only on phenotypic data.

biomimetic membrane *in vitro* reconstitution of a biological membrane, based on the self-organization of polar lipids in mono- or bilayers.

Biomphalaria alexandrina one intermediate snail host for *Schistosoma mansoni* parasite.

biopanning a technique used to enrich particles from phage-display libraries. To select ligands with desired binding specificity, a phage-display library is incubated with an immobilized receptor. Unbound particles are removed by washing and bound phage is eluted for amplification and further biopanning.

biopesticides pesticides (chemicals used to kill unwanted organisms such as insects) that are derived from plants, microbes, or other natural materials.

biopolymer a large biological molecule formed by the linking together of smaller subunit molecules. Examples are amino acids to proteins or monosaccharides to polysaccharides, etc.

biopsy removal of tissue from a living body for diagnostic purposes (e.g. lymph node biopsy).

bioremediation the use of biological organisms such as plants or microbes to aid in removing or detoxifying hazardous substances from contaminated areas.

bioscience the study of biology wherein all the applicable sciences, such as physics, chemistry, are applied.

biosecurity security from transmission of infectious diseases, parasites and pests.

biosequestration biologically mediated uptake and conversion of carbon dioxide to inert, long-lived, carbon-containing materials.

biosphere all living organisms.

biosynthesis the production of needed cellular constituents from other (usually simpler) molecules.

biotechnologists scientists who use biological processes to develop novel products.

biotechnology a broad term generally used to describe the use of molecular biological techniques in industrial processes, such as agriculture, brewing and drug development. The term also refers to the production of genetically modified organisms or the manufacture of products from genetically modified organisms.

biotic any living component that affects another organism. Biotic components include plants, animals, bacteria, fungi and parasite.

biotic factor any living component of an ecosystem.

biotic niche the collection of species in an ecosystem that either inhibit (e.g. predators, competitors) or facilitate (mutualists) the focal species. The focal species may be a pathogen, its vector(s), or its reservoir(s). Thus, the biotic niche implies a large cluster of connections that may change over time or be altered by human actions or climate.

biotreatment the treatment of a waste or hazardous substance using organisms such as bacteria, fungi and protozoa.

biotype biological or biochemical type of an organism. Organisms of the same biotype will display identical biological or biochemical characteristics. Certain key markers are used to define and recognize biotypes in tracing the spread of organisms in the environment and in epidemics or outbreaks.

biramous divided into two branches; typical of the terminal segments of the legs of Crustacea.

birefringent optical property; double refraction of light in a molecular ordered material, causing depolarization of polarized light.

bispecific antibody an artificially produced hybrid antibody in which each of the two antigen-binding arms is specific for a different antigenic epitope. Such antibodies, which can be produced either by chemical cross-linkage or by recombinant DNA techniques, can be used to link together two different antigens or cells (e.g. a cytotoxic T-cell and a tumour cell).

bistable system a system for which two alternative stable states exist.

biting midges the dipteran genus *Culicoides* may act as vectors of several important diseases, such as *Loa loa* (eyeworm in humans), arboviruses including African horse sickness virus, bluetongue virus and Schmallenberg virus of veterinary importance.

bivalent a pair of homologous chromosomes that synapse during the first meiotic division.

BK/Maxi-K/SLO-1 channels channels for potassium ions (K^+) activated by depolarization of the membrane potential and increased intracellular Ca^{2+} concentrations. These channels show a large K^+ conductance [100–300 pS, therefore termed big K^+ (BK) channels] and are typically involved in rapid repolarization of the membrane after excitation.

blackwater fever a rare complication of malaria, usually *Plasmodium falciparum* malaria. Massive intravascular haemolysis causes haemoglobinuria, with a blackening of the urine, due to a high methaemoglobin level in the urine.

bladder worm infective stage of *Taenia* tapeworm (*Cysticercus*). The name refers to the fluid-filled bladder which surrounds the larval scolex (*Taenia*) or scolices (*Coenurus*). Bladder worm is also a common name of *Capillaria plica*, found in the dog urinary bladder.

BLAST (basic local alignment search tool) a computer program that, when given any nucleotide or amino acid (protein) sequence, returns similar sequences retrieved from a chosen query database, usually the non-redundant database which houses all submitted DNA or protein sequences without duplication of a given sequence.

blast cells cells prior to division when they have large amounts of cytoplasm.

BLAST E-value a statistic describing the significance of the scored protein sequence similarity between two sequences aligned using the sequence comparison program BLAST. The expectation value (E-value) is the number of different alignments expected to have an equal or better score in the current database of sequences. Smaller E-values correspond to better alignments.

blastocyst after a mammalian ovum is fertilized, it begins to divide. The blastocyst, an early stage of this process, consists of a fluid-filled sphere surrounded by a layer of cells. A mass of cells at one side will become the embryo. The blastocyst is formed before implantation in the uterus.

Blastocystis an intestinal protist that is commonly found worldwide in a diverse range of species including humans, non-human primates, other mammals and birds. It is genetically polymorphic and molecular analysis of the small subunit ribosomal RNA (SSU rRNA) gene has enabled division of this organism into 17 distinct subtypes found in a wide range of species.

blastomere any of the cells in the early embryo produced as the result of cell division in the fertilized egg. A blastocyst is made up of many blastomeres.

blebbing the formation of irregular bulges in the plasma membrane of a cell, usually brought about when the cytoskeleton separates from the plasma membrane. Blebbing is a hallmark of cells undergoing apoptosis.

blepharitis inflammation of eyelids.

blepharoplasty a small granule-like body, usually appearing in the cytoplasm, from which an axoneme arises. Axonemes may form rod-like structures in the cytoplasm, cilia or flagella.

blind assessment the evaluation of an outcome that is made without the evaluator knowing which results are from the test under study and which are from the control or 'gold standard'.

blind(ed) study or masked study a study in which observer(s) and/or subjects are kept ignorant of the group to which the subjects are assigned, as in an experimental study, or of the population from which the subjects come, as in a non-experimental or observational study. Where both observer and subjects are kept ignorant, the study is termed a double-blind study. If the statistical analysis is also done in ignorance of the group to which subjects belong, the study is sometimes described as triple blind. The purpose of 'blinding' is to eliminate sources of bias.

blocking antibodies antibodies that by binding to a target cell, serve to protect it from immune destruction.

blood cycle asexual cycle of the parasite that occurs in the bloodstream.

blood flukes a clade of digenetic trematode flatworms that share a feature of living in the venous or arterial system of their vertebrate hosts. There are three families, Sanguinicolidae in fishes, Spirorchiidae in turtles and crocodiles, and Schistosomatidae in birds and mammals.

blood groups antigens found on the surface of red blood cells. Their expression is inherited.

blood sampling laboratory examination of samples of blood. These are usually collected from an appropriate vein, but arterial samples are required for some special techniques. Collection is done by a closed method, either a hypodermic syringe or a vacuumized container. The site of collection varies with the species and the circumstances. The sample may be clotted, for serum estimations, or collected with an anticoagulant for examinations on plasma or whole blood.

blood–brain barrier (BBB) a highly selective diffusion barrier that separates blood from the CNS. It is formed by tight junctions interconnected endothelial cells that surround all blood vessels in the brain. The BBB endothelial cells are underlined by a basal lamina and associated with pericytes and astrocyte endfeet.

blood–cerebrospinal fluid barrier (BCSFB) a diffusion barrier formed by tight junction interconnected epithelial cells of the choroid plexus localized in the ventricles of the brain.

blood–leptomeningeal barrier (BLMB) a diffusion barrier formed by the unique cellular and molecular characteristics of endothelial cells of the leptomeningeal microvessels at the surface of the brain and spinal cord. The BLMB forms a blood–CSF barrier.

blood-stage vaccines vaccines in this group are 'curative' because they aim to attenuate the severity of disease in infected individuals by preventing the cyclical invasion of red blood cells by merozoites.

bloodstream form a *Trypanosoma brucei* life cycle stage that is found in the bloodstream of the mammalian host and is commonly cultivated *in vitro*.

blowfly is one of the most widespread ectoparasite affecting sheep, usually caused by the greenbottle fly. Affected sheep can be treated by plunge dipping using an organophosphate compound but it is more usual to treat individual infested sheep with dip wash applied directly to the struck area after clipping away overlying wool.

bolus a dose of drug given in such a way as to rapidly achieve the desired effect, often in pulse fashion.

bone marrow stromal cells mixed population of mesenchymal cells derived from the non-blood forming fraction of bone marrow.

Bonferroni a multiple comparison adjustment approach that tests each individual hypothesis by dropping the threshold for declaring statistical significance by *n*-fold, when n hypotheses are being tested.

Bonferroni correction the simplest and perhaps most conservative method to control family error rate (α) by correcting for the number of independent hypothesis tests (*n*) when *n* is large; that is, the corrected threshold $P_{corrected} = \alpha/n$.

booster a second administration of immunogen to an individual primed months or years ago.

bootstrap analysis (in phylogenetic reconstruction) a procedure to evaluate the level of support provided by the data for a particular element of the tree topology. It involves omission of some of the data, with replacements drawn from the remaining data. Frequent appearance of a clade ('bootstrap support') in a majority consensus tree is often (over)interpreted as a confidence level in the tree topology.

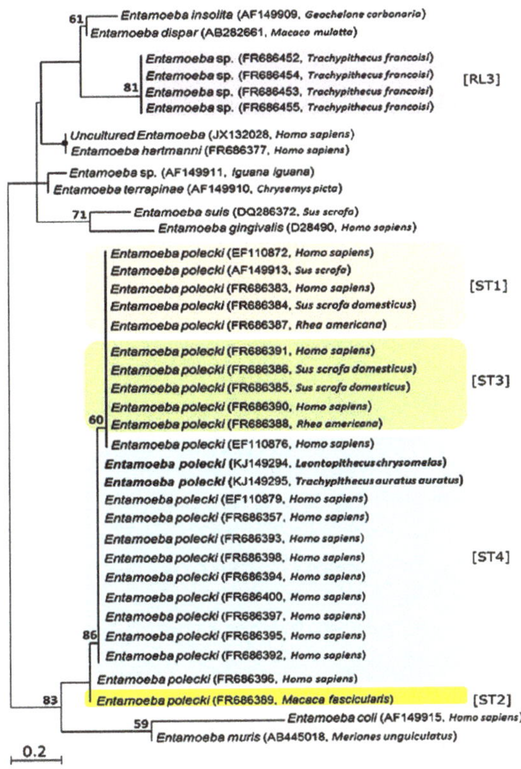

border testing a proactive ID tactic aimed at preventing cross-border transmission at checkpoints. Border testing may be preceded by fever screening followed by testing of all patients with a recent history of malaria symptoms.

bothria (singular **bothrium**) shallow longitudinal sucking grooves found on the scolex of the broad (or fish) tapeworm

Diphyllobothrium latum of the order Psuedophyllidea.

bots parasitic larvae of several fly species, particularly *Gastrophilus* (horse bot), *Oestrus* (sheep bot), *Dermatobia* and *Hypoderma* (affect cattle and other species).

bottle jaw fluid accumulation under the lower jaw (also known as submandibular oedema).

bottleneck reduction in population size that can have major influence on genetic variation because of the relationship between genetic drift and population size.

bovine anaplasmosis a disease caused by *Anaplasma marginale* and affects a wide variety of ruminant hosts with some of these species acting as reservoir hosts.

bovine lungworms *Dictyocaulus viviparous* is the only lungworm of cattle. Adults are up to 80 mm long. They live in the lumen of bronchioles. It has a direct life cycle, no intermediate host. Infective stage is L3 larvae ingested with contaminated grass. The L3 larvae penetrate the gut, moult to L4 stage and enter lymphatics. L4 migrate to lungs via circulation. L4 larvae moult to L5 (early adult stage) in the alveoli. L5 migrate up through the bronchial tree as they mature. Adults are found in the bronchi and bronchioles. At 21 days eggs containing L1 are shed by fertilized females, hatch almost immediately into L1, which coughed up and swallowed and pass in faeces. On pasture, the L1 larvae moult into L3 within 5 days. Clinical signs range from bronchitis to severe consolidating pneumonia.

Bp (base pair) two nitrogenous bases (adenine and thymine or guanin and cytosine) held together by weak bonds.Two strands of DNA are held together in the shape of a double helix by the bonds between bps. *See* **nucleotide**.

bradyzoite a long-lived, slow growing form of *Toxoplasma gondii* found during the chronic phase of toxoplasmosis. The intracellular bradyzoites form cysts in tissues, such as muscles and brain where they can evade the immune response.

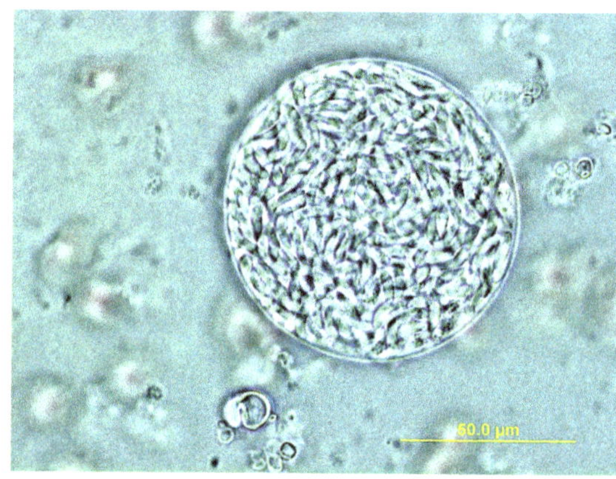

brain parenchyma the brain parenchyma represents the entity of brain cells including neuronal and glial processes and synapses; it is completely surrounded by the glia limitans.

Branchiura subclass of exclusively ectoparasitic crustaceans; 'fish lice' dorsoventrally flattened, oval body; oviparous; most species parasitize fish.

branchpoint a sequence as part of the 3′ splice site that is recognized by the spliceosome and that reacts with the 5′ splice site in the first step of the splicing reaction to form the lariat.

breakpoint worm burden an unstable parasite density threshold occurring in the human host in helminth infections, below which initial values or fluctuations in mean worm burdens lead to parasite elimination, and above which the system tends to settle to an endemic mean worm burden.

BRENDA database of functional and molecular information (e.g. pathway, reaction specificity, structure, methods for isolation and purification) for enzymes based on primary literature. Enzymes are classified according to the Enzyme Commission system, and entries are cross-referenced to other databases.

brightfield microscope the standard bench conventional microscope in which the object to be viewed is illuminated from below.

broad-sense heritability (*H^2*) the proportion of variation due to all genetic effects (i.e. both additive and non-additive, including dominance and epistasis).

broken pairs paired-end reads that do not map as proper pairs.

bronchial lavage similar to bronchial washings but this term implies instillation of a larger volume of fluid before aspiration.

bronchial washings fluid that may be aspirated from bronchial tree during bronchoscopy.

bronchiolitis inflammation of the bronchioles.

bronchitis inflammation of mucous membranes of bronchi; often caused by infectious agents.

bronchoscopy examination of bronchi through a bronchoscope, a tubular illuminated instrument introduced through the trachea.

BSFs (B cell stimulating factors) earlier terms for interleukin 4 (BSF-1) and interleukin 6 (BSF-2).

buccal cavity/capsule in nematodes, the mouth chamber that joins the mouth opening with the oesophagus.

buccal teeth/plates tooth-like structure in the mouth cavity of the adult hookworms of the genus *Ancylostoma* (compared with buccal plates found in the mouth cavity of the genus *Necator*).

budding reproduction by which a new individual unlike the parent is produced. The new individual formed may grow and develop to be the same as the parent as occurs in some free-living protozoa or may develop into a different morphological form in some species, e.g. in trematodes; cercariae develop from the germinal layer within a sporocyst and emerge from the sporocyst as free-living cercariae.

bulinids snails of the genus *Bulinus*, in the family Planorbidae.

bulk segregant analysis (BSA) analysis in which offspring from diverged parents are phenotyped and the DNA of individuals from opposing tails of the phenotypic distribution is combined. Causative variants are identified by contrasting allele frequency differences among the pools.

bunostomiasis infection with nematodes of the genus *Bunostomum* (hookworm) and related hookworms. Transmission is generally by skin penetration and is favoured by warm, humid conditions.

bursa an umbrella-like cuticular expansion at the posterior end of males of the nematode suborder Strongylida. It is supported by muscular rays and aids in copulation (hookworms, trichostrongyles) and is useful in nematode taxonomy and species identification. *See* **copulatory bursa**.

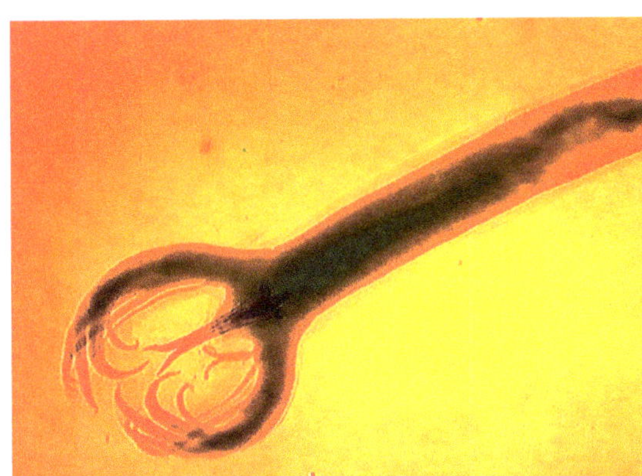

Bursa of Fabricius a primary lymphoid organ in avian species and is located at the cloacal-hind gut junction. It is the site of B lymphocyte maturation. B cells were named for this organ where they were originally discovered Bursal cells or B cells.

bursitis inflammation of a bursa, which is a small sac lined with synovial membrane and filled with fluid interposed between parts that move on each other.

by-product mechanism a mechanism of speciation in which premating and/or postmating reproductive isolation evolves incidentally as a consequence of evolution of other traits.

BZ-resistance mechanisms nucleotide point mutations have been localized at codons 167 and 200 (TTC to TAC) and 198 (GAA to GCA) in parasite populations in isotype 1 of the β-tubulin gene, the target of benzimidazoles derivatives.

A B **C** D E F G H I J K L
M N O P Q R S T U V W
X Y Z

C–C chemokine receptor 2 (CCR2) receptor mainly expressed by myeloid cells with the main chemokine ligand CCL2. It causes emigration of Ly6C^{high} monocytes from the bone marrow, thus exerting a central role in the trafficking of these monocytes to inflammation sites.

C1 first component of the classical complement activation pathway. It is composed of C1q and the C1r and C1s serine proteases. C1 binds to antigen-bound IgG or IgM via the C1q globular heads.

C2, C3 and C4 complement proteins that, after activation, can form C3 (C4b,2b) and C5 (C4b,2b,3b) convertases. These are protein enzymatic complexes that can activate C3 and C5, respectively.

C3a, C4a, C5a are anaphylotoxins and are structurally similar to chemokines. When produced in large amounts or injected systemically, they induce a generalized circulatory collapse and shock-like syndrome similar to that seen in a systemic allergic reaction involving IgE antibodies.

C3b and C4b opsonic (phagocytosis-stimulating) protein fragments generated during classical activation or lectin complement activation. C3b is also generated via the alternative pathway.

C5b-9 membrane attack complex (MAC) composed of complement components C5b, C6, C7, C8 and C9.

Caco-2/TC7 epithelial cell line derived from colorectal adenocarcinoma.

cadherins calcium-dependent cell–cell adhesion molecules. Cadherins form homodimers that link cells together.

caecum (plural **caeca**) sac-like extension of the intestine that is open only at one end. Trematoes exhibit caeca rather than complete intestinal tracts.

calabar (fugitive) swelling localized patches of subcutaneous oedema seen in individuals with the African eyeworm, *Loa loa*.

calcium imaging genetically encoded calcium indicators can be expressed in subsets of neurons to measure Ca^{2+} concentrations. When neurons are activated, the concentration of intracellular calcium increases, and fluorescence of these indicators can be used as a proxy for neuronal activity in model organism, such as live *Caenorhabditis elegans*.

calibration point a node in a phylogenetic tree for which there is independent age evidence. This can be fossil evidence or geological data, such as the timing of a continental break-up or the formation of volcanic islands.

CALP 1.3 calpain-like protein found at the distal tip of the flagellum in *T. brucei*.

Calvin cycle a series of photosynthetic chemical reactions that do not require light to occur. The Calvin cycle uses energy produced by light-dependent reactions of photosynthesis to incorporate carbon from carbon dioxide into organic compounds used to make sugars, starches, and other biological molecules.

camelid-like antibodies single domain antibodies that are derived from the heavy-chain antibodies of camelids. They are considered a new generation of therapeutic agents and immunoreagents.

camera lucida an optical device attached to the eye piece of a microscope. It operates by superimposing an image of the subject being viewed upon a drawing surface. Such drawings were often used in the past to record the morphology of protists in lieu of microphotography.

cancer an abnormal, uncontrolled and rapid growth of cells that invade and destroy surrounding tissues. Cancer is a broad term for more than 100 diseases characterized by this growth. Cells from the tumour can break away (metastasise) and spread through the bloodstream or lymph system to other parts of the body, creating new tumours.

candidate gene a gene selected for study because it is either known to be located by linkage analysis in a region of interest or existing data implicates the gene or its variants in a plausible disease model.

candidate phyla uncultivated microbial groups that branch independently from known sequences near the base of the bacterial clade.

canine filariases skin dwelling microfilariae, such as those within the genus *Cercopithifilaria* (Spirurida, Onchocercidae). Dogs may become infected with least three species of *Cercopithifilaria*, namely *Cercopithifilaria grassii*, *Cercopithifilaria bainae* and *Cercopithifilaria* sp. II sensu. The distribution of canine *Cercopithifilaria* spp. is strictly associated with that of their main arthropod vectors, *Rhipicephalus sanguineus* sensu lato ticks.

canine heartworm a disease due to *Dirofilaria immitis*, which can cause cardiopulmonary problems and even death in dogs. The disease is more prevalent in warm tropical and subtropical regions than in other areas of the world because these conditions are favourable for mosquito vectors of *D. immitis*, including *Culex* spp., *Aedes* spp., *Anopheles* spp., *Ochlerotatus* spp.

canonical histones histones synthesized solely during the S phase of the cell cycle, constituting the 'standard' octameric nucleosome core.

CAP5.5 calpain-related cytoskeleton-associated protein in the pellicular membrane of procyclic form *T. brucei*; essential for cell morphogenesis; homologue in bloodstream form is CAP5.5 v.

capillariases a range of diseases caused by nematode species of the genus *Capillaria*. Feline lung capillariasis caused by *Capillaria aerophila* (also known as *Eucoleus aerophilus*) is a subclinical to severe, potentially fatal disease, with the following clinical signs: general respiratory distress, bronchovesicular breathing sounds, sneezing, wheezing, and chronic dry or moist and productive cough, complicated bronchopneumonia and respiratory failure.

capillary array gel-filled silica capillaries used to separate fragments for DNA sequencing. The small diameter of the capillaries permit the application of higher electric fields, providing high speed, high throughput separations that are significantly faster than traditional slab gels.

capitulum pseudo-head of a tick; bears the mouthparts and the probing and sensory structures. Its shape is characteristic of the tick species.

capping the active clustering of surface structures such as antigens or receptors in a small area on the cell surface.

capture-marking-recapture a method commonly used in ecology to estimate population size, for example. It consists of marking individuals and following the presence or absence of those marked individuals in a given study area.

carbamates the most commonly used carbamates are bendiocarb, carbaril insecticides inhibit acetylcholine (Ach) breakdown by inhibiting acetylcholine esterase (AChE) via carbamylation. They bind to the enzyme that is normally responsible for inactivating ACh in the synapse. When an insect has been poisoned by a AChE inhibitor, the AChE is not available to help break down the ACh, and the neurotransmitter continues to cause the neuron to 'fire', or send its electrical charge. This causes overstimulation of the nervous system, and the insect dies.

carbamoylcholine chloride a cholinergic agonist that stimulates both muscarinic and nicotinic subtypes of acetylcholine receptors and has been shown to stimulate the activity of the feeding tube or pharynx (pump) that facilitates nematode feeding.

carbohydrate a chemical compound that contains only carbon (C), hydrogen (H), and oxygen (O) and has the general formula Cx (H_2O)y. Examples include mono, di, tri and polysaccharides. Plant carbohydrates constitute a major food class (starches), cellulose and chitin are often structural polysaccharides in cell walls and arthropod exoskeletons and the monosaccharide glucose is a basic source of energy for all animals.

carbon cycle the complex carbon flows and transformations among major earth system components (atmosphere, oceans, and terrestrial systems). The global flow of carbon from one reservoir (carbon sink) to another. Each carbon exchange among reservoirs is mediated by a variety of physical, biogeochemical, and human activities.

carbon dioxide (CO_2) gas that is an important part of the global carbon cycle. CO_2 is emitted from a variety of processes (e.g. cellular respiration, biomass decomposition, fossil fuel use) and taken up primarily by the photosynthesis of plants and micro-organisms. CO_2 is a greenhouse gas that absorbs infrared radiation and traps heat in the Earth's atmosphere.

carbon fixation conversion of inorganic carbon dioxide to organic compounds by photosynthesis.

carbon flux rate of carbon movement as it flows from one carbon reservoir to another in the global carbon cycle. For the global carbon budget, carbon flux is usually expressed in gigatons of carbon per year (GT C/year).

carbon sequestration biological or physical process that captures carbon dioxide and converts it into inert, long-lived, carbon-containing materials.

carbon sink a pool (reservoir) that absorbs or takes up released carbon from another part of the carbon cycle. For example, if the net exchange between the biosphere and the atmosphere is towards the atmosphere, the biosphere is the source, and the atmosphere is the sink.

carbon source a pool (reservoir) that releases carbon to another part of the carbon cycle.

carcinoembryonic antigen an antigen present in fetal tissue and is re-expressed on the surface of neoplastic cells.

carcinogen an agent with the capacity to cause cancer. Carcinogens may be natural, such as aflatoxin, which is produced by a fungus and sometimes found on stored grains, or man-made, such as asbestos or tobacco smoke. Carcinogens work by interacting with a cell's DNA and inducing genetic mutations.

cardiomyopathy deterioration of heart muscle function, usually leading to heart failure.

carrier **1** a distinct group of solute transport proteins that mediate transport at rates that are usually slower than through ion channels. Carriers require protein conformational changes for each transport cycle. Also known as solute transporters, they may couple the movement of two or more solutes (e.g. co- and counter-transporters). **2** a person or animal that harbours a specific infectious agent in the absence of clinical disease and serves as a potential source of infection. **3** an immunogenic macromolecule to which a hapten may be bound so making the hapten immunogenic. This part of the molecule is recognized by T cells. **4** an individual who carries one copy of a recessive gene for a hereditary condition.

carrier host a natural definitive host carrying the parasite without the appearance of any pathological lesions or disease symptoms and at the same time it can spread the infection to other hosts, e.g. *Tritrichomonas foetus* in bulls.

carrier screening a type of genetic testing performed on people who display no symptoms for a genetic disorder, but may be at risk for passing it on to their children. A carrier for a genetic disorder has inherited one normal and one abnormal allele for a gene associated with the disorder. A child must inherit two abnormal alleles in order for symptoms to appear. Prospective parents with a family history of a genetic disorders are candidates for carrier screening.

carrying capacity the maximum population size of a species that can be indefinitely sustained in an environment.

cartography the study and craft of map making.

cascade reaction a series of enzyme reactions in which the products of one reaction catalyse a second reaction, and so forth. Typical example is the complement cascade.

case–control analysis a study design based upon the comparison of cases and controls defined at the outset of study. Genotype or allele frequencies are compared between the affected cases and unaffected control group. A difference in the frequency between case and control groups is determined by statistical analyses and is generally reported by the odds ratio.

case–control study retrospective comparison of exposures of persons with disease (cases) with those of persons without the disease (controls).

case-series report of a number of cases of disease.

caspase-recruitment domain (CARD) a homotypic interaction motif present in proteins typically involved in inflammation and apoptosis that belongs to the death-domain superfamily.

caspases a family of cysteine proteases that cleave cellular targets and lead to a cascade of cellular changes as observed during apoptosis in higher eukaryotes.

castes specialized behavioural groups within a eusocial colony that often correspond to morphological features and that are generally considered to be a stable, if not permanent, characteristic of an individual. For example, members of the queen caste hatch as adults with wings and can reproduce, whereas members of the worker caste (or castes) are wingless and do not normally reproduce.

catabolism the degradation of complex molecules, such as nutrients, leading to the release of energy.

catarrhal exudate fluid formed by mucus and white blood cells, which is caused by an over-stimulation of respiratory, mucus-producing muciparous cells as a response to inflammatory processes.

CATCH-IT (covalent attachment of tagged histones to capture and identify turnover) a method for measuring genome-wide nucleosome turnover using metabolically labelled, newly synthesized histones for affinity-based chromatin capture.

cathepsin S an enzyme crucial for the loading of antigen onto MHC II molecules on antigen-presenting cells to present to $CD4^+$ T cells and elicit an adaptive immune response.

cationic antimicrobial substances present within granules of phagocytes.

causal variants genetic variants that directly modify a phenotype and/or cause a change of disease risk. Owing to the limited amount of variation interrogated by single-nucleotide polymorphism (SNP) genotyping microarrays, SNPs in genome-wide association studies typically merely tag the causal region rather than being the causal variants themselves.

causality the relating of causes to the effects they produce. Most of epidemiology concerns causality and several types of causes can be distinguished. It must be emphasized, however, that epidemiological evidence by itself is insufficient to establish causality, although it can provide circumstantial evidence.

caval syndrome a form of heart worm disease characterized by acute onset of depression, dyspnoea, anaemia, systolic murmur, jugular vein distension, haemoglobinuria, and hepatic and renal dysfunction. It is often associated with a poor prognosis. The caval syndrome develops as a result of retrograde migration of adult heart worms from the pulmonary arteries to the right ventricle, the right atrium and vena cavae; resulting in heart failure.

caveolae specific type of lipid rafts containing caveolin proteins, often involved in endocytosis.

caveolar endocytosis a clathrin-independent mechanism of endocytosis. Entry into cells is mediated by caveolae, which are flask-shaped invaginations on the cell surface that contain the protein caveolin.

caveolin proteins family of integral membrane proteins which play a role in receptor-independent endocytosis and may also act as scaffolding proteins.

CCCTC-binding factor (CTCF) a chromatin-binding factor with highly conserved zinc finger domains that control binding to consensus DNA target sequences. CTCF regulates transcription by binding to chromatin insulators and preventing interaction between the promoter and enhancers or silencers.

CD antigen cluster of differentiation designation assigned to leucocyte cell surface molecules which are identified by a given group of monoclonal antibodies.

CD160 a membrane protein found on normal natural killer and T cells. CD160 improves proliferation and cytotoxic activity in CD8 T cells. It has recently been recognized as an inhibitory receptor.

CD1d an immunoglobulin family protein with a specific role in lipid antigen presentation.

CD244 (2B4) a transmembrane protein belonging to the Ig superfamily, is a co-stimulatory receptor expressed by memory CD8 T-cells and natural killer cells. 2B4 can elicit both activating and inhibitory signals.

CD25 the IL-2 receptor α chain present on activated T-cells, B cells, NK cells, and monocytes. Additionally, CD25 along with FoxP3 is considered as one of the potent markers of T regulatory cells.

CD3 a trimeric complex of γ, δ and ε chains which together with a ζζ homodimer or ζη heterodimer acts as a signal transducing unit for the T-cell receptor.

CD4 cell surface glycoprotein, usually on helper T-cells that recognizes MHC class II molecules on antigen-presenting cells.

CD4$^+$ T-cells/T helper (TH) cells a T-cell subset characterized by the expression of the surface protein CD4. These cells express receptors that recognize specific antigens presented on MHC II molecules of antigen-presenting cells. They orchestrate adaptive immune responses by releasing cytokines and aiding other immune cells. Upon activation, TH-cells are polarized into different effector subsets, including TH1, TH12, and TH17 cells, which are each specialized to combat different classes of pathogen.

CD40 a member of the tumour necrosis factor superfamily, is a co-stimulatory molecule required for activation of antigen-presenting and T cells. With its ligand CD40L, CD40 is a critical contributor to the inflammatory process.

CD8 cell surface glycoprotein, usually on cytotoxic T-cells, that recognizes MHC class I molecules on target cells.

CD8$^+$ cytoxic T-cells T-cells that recognize peptides presented on MHCI, a complex that specializes in presenting peptides derived from intercellular compartments and pathogens.

CD80 (or B7.1) found on the surface of activated B- cells and monocytes and provides the dominant co-stimulatory pathways that regulate T- and B- cell responses. It can bind to either CD28 to provide a stimulatory signal to T cells or can combine with cytotoxic T lymphocyte antigen 4 (CTLA-4) to send a negative signal to T-cells.

CD86 (or B7.2) a protein expressed on antigen-presenting cells that provide co-stimulatory signals necessary for T-cell activation and survival.

Cdc13 cell division control protein 13, a G-strand telomere-binding protein that acts as a regulator of telomere length in yeast and co-ordinates C-strand re-synthesis by DNA polymerase alpha.

cDNA library a collection of DNA sequences that code for genes. The sequences are generated in the laboratory from mRNA sequences. *See* **messenger RNA**.

cell the basic building block of living things. All cells can be sorted into one of two groups: eukaryotes and prokaryotes. A eukaryote has a nucleus and membrane-bound organelles, while a prokaryote does not. Plants, fungi, protists and animals are made of eukaryotic cells, while many microbes, such as bacteria, consist of single prokaryotic cells. An adult human body is estimated to contain between 10 and 100 trillion cells.

cell cycle a series of events that occurs in a cell as it grows and divides. A cell spends most of its time in what is called interphase, and during this time it grows, replicates its chromosomes, and prepares for cell division. The cell then leaves interphase, undergoes mitosis, and completes its division. The resulting cells, known as daughter cells, each enter their own interphase and begin a new round of the cell cycle.

cell division the process by which cells split into two copies of the original. The DNA of the original cell is copied and one copy sent to each cell, ensuring that both have the correct amount of DNA.

cell line cells that grow and divide indefinitely outside the body, and are originally derived from one cell of a specific type.

cell membrane also called the plasma membrane, is found in all cells and separates the interior of the cell from the outside environment. The cell membrane consists of a lipid bilayer with integrated proteins that is semipermeable. The cell membrane regulates the transport of materials entering and exiting the cell.

cell migration the movement of a cell towards or away from a stimulus, often based on a chemical gradient.

cell segmentation automated detection and delineation of the outside of single cells and nuclei in microscope images.

cell spreading certain insect haemocyte types have the ability to strongly adhere to foreign surfaces and spread to phagocytose a small particle or encapsulate a larger foreign object.

cell traversal process of sporozoite migration through cells with disruption of their plasma membrane and without formation of a parasitophorous vacuole.

cell-based therapies involve transplanting stem cells into damaged tissues to regenerate the various cell types of that tissue. For example, bone marrow transplants are a form of cell-based therapy that have been used to treat leukaemia for more than 30 years. New stem cell research may lead to cell-based therapies to treat a range of conditions, including heart disease, spinal injuries, diabetes and Parkinson disease.

cell-mediated immunity (CMI) T-helper (Th) lymphocytes are the principal orchestrators of the immune response because they are needed for activation of important effector cells. Effector cells include CTL, B- cells and antibody producing, activated B-cells called plasma cells. However, some Th cell types (Th1 cells) also stimulate innate immunity by increasing phagocytosis and the magnitude of biochemical killing pathways within these phagocytes. The activation of Th cells occurs early in the immune pathway and requires at least 2 signals. One signal is provided by binding of the T-cell antigen receptor to the class II MHC–antigen complex on the APC. The second signal is provided by binding of co-stimulatory molecules such as the B7 family (CD80 and CD86) on the APC surface to CD28 on the T-cell membrane. Subsequent activation of CTL and B lymphocytes is dependent upon cytokines, which are secreted by Th lymphocytes early in infection. The major function of the CTL is to kill cells that express foreign antigens in association with MHCI. This interaction of CTL with infected target cell includes the binding of Fas ligand (FasL) on the surface of the CTL with Fas on the surface of the target cell. Fas/FasL interaction induces production of a protein (Perforin) by the CTL which creates pores in the target cell membrane. The CTL then produces another enzyme (Granzyme) which enters the pores and activates caspase pathways within the target cell. Activation of these caspase pathways induces apoptosis of the target cell. Natural killer cells also kill target cells by this mechanism. However, although NK cells are lymphoid in origin they are not MHC restricted.

cell-to-cell variability the phenomenon that individual cells in a population of genetically identical cells display variable activities and behaviours.

cellular automata models consisting of a grid of subunits that change between discrete states over time according to fixed rules. Subunits could represent cells, individual organisms or geographical areas.

cellular capsule a multicellular aggregate of haemocytes that isolates a foreign object. In some insects the capsule is composed of melanin.

cellular heterogeneity similar to cell-to-cell variability. Sometimes, ‘heterogeneity’ is used to indicate multiple discrete phenotypes within a population, while ‘variability’ is used to indicate variation around a single phenotype. There is no consensus on which term to use in which occasion, and both terms are interchangeable.

cellular immunity the arm of the immune response that is based upon the activation of T-cells, natural killer cells and macrophages.

cellular states a quantitative description of the physiological states of single cells reflected in, for instance, their sizes, shapes, cell cycle phases, senescence or other detectable readouts such as metabolic states.

cellulose linear polysaccharide polymer with many β1,4 glycosidically linked glucose monosaccharide units. Cellulose is the major component of plant cell walls and the most abundant biological material on Earth.

centimorgan a unit used to measure genetic linkage. One centimorgan equals a 1% chance that a marker on a chromosome will become separated from a second marker on the same chromosome due to crossing over in a single generation. It translates to approximately one million bps of DNA sequence in the human genome.

central carbon metabolism a core network of about 50 enzymatic reactions that convert carbon nutrients into 'building blocks'.

central memory immunological memory that is dependent on $CCR7^+$ T-cells which, under the influence of chemokines, travel to secondary lymphoid organs where they give rise to $CCR7^+$ effector memory T-cells.

central memory T-cell (TCM cell) an antigen-experienced resting T-cell that expresses cell surface receptors that are required for homing to secondary lymphoid organs. These cells are generally long-lived and can function as the precursors for effector T-cells during recall responses.

central tolerance specific immunological tolerance due to the induction of lymphocyte apoptosis or anergy within the primary lymphoid organs (bone marrow in the case of B-cell tolerance and the thymus for T-cells).

centrifuge a laboratory instrument used for the separation of substances by sedimentation through rotation at high speeds, e.g. the separation of components of cells or blood.

centrioles are paired barrel-shaped organelles located in the cytoplasm of animal cells near the nuclear envelope. Centrioles play a role in organizing microtubules that serve as the cell's skeletal system. They help determine the locations of the nucleus and other organelles within the cell.

centrocone a specialized structure of the apicomplexan nucleus associated with the intra-nuclear mitotic spindle.

centromere **1** a specialized region of the eukaryotic chromosome that serves as the point of attachment of the mitotic spindle during cell division. It is usually located near the middle of the chromosome and is characteristically heterochromatic. By convention, the centromere is the separation point for the two 'arms' of the chromosome. These are referred to as a short 'p' arm and a longer 'q' arm. **2** a constricted region of a chromosome that separates it into a short arm (p) and a long arm (q). During cell division, the chromosomes first replicate so that each daughter cell receives a complete set of chromosomes. Following DNA replication, the chromosome consists of two identical structures called sister chromatids, which are joined at the centromere. **3** the most condensed and constricted region of a chromosome. Joins the two chromatids of the chromosome and is also the attachment point of spindle fibres during cell division when the two chromatids separate.

centrosome a cellular structure involved in the process of cell division. Before cell division, the centrosome duplicates and then, as division begins, the two centrosomes move to opposite ends of the cell. Proteins called microtubules assemble into a spindle between the two centrosomes and help separate the replicated chromosomes into the daughter cells.

cephalic gland in trematodes, the gland in a miracidium that produces a fluid which enables the miracidium to penetrate the tissues of its snail host.

cephalic gland duct the duct that carries penetration fluid from the gland to the tissue of the host.

cercaria (plural **cercariae**) a free-living larval stage of digenetic trematodes, which develops from a sporocyst or redia in snail intermediate hosts. The free-swimming larva of a trematode (usually possessing a tail) develops asexually from the sporocyst or redia generation within the intermediate, molluscan host and to later become the metacercaria if the trematode has a metacercarial stage.

cercarial dermatitis (or swimmers' itch) skin hypersensitivity reaction to the penetration of cercariae of certain members of the non-human schistosomes.

cercomer a tail-like appendage, which often retains the hooks of the hexacanth embryo, on procercoid and cysticercoid larvae of certain tapeworms.

cerebrospinal fluid (CSF) fluid that fills the brain cisterns and where antibodies are secreted to protect the brain.

cestodal cysts the larval (metacestode) stage of cestodes in mammal hosts, e.g. *Echinococcus* spp. cysts in humans, *Cyticercus tenuicollis* cysts in sheep.

cestodes a class of parasitic flatworms with a complex life cycle, commonly called tapeworms, including 14 orders. Most often intermediate stages of cestodes exploit invertebrates as intermediate hosts and vertebrates as final hosts, although in some species both hosts are vertebrates. They are flat, ribbon-like worms, with no digestive tract; nutrition is absorbed through the syncytial tegument. All are hermaphroditic. The tapeworm has a head, called a scolex, and a body made of many segments (band-like) called proglottids.

CFU (Colony Forming Unit) a viable bacterium in a culture from which a colony grows. The formula that gives the total number of viable bacteria per unit volume is viable bacteria = CFUs/(amount plated) × (dilution factor).

CH50 unit the quantity or dilution of the serum which required to lyse 50% of the red cells in a standard haemolytic complement assay.

chabertiasis is caused by *Chabertia ovina* (large-mouthed bowel worm) a worm 1–2 cm in length, which inhabits the colon. The adult worm has a characteristic, curved, bell-shaped buccal capsule, which lacks teeth. It is more common in sheep and goats than in cattle. It has a worldwide distribution, but it tends to be more common in temperate regions. The life cycle is direct. After ingestion by the final host, L3 larvae undergo a period of development in the wall of the small intestine before passing to the caecum and then to the colon. The prepatent period is approximately 6 weeks. Pathogenic effects are caused by the feeding adults which become attached to the mucosa and draw a plug of mucosa into the buccal capsule which is then digested. This results in area of mucosal ulceration and local haemorrhage with protein loss into the gut through these lesions.

Chagas disease is named after the Brazilian physician Carlos Chagas, who discovered the disease in 1909. It is caused by the parasite *T. cruzi*, which is transmitted to animals and people by insect vectors in the family Reduviidae and is found only in the Americas. Chagas disease is also referred to as American trypanosomiasis. In the United States, Chagas disease is considered one of the neglected parasitic infections, a group of five parasitic diseases that have been targeted by CDC for public health action.

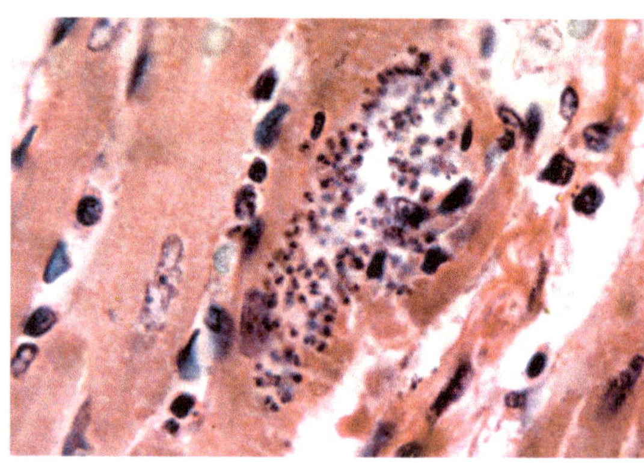

chalimus (plural **chalimi**) the four stages of immature salmon louse fixed to the host by a frontal filament around which they feed on host mucus and skin.

challenge infection the experimental human malaria infection procedure that uses the bites of laboratory-reared mosquitoes that are infected with *P. falciparum* to deliver sporozoite inocula to assess the efficacy of immunization.

Charcot–Leyden crystals a slender crystal, pointed at both ends, formed from the breakdown products of eosinophils usually following some type of immune response.

cheats individual bacteria that benefit from the co-operative behaviour of others, but do not contribute.

chelicerae pair of moveable oral appendages adapted for cutting, carried by all members of the Chelicerata, including ticks.

chem-seq (chemical affinity capture and massively parallel DNA sequencing) an unbiased molecular biology-based protocol designed to identify sites of interaction of small molecules with the genome. This method involves the use of small molecules of interest to isolate bound DNA targets, which are then analysed by sequencing.

chemical ligaturing the use of a chemical compound that paralyses the feeding tube or pharynx (pump) and so prevents the nematode from feeding.

chemoautotroph an organism that biochemically synthesizes its own organic materials from inorganic compounds using chemical energy.

chemokine a group of cytokines produced by cells present at the site of inflammation. They have the ability to induce chemotaxis in various leucocytes.

chemokine-binding proteins secreted proteins that selectively bind to and inhibit the action of specific chemokines.

chemoprophylaxis the preventative use of antimalarial drugs in a regimen that achieves sustained blood concentrations above the level that inhibits the multiplication of the malaria parasite for the whole of a period at risk.

chemosensitizer substance that blocks the effect of a drug resistance system.

chemosensory olfactory and gustatory receptors proteins capable of responding to volatile (olfactory) or water-soluble (gustatory) chemical stimuli.

chemostat a closed system for microbial growth that is used to control the microbial replication rate and/or metabolism by the use of chemically defined media. Chemostat keeps bacterial cultures in an optimal growth state by continually adding medium and removing old cells.

chemotaxis migration of cells or organism towards increasing concentrations of gradient of chemotactic factors.

chemotherapy the use of a chemical that has a specific action in removing or killing a microbial agent, or in controlling the growth of cells that form a cancer.

chimera (plural **chimeras**) **1** an organism that contains cells or tissues with a different genotype. These can be mutated cells of the host organism or cells from a different organism or species. **2** a recombinant molecule of DNA composed of segments from more than one source. MDA can generate chimaeras that are predominantly inversions through its branching mechanism of DNA replication.

chimeraplasty an experimental targeted repair process in which a desirable sequence of DNA is combined with RNA to form a chimeraplast. These molecules bind selectively to the target DNA. Once bound, the chimeraplast activates a naturally occurring gene-correcting mechanism. *See* **cloning vector**.

chimeric composite of genetically distinct individuals, e.g. following an allogeneic bone marrow graft.

Chinese hamster ovary (CHO) cells epithelial-like cell line.

ChIP (chromatin immunoprecipitation) proteins associated with DNA are immunoprecipitated as a protein–DNA complex, and the abundance of associated DNA is compared to input DNA. X-ChIP, which involves cross-linking of DNA–protein complexes with formaldehyde, or N-ChIP, which involves native immunoprecipitation conditions (frequently after micrococcal nuclease digestion), are common approaches.

ChIP–seq (Chromatin immunoprecipitation followed by sequencing) an unbiased molecular biology-based protocol designed to identify interaction sites of chromatin-binding proteins with the genome. This method uses specific antibodies to precipitate the protein of interest, which isolates bound DNA fragments that are then subjected to sequencing.

chitin long chain polymers of β(1,4)-N-acetylglucosamine residues found in many species, including some fungi, some protozoan cyst walls (*Entamoeba*) and virtually every arthropod in their exoskeleton.

chitin synthase the transferase that adds successive N-acetylglucosamine residues to the growing chitin polymer with a β1,4 glycosidic linkage.

chitin synthesis inhibitors exert an effect on fecundity and prolificacy of insect females and inhibit egg hatching and larval moulting (e.g. lufenuron).

chitinous shell the hard shell of nematode eggs lined with the vitelline membrane which encases the embryo.

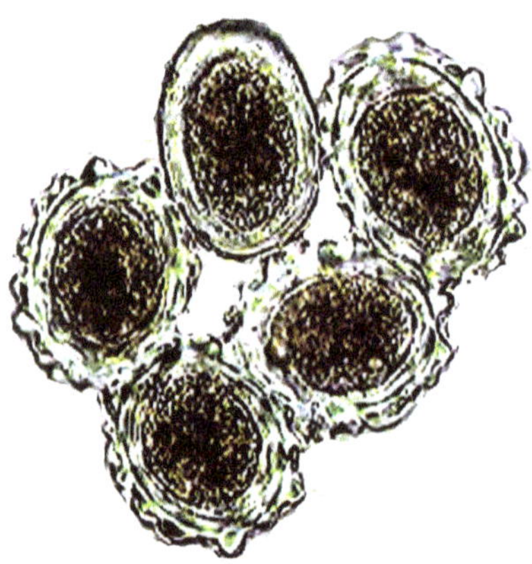

chlorophyll a type of green pigment used to harness light energy in the chloroplasts of plants and other photosynthetic organisms.

chloroplast an organelle in the cells of green plants. It contains chlorophyll and functions in photosynthesis and protein synthesis.

chloroplast chromosome circular DNA found in the photosynthesizing organelle (chloroplast) of plants instead of the cell nucleus where most genetic material is located.

chloroquine (CQ) the most widely used antimalarial drug to date, a 4-aminoquinoline derivative of the natural product quinine (QN).

chloroquine resistance (CQR) the ability of parasites to proliferate in the presence of CQ at concentrations that are inhibitory to sensitive strains.

chloroquine sensitive (CQS) the natural sensitivity to chloroquine for wild type strains of *P. falciparum*.

choanomastigote a morphotype characterized by barley-shape cells with a wide flagellar pocket and kDNA pre-nuclear or adjacent to the nucleus.

cholera toxin B secreted toxin that binds to GM1 gangliosides; fluorescently labelled form used to identify lipid rafts, which contain GM1 gangliosides.

cholesterol sterol that is absorbed from food in the intestine or produced in the liver. It is the most abundant sterol in mammalian cells.

cholinergic anthelmintic antinematodal drugs which interact with acetylcholine receptors.

cholinergic nerve terminals axonal terminals that use acetylcholine for neurotransmission.

cholinomimetics drugs which mimic the actions of acetylcholine and activate acetylcholine receptors.

chondroitin sulfate A (CSA) belongs to a family of glycosaminoglycans that are present in a variety of tissues. Its functions range from structural stability to surface receptors and signalling molecules, or even essential components of embryonic development. In the placenta, CSA acts as a receptor for IE expressing VAR2CSA.

chorioptic mange chorioptic mange occurs commonly in cattle, sheep and more rarely in goats. In cattle two species have been described: *Chorioptes ovis* and *Chorioptes texanus*; *C. bovis* is the species present on sheep and goats. Generally, the condition is less severe than psoroptic or sarcoptic mange. These mites have a predilection for the perineum, udder, caudal areas of thigh, rump and feet. Lesions consist of alopecia, erythema, excoriations and (small) crusts associated with pruritus.

chromatid one of two identical halves of a replicated chromosome. During cell division, the chromosomes first replicate so that each daughter cell receives a complete set of chromosomes. Following DNA replication, the chromosome consists of two identical structures called sister chromatids, which are joined at the centromere.

chromatin a DNA–protein–RNA complex, which forms the basis of DNA compaction and genome organization. The major proteins in chromatin are histones, which help package the DNA in a compact form that fits in the cell nucleus. Changes in chromatin structure are associated with DNA replication and gene expression.

chromatin immunoprecipitation (ChIP) a method used to determine the location in a genome of DNA binding sites recognized by a particular protein of interest.

chromatin looping a mechanism allowing remote regulatory elements to participate in the formation of a transcription initiation complex.

chromatin loops higher-order chromatin architecture that brings into close physical proximity DNA regions separated by a hundred or more nucleotides.

chromatin remodelling the dynamic disruption and formation of nucleosome complexes that constitute chromatin. Chromatin remodelling involves the shifting of nucleosome cores along the DNA molecule. These changes can affect the state of chromatin and subsequently alter spatial and temporal patterns of gene expression.

chromatin states represent different spatial combinations of histone modifications. Distinct chromatin states can define different functional units of the genome.

chromatography a variety of techniques useful for the separation of various compounds including but not limited to lipids, carbohydrates and proteins.

chromatoid bar or BodyA bar, rod or splinter-shaped body in the cytoplasm of an amoeba that stains darkly and resembles chromatin.

chromatoid basal rod the rod-like structure that forms the base of the undulating membrane of flagellates.

chromomere one of the serially aligned beads or granules of a eukaryotic chromosome, resulting from local coiling of a continuous DNA thread.

chromosomal deletion the loss of part of a chromosome's DNA.

chromosomal fission breakage of chromosome segments.

chromosomal fusion reunion of previously independent chromosome segments.

chromosomal inversion chromosome segments that have been turned 180 degrees. The gene sequence for the segment is reversed with respect to the rest of the chromosome.

chromosomal position effects alteration in transgene expression patterns due to the integration site in the host genome. Position effects are caused by the interference of neighbouring genomic sequences associated with the transgene and its expression.

chromosome the self-replicating genetic structure of cells containing the cellular DNA that bears in its nucleotide sequence the linear array of genes. In prokaryotes, chromosomal DNA is circular, and the entire genome is carried on one chromosome. Eukaryotic genomes consist of a number of chromosomes whose DNA is associated with different kinds of proteins.

chromosome conformation capture (3C) and 4C (3C on Chip) a foundational, high-throughput technology for the identification and mapping of interactions between specific, distant chromosomal loci. Interacting chromosomal loci are fixed via cross-linking and the bound DNA processed via restriction digestion, ligation and quantitatively amplified via PCR or hybridization to an array chip.

chromosome painting attachment of certain fluorescent dyes to targeted parts of the chromosome.

chromosome region p a designation for the short arm of a chromosome.

chromosome region q a designation for the long arm of a chromosome.

chronic fluke disease is the most common and widespread form of liver fluke disease (fasciolosis) in both sheep and cattle and can be associated with either 'summer' (after the main peak of shedding) or 'winter' infection of snails. It is associated with a prolonged intake of low to moderate numbers of metacercariae from herbage and results in a progressive loss of body condition associated with the accumulation of adult flukes in the bile ducts of the liver. Anaemia is often severe in undernourished sheep and they may also exhibit submandibular oedema ('bottle jaw'), an accumulation of fluid caused by low blood protein (or hypoalbuminaemia). Deaths are uncommon in well-nourished sheep but chronic fascioliasis is often exacerbated by poor nutrition. *See* **fasciolosis**.

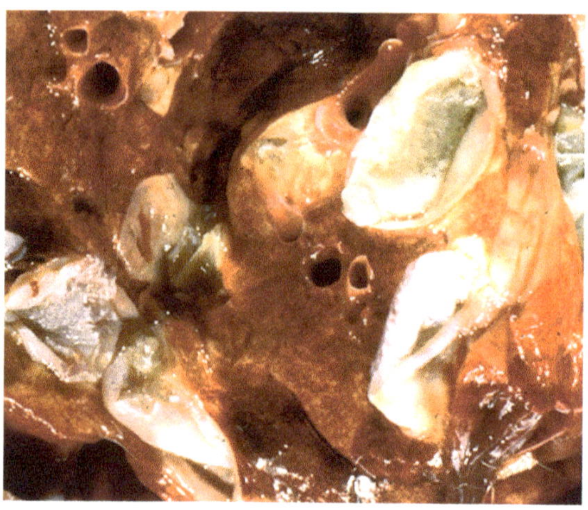

chronic infection is usually caused when the host is unable to fully clear the pathogen. This may be a consequence of host immune dysfunction or immune evasion by the pathogen or even both.

chronic malaria (or *Plasmodium falciparum* infection) *P. falciparum* can persist in the human host for several months. Frequently, such infections are asymptomatic or cause mild (recurrent) symptoms. Genotyping is necessary to differentiate chronic infections from repeated new infections.

chronic scleroderma stage chronic phase of bovine besnoitiosis, characterized by skin lesions such as hyperkeratosis and alopecia that occur as a consequence of development of tissue cysts in the connective tissues.

chronospecies a species that is part of the continuum of a lineage documented in the fossil record, distinguished from the chronospecies below and above by the accumulation of sufficient morphological change over time.

cilia (singular cilium) minute, axial fibrils that arise from granules in the ectoplasm of Ciliata (ciliates) and function in locomotion.

ciliata a class of protozoa of the subphylum Ciliophora, whose members possess cilia during the life cycle and of which a few species are parasitic.

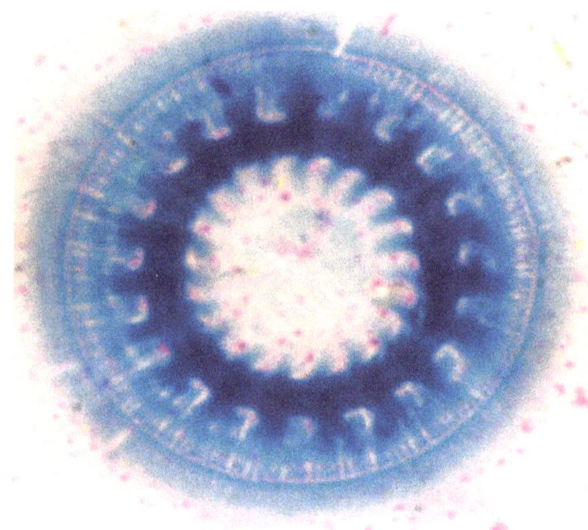

Ciliophora the phylum consists of the ciliates, which include a variety of free-living and symbiotic species. Ciliate locomotion involves the co-ordinated movement of rows of hair-like structures, or cilia. Cilia are structurally similar to flagella but are usually shorter, more numerous and more rounded.

circulating anodic antigen (CAA) gut-associated proteoglycan schistosome compounds.

cirrus the retractile muscular organ at the outer end of the male reproductive system of species of Platyhelminthes.

cirrus pore a term used in relation to trematodes. The opening through which the cirrus is protruded.

cirrus pouch a term used in relation to trematodes. A hollow organ surrounding the introverted cirrus.

***cis*- and *trans*-acting regulatory mechanisms** mechanisms that regulate gene expression. *Cis*-acting regulatory elements (such as promoters or enhancers) are DNA sequences located on the same chromosome and generally in the vicinity of the target gene, and they recruit molecules that can modulate the expression of the target gene. *Trans*-acting factors (such as transcription factors or microRNAs) control gene expression by binding to *cis*-acting elements; their genomic localization is independent of that of the regulated genes.

***cis* conformation** the conformation of a glycosidic linkage of a receptor analogue, in which the two carbon atoms at either end of the glycosidic bond are oriented in the same direction.

***cis*-acting RNA elements** RNA sequences or structures that regulate the activity of the RNA in which they reside.

citrullination the enzymatic conversion, by peptidyl arginine deiminase, of an arginine in a protein to a cittruline.

clade or monophyletic group a taxonomic group with a single common ancestor.

cladistic analysis a specific method of phylogenetic analysis that is based on the polarization of characters that are shared between ancestral (plesiomorphic) and derived (apomorphic) characters. Only those apomorphic characters shared by all members of a given clade (synapomorphic characters) are considered to have a phylogenetic value.

cladistics method of drawing trees of organismal relationships founded on shared derived characters.

cladogenesis the splitting of a species lineage.

cladogram a diagram, in the form of a stylized tree, showing inferred historical branching patterns among taxa.

clag genes a multigene family specific to *Plasmodium* spp., named for early evidence linking these genes to cytoadherence.

class I ligase ribozyme a ribozyme selected from a random pool of RNAs that can catalyse the ligation of an exogenous fragment of RNA to its own 5′ end.

class I MHC protein heterodimeric surface glycoproteins encoded by the A,B and C locus of the major histocompatibility complex, which functions mainly in antigen presentation to $CD8^+$ T-cells.

class switch the process by which a B-cell changes the class but not specificity of a given antibody it produces, e.g. switching from an IgM to an IgG antibody.

class switch recombination the recombination of immunoglobulin heavy chain constant region gene segments, e.g. switching from Cμ and Cδ to Cγ1 to convert an IgM (and IgD) antibody into an IgG1 antibody.

classical inbred strain a strain that has been maintained for >20 generations of brother–sister mating. It is essentially homozygous at all genetic loci, except for mutations arising spontaneously. Many strains have been bred for ≥150 generations and thus are essentially homozygous at all loci.

classical pathway (of complement activation) activation pathway involving complement components C1, C2 and C4 which, following fixation of C1q, e.g. by antigen–antibody complexes, produces the classical pathway C3 convertase C 4 b 2a.

classically activated macrophages (CAM) activated by Th1 cytokines such as IFN-γ, TNF and IL-12 to induce the production of nitric oxide that leads to inflammation and tissue damage.

classification the taxonomic placement of organisms into groups according to similarities in structure, physiology, biochemistry, and other characteristics. The smallest group is the species. Similar species are placed in a genus, similar genera are grouped into families, families into orders, orders into classes, classes into phyla (or divisions in plants), and phyla into kingdoms. Modern classification is usually intended to reflect degrees of evolutionary relationship, although not all experts agree on single classification schemes for animals or for plants.

clathrin-mediated endocytosis a process by which cells internalize molecules in a receptor-mediated uptake pattern. Also, it is called receptor-mediated endocytosis.

CLAVATA3/ESR (CLE)-like proteins a family of secreted peptide ligands that play important roles in plant growth and development. They share a putative N-terminal signal peptide and a conserved C-terminal 14-amino acid CLE motif.

click chemistry selective, high-yielding and biocompatible chemical reactions used to covalently link two or more molecules. The copper-catalysed azide–alkyne cycloaddition (CuAAC) and its copper-free version are the most commonly used.

climate average weather conditions over a time period, usually several decades. Climate is largely determined by local geographical features, latitude, altitude, land- and sea-masses, and atmospheric circulation patterns.

clinal patterns genotypic or phenotypic gradations along a continuous sampling area, for example with increasing latitude or altitude.

cline the gradual change in phenotypes or allele frequencies along a geographical or environmental gradient.

clinical attack rate the cumulative proportion of the population that experiences an episode of a disease over a defined time period. The term 'clinical' indicates that the episode was symptomatic and that the diagnosis was established presumptively, without parasitological confirmation.

clinical cure the disappearance of all clinical signs, but does not necessarily mean complete elimination of the causative organism. On the contrary, sterile cure or radical cure is complete elimination of the causative organism.

clinical decision support patient-specific information that is intelligently filtered or presented at appropriate times to enhance health and health care. These tools include alerts and reminders to clinicians or patients and access to clinical guidelines, condition-specific order sets, and data summaries within the health care work flow.

clinical outcome measures patient health or well-being. Ideally it should be credible, comprehensive, sensitive to change, accurate, sensible, and biologically sensible.

clinicopathological pertaining to both signs of disease and to its pathology.

cloacal drop a reverse peristalsis of the cloaca that enables the outside environment to be sampled by the bursa to aid in the diversity of antibody production in young birds.

clonal cell population a population of cells derived from a single mother cell that share identical genomic contents (i.e. without mutations).

clonal deletion a process by which contact with antigen (e.g. self antigen) at an early stage of lymphocyte differentiation leads to cell death by apoptosis.

clonal selection the selection and activation by antigen of a lymphocyte bearing a complementary receptor, which then proliferates to form an expanded clone.

clonal selection theory the concept that specificity and diversity of an immune response are the result of selection by antigen of specifically reactive clones from a large repertoire of preformed lymphocytes, each with individual specificities.

clonality absent or severely restrained genetic recombination.

clone an exact copy made of biological material such as a DNA segment (e.g. a gene or other region), a whole cell, or a complete organism.

cloning (also nuclear transfer) the process of producing a group of cells (clones), all genetically identical, from a single ancestor cell. In recombinant DNA technology, the use of DNA manipulation procedures to produce multiple copies of a single gene or segment of DNA is referred to as cloning DNA. Recently, 'cloning' has been used to describe the asexual production of whole organisms from a single somatic diploid cell (i.e. not a sperm or ovum but a cell that has a complete set of chromosome), e.g. 'Dolly the sheep'.

cloning vector DNA molecule originating from a virus, a plasmid, or the cell of a higher organism into which another DNA fragment of appropriate size can be integrated without loss of the vector's capacity for self-replication. This vector introduces foreign DNA into host cells, where the DNA can be reproduced in large quantities. Examples are plasmids, cosmids, and yeast artificial chromosomes. Vectors are often recombinant molecules containing DNA sequences from several sources.

clonorchiasis a fish-borne trematode infection which is considered one of the major parasitic zoonoses in many countries. Clonorchiasis can cause a series of hepatic diseases such as periductal inflammation, fibrosis, pyogenic cholangitis, cholecystitis, cholelithiasis and liver cirrhosis, and *C. sinensis* is currently classified as carcinogenic to humans for cholangiocarcinoma. *C. sinensis* is actively transmitted along the rivers, especially among the residents who have the habit of eating raw fish. Humans, cats, dogs, pigs and some wild animals are definitive hosts for *C. sinensis*. Some aquatic snails, freshwater fishes and shrimp act as the first and the second intermediate hosts, respectively.

clorsulon a benzosulfonamide used against *Fasciola hepatica* in beef and dairy cattle, and is effective against both mature and immature flukes. However, its activity against *F. magna* is fair to poor. It is not effective against rumen flukes or lung flukes. Clorsulon is safe in pregnant and breeding animals.

cluster a genetic grouping of individuals, from a single or several nearby sampling sites, that share a similar multilocus genotypic profile.

cluster analysis a set of statistical methods used in many scientific disciplines to group objects within a data set based on similarity.

cluster determinant (CD) cluster of antigens with which antibodies react that characterize a cell surface marker.

cluster of differentiation a designation to one or more cell surface proteins.

cluster randomized trial (CRT) a type of randomized controlled trial in which groups of subjects (as opposed to individual subjects) are randomized.

clustered regularly interspaced short palindromic repeats (CRISPR) DNA elements containing short palindromic elements and internal 'spacer' DNA. In genome editing use, delivery of CRISPRs with spacer DNA that is homologous to a targeted genomic locus, together with a Cas9 protein, can result in double-strand DNA breaks.

clustered regularly interspaced short palindromic repeats/CRISPR-associated 9 system (CRISPR/Cas9 system) a system used by prokaryotes for immunization processes by integrating exogenous DNA to keep track of viral attacks and to defend their organism from invaders with similar DNA. New technologies now allow for the transfer of this system to eukaryotes. Customized Cas9 nucleases are guided by small RNAs to the target site in the genome to manipulate specific genes.

C_{max} maximum or peak concentration of drug achieved after administration.

CMG complex a higher order structure composed of CDC45, the MCM helicase complex, and the GINS complex. The

MCM helicase acts to unwind DNA during replication, but displays optimal activity only once it is associated with CDC45 and GINS.

co-adaptation selection by which harmoniously interacting genes accumulate in the gene pool of a population. Selective forces acting on one of the genes will also affect its interacting partners.

coagulation cascade a sequence of biochemical activities involving fibrin, platelets and endothelial cells that halt bleeding by forming a clot. Two pathways (cascades of proteases) lead to the conversion of fibrinogen to fibrin and platelet activation. The damaged wall is then covered by a platelet and fibrin-containing clot to stop bleeding and begin repair of the damaged vessel.

coalescent theory all alleles in a sample are assumed to derive from a common ancestor. Thus, it is possible to estimate the coalescent time of two alleles into a common ancestor. After one coalescent event, a sample of *N* alleles is reduced to *N* – 1. Time is considered to move backwards in the model until the time to a common ancestor for the entire sample is estimated. By modelling this stochastic process, a distribution of genealogies emerges, enabling the estimation of parameters of interest, such as the TMRCA.

coalescent tree each leaf of the tree corresponds to an observed allele, and the root represents the most recent common ancestor (MRCA) of all observed alleles. The internal nodes (branching points) represent the MRCA of the alleles at the leaves connected to that node (without passing the root). Distances along branches represent time, measured in generations.

coancestry (θ) a kinship coefficient defined as the probability that two homologous alleles, one drawn from each of two individuals, are identical-by-descent.

coccidia a group of intracellular parasites of the gut and other organs of vertebrates, with alternating asexual and sexual developmental phases resulting in the production of an environmentally resistant oocyst (usually in the faeces of the definitive host).

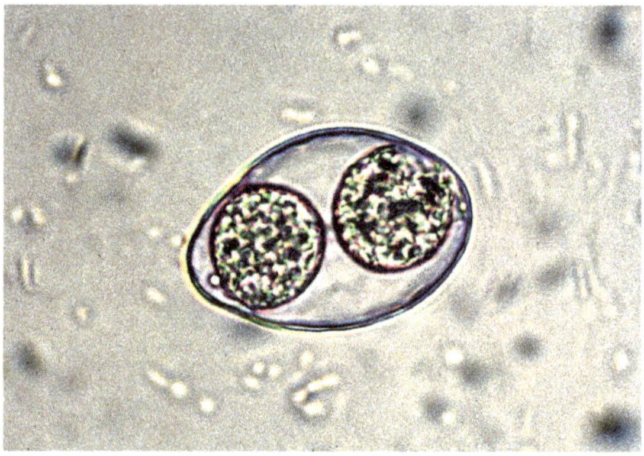

code *See* **genetic code**.

coding sequences DNA sequences that are translated into proteins. In conventional usage, all other sequences are 'non-coding'.

co-dominance a relationship between two versions of a gene. Individuals receive one version of a gene, called an allele, from each parent. If the alleles are different, the dominant allele usually will be expressed, while the effect of the other allele, called recessive, is masked. In co-dominance, however, neither allele is recessive and the phenotypes of both alleles are expressed. *See* **recessive gene**.

co-dominant expression of heterozygote phenotypes that differ from either homozygote phenotype. Microsatellites are co-dominant genetic markers, because one can distinguish a heterozygote (two bands) from each of the homozygotes (single band).

codon a trinucleotide sequence of DNA or RNA that corresponds to a specific amino acid. The genetic code describes the relationship between the sequence of DNA bases (A, C, G, and T) in a gene and the corresponding protein sequence that it encodes. The cell reads the sequence of the gene in groups of three bases. There are 64 different codons: 61 specify amino acids while the remaining three are used as stop signals. *See* **genetic code**.

codon harmonization during the translation in the ribosomes, low-frequency codons can cause translational pausing, favouring the correct folding of the proteins. Codon harmonization consists of matching the usage frequencies, particularly the low usage frequencies, of the native codons from the gene within the native organism as near as possible to codons recognized as low usage frequency codons by the surrogate host. This ensures that the positional codon frequency of low/intermediate and high usage codons remains unaltered in the surrogate host, allowing the needed halts in the translational processes to match that of the organism of interest (natural host). Differences in synonymous codon usage between the natural host and surrogate cells can lead to low protein expression and the formation of insoluble aggregates.

codon optimization most of the 20 genetically determined amino acids are encoded by multiple codons, which each species use at a distinct frequency. Protein synthesis in protozoan parasites relies on codons that are rarely employed by the surrogate organism in a heterologous expression system. Codon optimization consists of matching the frequencies at which different codons are used in the natural host (donor) and the surrogate organism.

codon usage bias the frequency of occurrence of synonymous codons in a coding sequence.

coefficient of relatedness (*r*) a measure of the degree of relatedness between individuals, ranging from −1.0 (no genes in common, at least over the genetic markers assayed) to +1.0 (identical twins or clones). In an outbred diploid population, siblings should have $r = 0.5$, individuals chosen at random should have $r = 0.0$.

coefficient of variation standard deviation divided by the mean.

coelozoic inhabiting the intestinal canal of the body.

coenurosis ('gid' or 'sturdy') is a disease of the brain and spinal cord caused by *Coenurus cerebralis*, the intermediate stage of *Taenia multiceps*, which inhabits the intestine of dogs, cats and wild carnivores. The clinical disease occurs in sheep and rarely in cattle. *C. cerebralis* causes cortical meningitis and encephalitis. An affected sheep holds its head to one side and turns in circles. There may be blindness, paralysis of limbs, and pressure atrophy of the brain and bones of the skull adjacent to the cyst.

Coenurus larva of *Taenia multiceps* which is a fluid-filled cyst containing many invaginated scolicies. It has a non-laminated wall that produces protoscolices, but without brood capsules.

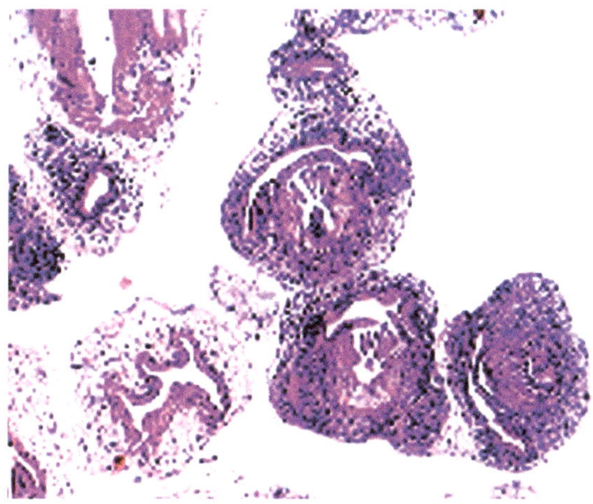

co-evolutionary arms race iterative, antagonistic cycles of evolution between pathogen, parasite or predator and host or prey, in which one continually drives the evolution of the other to better their chances of surviving or benefiting from the interaction. In the context of phytopathogens, the arms race often manifests in the evolution of new pathogen effectors and host nucleotide-binding domain and leucine-rich repeat-containing proteins (NLRs), which recognize and/or respond to the effectors.

cofactor an organic or inorganic substance required by an enzyme to function.

cohort study a study that begins with the gathering of two matched groups (the cohorts), one which has been exposed to a prognostic factor, risk factor or intervention and one which has not. The groups are then followed forward in time (prospective) to measure the development of different outcomes. In a retrospective cohort study, cohorts are identified at a point of time in the past and information is collected on their subsequent outcomes.

co-infection simultaneous infection of a host by two or more parasite species.

co-interventions interventions other than the treatment under study that may have been applied differently to the study and control groups. Co-intervention is a serious problem when double-blinding is absent or when the use of very effective non-study treatments is permitted.

co-isogenic or congenic nearly identical strains of an organism. They vary at only a single locus.

colicins a family of bactericidal proteins produced by some strains of *Escherichia coli*. They are composed of three domains, one of which causes death by nuclease or pore-forming activity.

co-linearity when the order of genes in the genome is the same as that of successive steps in the biosynthesis of a metabolite.

colitis inflammation of the large intestine.

collagen long fibrous structural protein composed predominantly of glycine-X-Y triplet amino acid repeats.

collinearity a specific form of synteny in which the same genetic loci are present on corresponding chromosomes and in corresponding order in different genomes.

colony stimulating factors (CSF) factors that permit the proliferation and differentiation of haematopoietic cells.

colpodellids several genera of predatory biflagellates that diverge near the phylogenetic origin of apicomplexan parasites, such as *Voromonas* and *Colpodella*. These flagellates share several homologous characteristics with apicomplexans, such as myzocytosis and an apical complex consisting of micronemes, rhoptries and a conoid.

combination B a multiprotein vaccine containing N-terminal sequences of MSP1, the 3D7 allele of merozoite surface protein 2 (MSP2) and RESA.

combination therapy the use of two or more classes of antimalarial drugs/molecules in the treatment of malaria that have independent modes of action.

combinatorial diversity that component of antibody and T-cell receptor (TCR) diversity that is generated by the recombination of variable (V), diversity (D, for immunoglobulin heavy chains, and for TCR β and δ chains) and joining (J) gene segments.

combinatorial joining the joining of segments of DNA to generate essentially new genetic information, as occurs with Ig genes during the development of B-cells. Combinatorial joining allows multiple opportunities for two sets of genes to combine in different ways.

commensal pertaining to a relationship between two organisms in which one organism benefits without adversely affecting the other.

commensalism (com = together, mensa = table) a symbiotic relationship between organisms of different species in which one, the commensal, is obligated to derive its nourishment from the other, its host, but in no way does it deprive or damage its host.

communicable disease a disease capable of being transmitted from an infected person or species to a susceptible host, either directly or indirectly.

community a group of different species of organisms living together and interacting in a particular ecosystem.

community effectiveness the ability of a particular intervention to alter the natural history of a particular disease for the better, under actual conditions of practice and use.

co-morbidity co-existence of a disease or diseases in a study participant in addition to the index condition that is the subject of study.

Companion Animal Parasite Council (CAPC) an independent council established to create guidelines for the optimal control of internal and external parasites that threaten the health of pets and people, brings together broad expertise in parasitology, internal medicine, public health, veterinary law, private practice, and association leadership.

comparative genomics the study of genetics by comparisons with other organisms.

comparison group any group to which the index group is compared. Usually synonymous with control group.

compatibility characteristic of a host–parasite system where the parasite species is capable of establishing infection and achieving transmission using this host species.

compatible vector the ability of an insect vector to support the development of a pathogen and transmit the disease.

compensatory mutations in highly integrated genetic systems, mutations in any one component can have a deleterious effect on fitness. However, mutations in several components could lead to a recovery of the fitness reduction or even lead to an overall increase in fitness compared with the current state.

competence a transient physiological state in which bacteria are proficient in the uptake of extracellular DNA.

competitive ability the ability of an organism to compete with others for access to limited resources, including food, territory, or mates.

competitive CSH cross-species co-hybridization of two RNA samples derived from two different target species to a microarray that represents either one or none of the target species.

complement a series of serum proteins involved in the mediation of immune reactions. The complement cascade is triggered classically by the interaction of complement-fixing antibody with specific antigen.

complement components an enzymatic system of serum proteins triggered by the classical and alternative pathways, and resulting in target cell lysis, phagocytosis, opsonization and chemotaxis.

complement factor C3b plasma protein derived from factor C3 that binds to antigens on pathogens to label them for destruction.

complement fixation (CF) test an immunological test to measure whether antibody–antigen complexes can deplete a known concentration of standard complement proteins and protect added red blood cells from lysis.

complement receptor a structure found on erythrocytes, lymphocytes, neutrophils, monocytes and macrophages that binds C3 fragments.

complement system a group of serum proteins, some of which act in an enzymatic cascade, that is activated by factors such as the combination of antigen and antibody and results in a variety of biological consequences including cell lysis and opsonization.

complementarity-determining regions (CDRs) sequences within the variable region of Ig receptors that make contact with antigen and determine B-cell specificity. The antigen-binding site of the antibody consists of six CDRs comprising three hypervariable loops each from the heavy- and light-chain variable region.

complementary DNA (cDNA) DNA that is synthesized in the laboratory from a messenger RNA template.

complementary RNA (cRNA) synthetic transcripts of a specific DNA molecule or fragment made by an *in vitro* transcription system.

complementary sequence nucleic acid base sequence that can form a double-stranded structure with another DNA fragment by following base-pairing rules (A pairs with T and C with G). The complementary sequence to GTAC for example, is CATG.

complement-like immune system a part of the insect immune system, analogous to the vertebrate complement system. It includes the thioester-containing protein 1 (TEP1) and leucine-rich repeat proteins LRIM1 and APL1.

complete genome sequence a high-quality, low error, gap-free DNA sequence of an entire genome of an organism.

complex disease is caused by the interaction of multiple genes and environmental factors. Complex diseases are also called multifactorial. Examples of complex diseases include cancer and heart disease.

complex genetics a term used in contrast to Mendelian patterns of inheritance, whereby instead of being dominantly or recessively inherited a phenotype is caused by multiple genes that could interact with each other and/or the environment.

complex trait a trait that has a genetic component that does not follow strict Mendelian inheritance. May involve the interaction of two or more genes or gene–environment interactions.

component populations all of the individuals of a parasitic species from one individual host at a particular time (infra-populations) from a population of hosts.

computational biology development and application of data-analysis and theoretical methods, mathematical modelling, and computational simulation techniques to the study of biological systems. *See* **bioinformatics**.

computer vision a field that processes, analyses and interprets images in order to produce numerical information.

concentration method a procedure carried out on a faecal specimen that increases the number of organisms found in a given unit over examination of a similar size unit before concentration.

concomitant immunity the phenomenon in which active infection with an organism prevents additional infections by the same species of organism. In filariasis, active infection with adult filarial worms confers partial protection against further infections by invading L3 larvae.

concordance the rate at which a second individual has the same phenotype as the first, for example in the second sibling of a pair, or the second twin.

concurrent disinfection immediate disinfection and disposal of discharges and infective matter all through the course of a disease.

condensed tannins polyphenols of high molecular weight that consist mainly of oligomers or polymers of monomeric units of flavan-3-ols (catechin, epicatechin, and so on). They are also described as proanthocyanidins. Depending on the chemical structure of the monomeric unit, in particular the number of hydroxyl radicals, they are classified into four groups, the two commonest being procyanidins and prodelphinidins.

conditional dependence two tests are conditionally dependent when the sensitivity (or specificity) of the second test depends on whether the results of the first test are positive or negative among infected (or non-infected) individuals.

conditional logistic regression analysis a variation on logistic regression that permits conditioning on having an affected offspring in every case parent trio, necessary for family data.

conductance an electrical measure of the rate of ion transport through an ion channel.

confidence interval (CI) a measure of the precision (or uncertainty) of study results for making inferences about the population of all such patients. The 95% CI is the range of values within which we can be 95% sure that the true value lies for the whole population of patients from whom the study patients were selected. Wide confidence intervals indicate less precise estimates of effect. CI is affected by sample size and by variability among subjects.

conformational (or discontinuous) epitopes formed through folding of a protein antigen bringing together discontinuous amino acid sequences to form the epitope.

confounded experimental factors are confounded when their potential effects cannot be distinguished. For example, if control and experimental measurements are obtained on different days, treatment and day effects are confounded.

confounding variable (confounder) a characteristic that may be distributed differently between the study and control groups and that can effect the outcome being assessed. Confounding may be due to chance or bias.

congenic animals which only differ at a single genetic locus.

congenital conditions are those present from birth. Birth defects are described as being congenital. They can be caused by a genetic mutation, an unfavourable environment in the uterus, or a combination of both factors.

congenital hypothyroidism an inherited trait that results in reduced activity of the thyroid gland, generally due to reduced production of thyroid stimulating hormone. The trait reduces the base rate of the body's chemical reactions, and results in tissue swelling and weight gain. It can cause neurological and development problems.

congruence agreement among or within phylogenetic data sets.

conjugate covalently linked complex of two or more molecules (e.g. fluorescein conjugated to antibody).

conjugation a temporary union of two ciliated protozoans and some bacteria for the exchange of nuclear material, which in bacteria is unidirectional.

conoid a cone-shaped spirally coiled filaments in the anterior tip of the zoite of certain apicomplexan protozoa. It functions as scaffolding for rhoptries of the apical complex. Conoids can be completely closed, as in many apicomplexans, or open-sided, as in colpodellids and perkinsids.

conserved hypothetical proteins the (often large) fraction of genes in sequenced genomes encoding proteins that are found in organisms from several phylogenetic lineages but have not been functionally characterized and described at the protein chemical level. These structures may represent up to half of the potential protein coding regions of a genome.

conserved sequence a base sequence in a DNA molecule (or an amino acid sequence in a protein) that has remained essentially unchanged throughout evolution.

Conserved Structural Motif Search Tool (COSMOS) a program that identifies RNA motifs based on the assumption that the regulatory network of RREs is conserved across different species. This alignment-free method searches for linear and structural motifs in the 3′-UTRs of multiple genomes.

consomic and congenic strains by repeatedly backcrossing one strain with another, it is possible to create animals that have a particular genomic region from one strain and the remainder of their genome from the other. Subsequent intercrossing makes the genomic segment homozygous and the mouse fully inbred. Thus, consomic and congenic mouse strains are created by replacing either a single chromosome (in a consomic or chromosome substitution strain) or a small chromosomal segment (in a congenic strain) with one from another strain.

conspecific an animal or organisms belonging to the same species.

constant region (C region) the invariant carboxyl-terminal portion of an antibody molecule, as distinct from the variable region which is at the amino-terminal of the chain.

constitutive ablation gene expression that results in cell death.

constraint-based analysis a methodology for analysis of metabolic networks that uses various constraints (physicochemical, topological, environmental or regulatory) to limit the range of achievable functional states that are accessible to the organism.

contact a person or animal that has associated with an infected person or animal that might provide an opportunity to acquire the infection.

contact rates the number of potentially infectious contacts made per infected host per unit time. In the context of vectors transmitting zoonotic pathogens, a potentially infectious contact is one that is capable of resulting in infection if contact occurs with an infected vector.

contig a set of overlapping DNA segments derived from a single genetic source that can be used to deduce the original DNA sequence.

contig map a map depicting the relative order of a linked library of overlapping clones representing a complete chromosomal segment.

contractile vacuole in *Balantidium coli* and many free-living protozoa, especially ciliates, a vacuole that is associated with removing liquid wastes from the body of the organism. In living ciliates, the contractile vacuole near the surface of the protozoa can be seen to fill with clear fluid, then by sudden contraction, discharge the fluid through the surface of the body then to disappear, only to reappear in the same place as a small vacuole that begins to fill again.

control reduction of disease incidence, prevalence, morbidity or mortality to a locally acceptable level as a result of deliberate efforts that comprise therapy and prevention (prophylaxis).

control event rate (CER) the percentage of the control/non-exposed group who experience outcome.

control measures actions or measures (e.g. elimination of reservoirs and vectors) adopted to restrain or reduce the prevalence of individual disease.

controlled reproductive number (*R*c) a quantitative description of whether and to what extent (in the presence of local interventions such as long-lasting insecticidal nets, indoor residual spraying, and detect-and-treat tactics) a single index case is likely to result in onward transmission. High R_c indicates a highly endemic area where control measures have not resulted in preventing significant onward transmission and where a single index case represents a risk of an outbreak. $R_c < 1$ indicates sustained onward transmission is unlikely.

controlled vocabulary (CV) a restricted list of terms relevant to a topic.

convergent evolution process by which a similar character evolves independently from different ancestors (also known as convergence).

cookie a packet of data passed from one computer application to another, which is stored and can be retrieved at a later date. Cookies often store your personal information and preferences so that you can sign on to a system and your prior preferences are remembered by the system.

Coombs' test diagnostic test that uses anti-immunoglobulin to agglutinate antibody-coated erythrocytes.

cooperation a phenomenon whereby two or more entities interact to provide benefits for themselves that are greater than those possible by the operations of the entities in isolation.

Cooperia adult nematode worms measure 5–8 mm in length and males have a prominent bursa in relation to their size. *Cooperia* are generally considered to be mild pathogens. They contribute secondary effects to the primary pathogens *Ostertagia* and *Haemonchus* in parasitic gastroenteritis.

coproantigens proteins released by parasites in the gastrointestinal tract that can be detected in the faeces for diagnostic purposes.

coproculture culture of the faeces for the purpose of hatching parasite eggs and obtaining larvae for morphological identification.

coprophagous a coprophagous organism is one that eats the faeces/excrement of another animal. Many insect species are coprophagous and often specialize in the consumption of faeces from large herbivores. Some of the most well-known coprophagous insects are the dung beetles.

coprophagy the consumption of faeces.

coproscopy parasitological technique by microscopy using faecal samples.

copulatory bursa the male bursa that embraces the female nematode during copulation. The morphological features of the bursa is useful for the identification of some species of nematodes.

copy number difference (CND) loss or gain of genomic material between individuals of different species.

copy number variant (CNV) variable number of copies of one or more segments of the DNA in a given organism. CNVs correspond to relatively large regions of the genome that have been deleted or duplicated.

coracidia (singular **coracidium)** actively swimming, free living larvae of diphyllobothriid (pseudophyllidean) tapeworms with aquatic life cycles.

core genome the set of genes shared by all members of a group of strains being examined. It is an operational definition that changes depending on the set of strains being examined, so the core genome of an entire parasitic species will be smaller than the core genome of a specific lineage of strains.

correlates of protection measurable signs that a host is immune and protected against becoming infected by a parasite and/or developing disease.

correlation study a study aiming to reveal a statistical association between two factors (e.g. toxoplasmosis and schizophrenia) by comparison of their occurrence in various populations.

cortex an outside (peripheral) layer of an organ, e.g. cortex of the kidney.

cortisol stress response cortisol is a stress hormone released by the interrenal cells of the kidney within minutes of an animal undergoing a stressful event. This hormone can be maintained at high levels by chronic or acute stressors. If maintained at high levels for extended periods of time, cortisol can have deleterious effects on host reproduction, immunology, ionic balance and many other processes.

cosmid a plasmid that contains the lambda phage cos sequence, which allows it to be packaged and amplified in a phage capsid. These hybrid plasmids have a much higher maximum size than traditional bacterial plasmids. The larger inserts of cosmids allow construction of parasite transfection vectors that efficiently target genes with very long flanks of homologous DNA to mediate homologous recombination.

cost–benefit analysis measures the difference between the costs of a control strategy and the expected reduction in economic loss from the disease. These are typically carried out to compare the expected economic effects of different control strategies.

costa a striated, rod-like structure that lies just under the recurrent flagellum of certain protozoa of the order Trichomonadida.

cost-effectiveness economic evaluation to assess what kind of results are produced for the amount of money spent.

co-transporter a molecule that enables the simultaneous and coupled movement of two substances across a cell membrane in the same direction.

counter-transport coupled movement of two substances across a cell membrane in opposite directions.

covariates variables that may confound the outcome variable of a statistical model; for example, age is a covariate of human weight.

coverage the number of reads that span a given genomic position.

CpG island an intergenic genomic region that contains a greater density of unmethylated CG dinucleotides than expected compared with genome-wide density. CpG islands were originally defined as regulatory regions in mammals. As insects largely lack intergenic DNA methylation, it remains unclear whether CpG island-like sequences are functional in insects.

CpG sites genomic regions of several hundred bps with a high GC content and many unmethylated CpG dinucleotides.

CR1 (first complement receptor) a receptor on the surface of phagocytes, including macrophages and neutrophils, that bind predominantly to C3b and C4b proteins of the complement cascade, facilitating clearance of particles opsonized by these proteins via phagocytosis.

C-reactive protein (CRP) an acute phase protein which is able to bind to the surface of microorganisms where it functions as a stimulator of the classical pathway of complement activation, and as an opsonin for phagocytosis. This serum molecule increases rapidly at the onset of infection.

crenarchaea a phylum of archaea distinguished from other phyla based on ribosomal RNA sequence. Crenarchaea are the most abundant type of marine archaea.

crenosomatids the nematodes belonging to the Crenosomatidae family.

CRISPR–Cas (clustered regularly interspaced short palindromic repeat–CRISPR-associated protein system) a genome editing technique for generating a site-specific mutant or transgenic organisms using the Cas9 protein-guide RNA complex to generate mutations or to direct exogenous DNA to specific genomic regions where it is incorporated via homologous recombination.

critical photoperiod the number of hours of daylight that results in diapause entry for 50% of the insect population, excluding those that do not enter diapause under unambiguous short day-lengths.

cross-hybridization having transcripts of more than one gene hybridized to one spot.

crossing over the type of recombination in which chromosomes exchange segments resulting in reciprocal recombinant genotypes. For example, in the heterozygote Ab/aB a cross over between the a and b genes would produce the recombinant chronosomes AB and ab.

cross-platform hybridization aliquots from the same RNA sample, or RNA isolated from the same biological source, are hybridized to two or more different microarray platforms.

cross-presentation the process by which an exogenous antigen is processed and presented via major histocompatibility complex I.

cross-reaction the reaction of an antibody directed against one antigen, with a second antigen. This occurs because the two antigens possess epitopes in common or similar epitopes.

cross-reactivity the ability of an antibody, specific for one antigen, to react with a second antigen; a measure of relatedness between two different antigenic substances.

cross-strain (within-species) hybridization (CTH) hybridization of RNA from one (target) strain to a microarray that contains DNA (cDNA or oligomers) from another (reference) strain of the same species.

cross-transmission transmission of a parasite from one host species to another host species.

Crustacea a group of aquatic arthropods having a hard exoskeleton, jointed bodies and appendages, and gills (e.g. lobsters, shrimps, crabs, copepods). The latter includes species, which cause direct damage to the host, such as the crustacean copepod *Lambroglena* species.

cryoelectron microscopy a technique that preserves the native state of biological material in a frozen hydrated state, which is usually achieved by rapid freezing in liquid ethane. Specimens are examined in a transmission electron microscope at ultralow temperatures and can be visualized without the need for heavy-metal staining.

cryptic genetic variation genetic variation present in a population that is not phenotypically expressed under benign or ambient conditions, but which may be visible upon genetic or environmental perturbations.

cryptic species two or more distinct species that are erroneously classified under one species name.

cryptosporidiosis a parasitic disease caused by *Cryptosporidium* species. The organism primarily infects the gastrointestinal epithelium, producing acute watery diarrhoea. Cryptosporidiosis is self-limiting in healthy adults and children. However, in immunocompromised patients, it might cause chronic diarrhoea, for which there is no effective treatment.

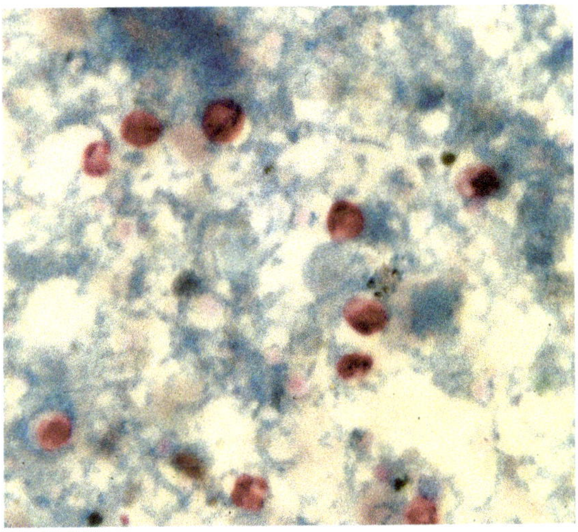

crypts the tubular invaginations of mucosa at the base of the intestinal.

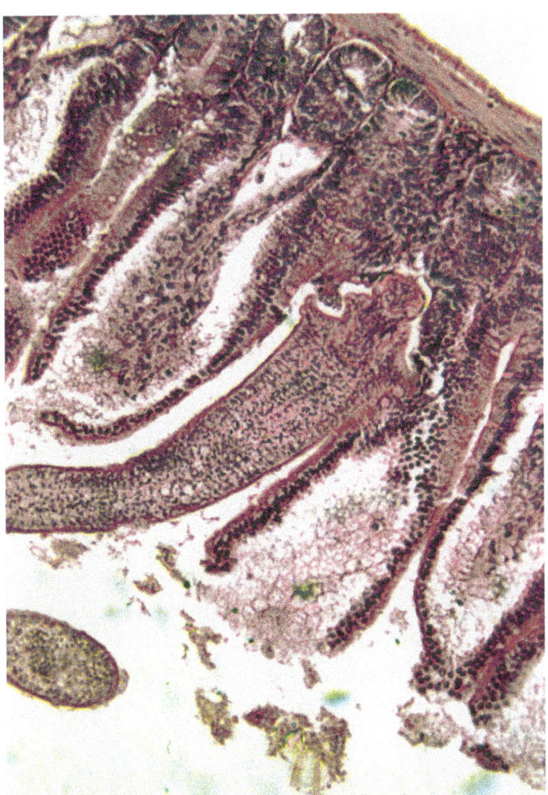

crystallinity degree of order presented by the molecules of a solid structure. The more organized the molecular structure is, the more crystalline is the solid.

CS antigen the major antigenic protein found on the surface of plasmodial sporozoites.

CSP the circumsporozoite protein of *P. falciparum*, an abundant, multifunctional and immunodominant protein that is present on the sporozoite surface.

CT (computed tomography) also known as 'CAT scanning' (Computerized Axial Tomography). This diagnostic method provides a form of imaging known as cross-sectional imaging, which is different from conventional X-ray imaging. It is used to create detailed images of internal organs, bones, soft tissue and blood vessels.

Ctenocephalides felis the main flea species infesting both dogs and cats, causing flea bite allergy dermatitis. *C. felis* can also transmit *Rickettsia felis* and some *Bartonella* species, such as *Bartonella henselae*. The *Ctenocephalides* fleas are also the primary intermediate host of *Dipylidium caninum*, the common pumpkin seed intestinal cestode of dogs and cats.

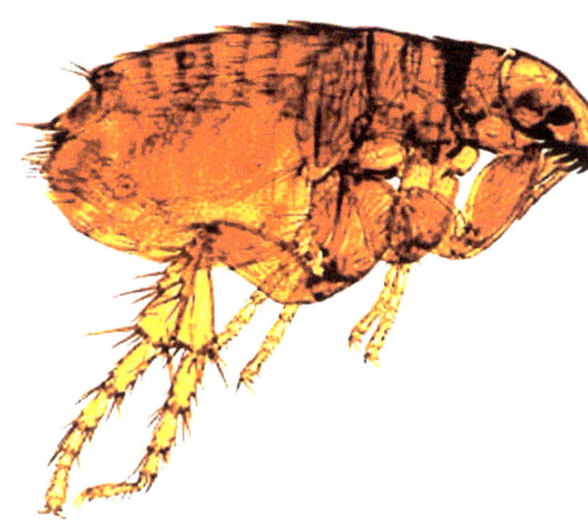

c-type cytochromes small haem-containing proteins that have an important role in respiratory electron transfer reactions.

culicine mosquito *See* **anopheline mosquito**.

cultured ELIspots peripheral blood mononuclear cells are cultured with antigen for 10 days and then restimulated for 18 hours in an ELIspot assay to detect primarily memory T-cells.

cumulative evolution the ongoing process by which parasites colonize new hosts and undergo speciation, increasing parasite diversity.

cure rate (CR) the proportion of individual hosts positive for parasites who become parasitologically negative after treatment.

cutaneous leishmaniasis (CL) an infection caused by vector-borne protozoa of the *Leishmania* species. Clinical signs range from small cutaneous nodules to gross mucosal tissue destruction. The disease is rarely fatal and can be subdivided into 'self-healing' or 'non-self-healing' forms.

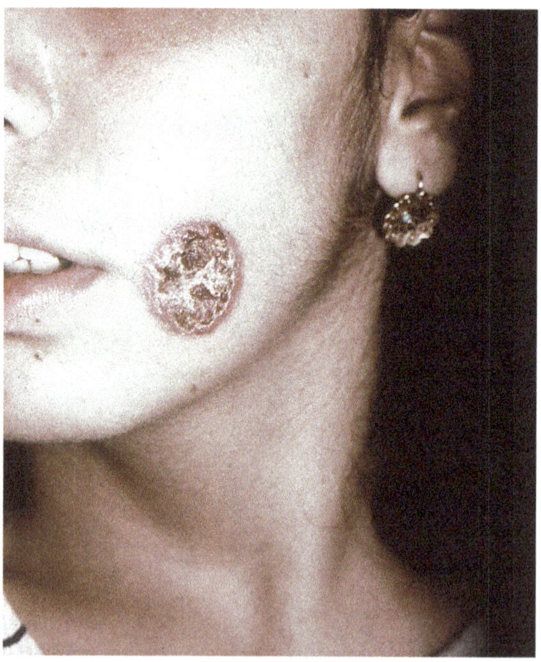

cutaneous T-cell antigen (CLA) a unique skin-homing receptor expressed by memory T-cells that infiltrate the skin. It facilitates targeting of T-cells to inflamed skin.

cuterebriasis infestation of skin of living animals with larvae of *Cuterebera* spp. Normally, the larvae crawl into the fur of the host, enter the body via a natural orifice, and then migrate to a subcutaneous site. Aberrant migration can occur, in unnatural host, to the central nervous system, trachea, nostrils, pharynx or intraocularly. In general, a disease seen in dogs, cats, and rabbits in warm climates in the late summer and autumn and imported animals in temperate climates. Animals are contaminated by picking up eggs laid by adult flies in the environment usually close to burrows or nests of rabbits or rodents, which are their natural hosts. Lesions usually found as solitary nodules on the head, neck and trunk. Nodular swelling is approximately 1 cm across with a central air hole and fistula. Other clinical signs can occur dependent on the site of aberrant migration. People are susceptible to wound myiasis. Blowfly maggots have been used clinically in people to clean wounds and promote formation of granulation tissue.

cuticle the outer covering layer secreted from the hypodermis or subcuticular layer in nematodes.

CXCR3 cell surface receptor for chemokines CXCL9, CXCL10, and CXCL11.

CXCR4 C-X-C chemokine receptor type 4 (CD184) a receptor for chemokines such as SDF-1 and is also named CXCL12.

cyanobacteria division of photosynthetic bacteria found in many environments, including oceans, fresh water, and soils. Cyanobacteria contain chlorophyll and other photosynthetic pigments in an intracellular system of membranes called thylakoids. Many cyanobacterial species also are capable of nitrogen fixation.

cyclin-dependent kinases (CDKs) serine/threonine kinases that are activated by association with cyclins to form CDK complexes which regulate cell-cycle progression.

cyclodevelopmental a relationship in which the parasite undergoes cyclic morphological change, but does not multiply.

cyclooctadepsipeptide (CODP) cyclic depsipeptides consisting of four alternating N-methyl amino acid moieties and four 2-hydroxy-carboxylic acid moieties.

cyclophosphamide cytotoxic drug used as an immunosuppressive.

cyclophyllidea (cyclophyllidian) an order of tapeworms. Tapeworms having four sucker discs encircling the scolex as in

Taenia saginata and sometimes also a rostellum with hooks as in *Taenia solium.*

cyclopoid copepods members of the order Cyclopoida, a major group of zooplanktonic organisms in both freshwater and marine ecosystems that often serve as intermediate hosts of fish parasites.

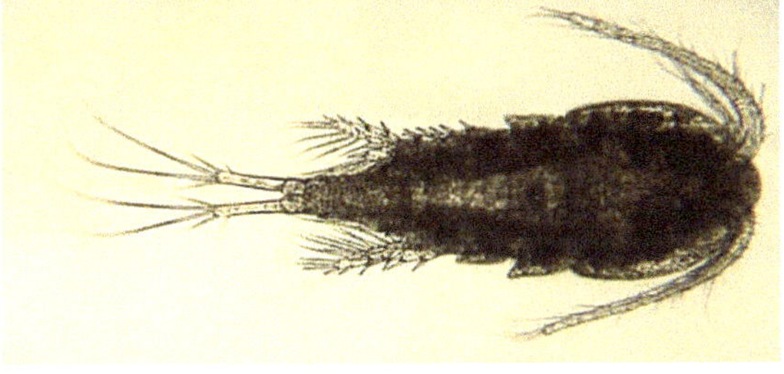

cyclopropagative a relationship in which the parasite undergoes cyclic, morphological changes and also multiplies.

cyclops or water fleas a genus of minute crustaceans with aquatic life cycles, some species of which act as hosts of *Diphyllobothrium* and *Dracunculus* spp.

cyclosporine A a T-cell specific immunosuppressive drug used to prevent graft rejection.

cyclozoonosis a zoonotic disease that requires at least two species of vertebrates as definitive and intermediate hosts. Examples include hydatid disease (*Echinococcus granulosus*) and trichinosis (*Trichinella spiralis*).

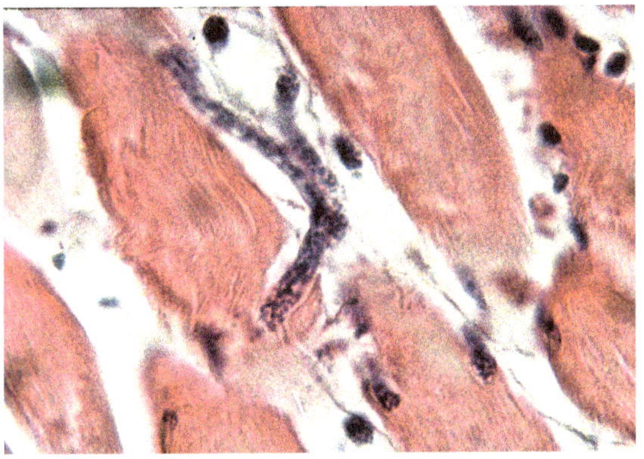

cys6 proteins possess characteristic domains with six conserved cysteines. Cys6 domains are unique to *Plasmodium* species.

cyst a cyst stage typically enables a parasite to survive outside the host, and is hence also the infective stage. It is usually characterized by a thick and resistant cyst wall. Cystic stage is a common resistant stage in protozoa. Excystation or hatching of cysts releases trophozoites.

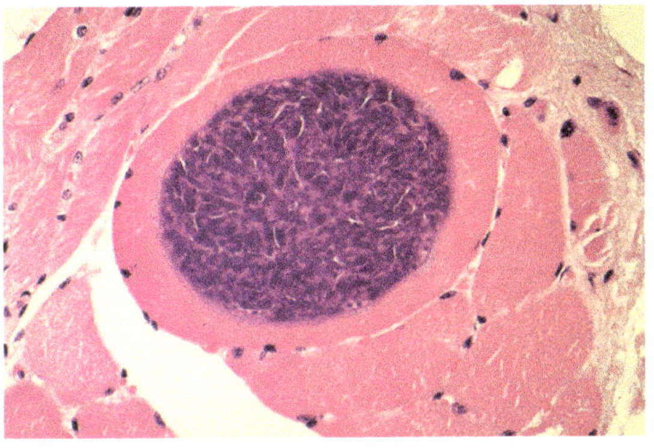

cyst nematodes nematodes belonging to the genus *Globodera*, including the soybean cyst nematode and potato cyst nematode that attack plant roots.

cyst wall the outermost part of the protective coating protecting an encysted protozoan.

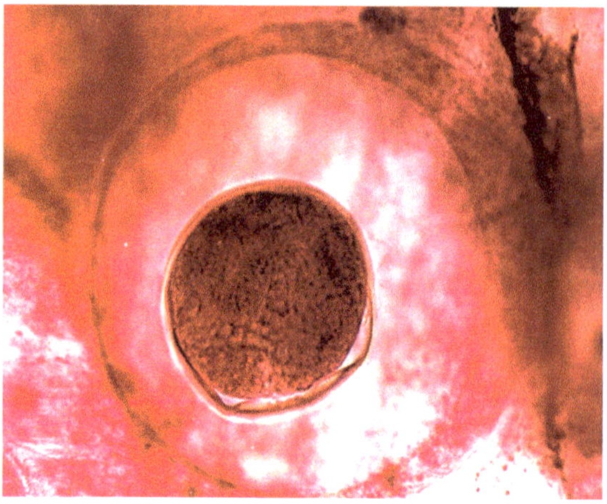

cyst wall synthase this is the name given to the recently described β1,3-N-acetylgalactosamine transferase, which is induced when *Giardia* trophozoites encyst and polymerizes the UDP-N-acetylgalactosamine into giardan. *See* **giardan**.

cystacanth the developmental stage of an acanthocephalan parasite at which it becomes infective to the final (or paratenic) host.

cysticercoid a larva of tapeworms in which the scolex is invaginated into a greatly reduced cystic cavity almost devoid of fluid. Examples, include tapeworm larval stage of *Hymenolepis* spp. and *Dipylidium caninum* that occurs in arthropod intermediate hosts.

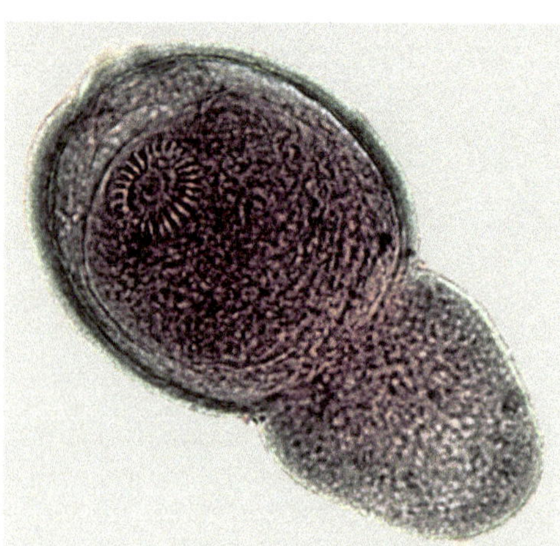

cysticercosis parasitic tissue infection caused by larval cysts of the tapeworm *Taenia solium*. These larval cysts infect brain, muscle, or other tissue, and are a major cause of adult onset seizures in most low-income countries. A person gets cysticercosis by swallowing eggs found in the faeces of a person who has an intestinal tapeworm. People do not get cysticercosis by eating undercooked pork. The highest rates of infection are found in areas of Latin America, Asia, and Africa that have poor sanitation and free-ranging pigs that have access to human faeces.

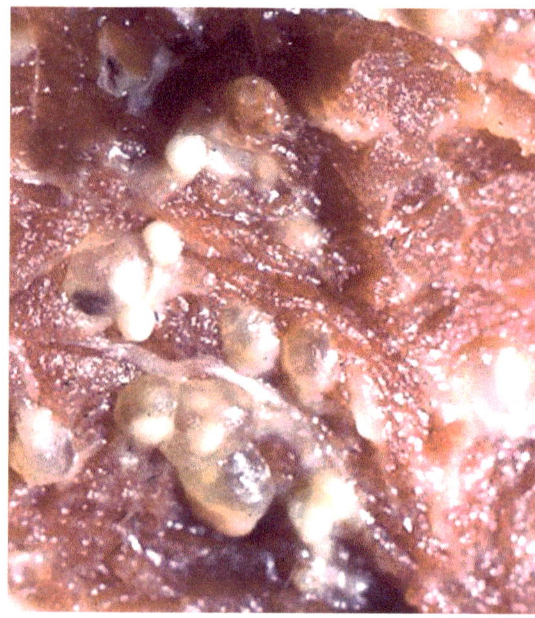

cysticercus a tapeworm larva in which the scolex develops in an inverted form and which has a fluid-filled bladder surrounding it. It is characteristic of the cyclophyllidean family Taeniidae. *See* **bladder worm**.

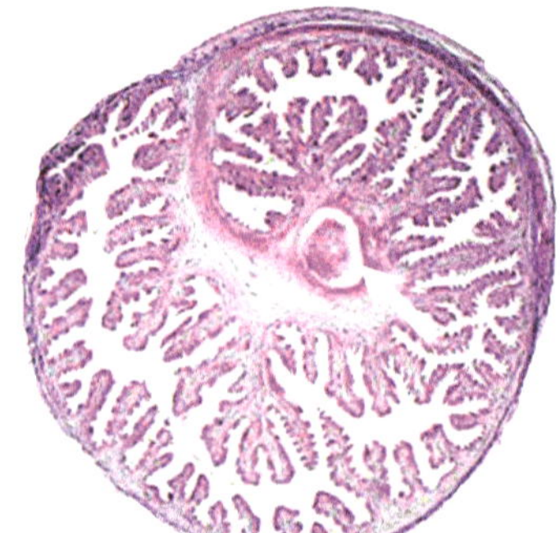

Cystoisospora (synonym ***Isospora***) ***suis*** the causative agent of neonatal porcine cystoisosporosis, a highly prevalent infection and leading cause of diarrhoeal disease in neonatal suckling piglets accompanied by diarrhoea and, in severe cases, weight loss and wasting. Ingestion of oocysts from the environment is considered the most important way of infection and usually occurs in contaminated farrowing crates within the first days after birth, resulting in oocyst shedding after a prepatent period of 4–6 days. Older pigs are generally less susceptible to *C. suis* and excrete few or no oocysts, without clinical signs of disease upon infection. Cystoisosporosis can be controlled by metaphylactic medication with toltrazuril which is applied in the first week of life as an individual single oral treatment in combination with hygiene measures.

cytoadherence the binding of parasite-infected erythrocytes to human host endothelial cells.

cytochrome c liberation protein of the intermembrane space of mitochondria, which translocates into the cytosol upon activation of a mitochondrial apoptotic pathway.

cytochrome P450 1A1 biotransforming enzyme of vertebrates, which is involved in the metabolism of xenobiotics such as benzo(a)pyrene and other polycyclic aromatic hydrocarbons.

cytochrome p450 27B1 (CYP27B1) a cytochrome P450 enzyme, also known as 5-hydroxyvitamin D_3 1α-hydroxylase, located on the inner mitochondrial membrane. It catalyses hydroxylation of 25-hydroxyvitamin D_3 at the 1α position to synthesize 1α,25-dihydroxyvitamin D_3, the active form of vitamin D_3, which then binds to the vitamin D receptor and regulates calcium metabolism.

cytochromes membrane-bound, haemgroup-containing electron-transfer proteins that are especially involved in chemiosmotic energy conservation in the aerobic and anaerobic respiratory chains or photosynthesis.

cytocidal chloroquine resistance (CQRCC) CQR phenotype as quantified by the shift in LD_{50} or cytocidal potency.

cytogeneticist a geneticist who specializes in the study of chromosomes and the structure and function of the cell.

cytogenetics the branch of genetics that studies the structure of DNA within the cell nucleus. This DNA is condensed during cell division and form chromosomes. The cytogenetic studies the number and morphology of chromosomes. *See* **karyotype**.

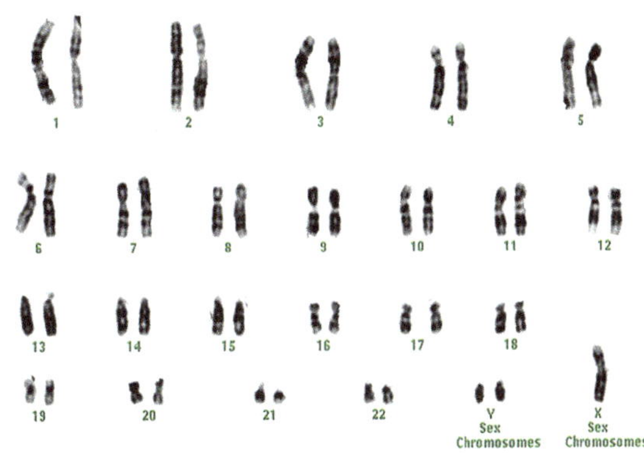

cytokine-inducible nitric oxide synthase (iNOS) cytokine-inducible enzymes that catalyse the production of nitric oxide.

cytokine-mediated immunoregulation Th ($CD4^+$) populations consist of different Th subsets which have different effects on the immune response; during an immune response naive (Th0) lymphocytes may differentiate into two distinct subsets (Th1 or Th2) which can be defined by the patterns of cytokines they produce and the immunological effect they orchestrate.

cytokines soluble low molecular weight proteins produced by cells in response to various stimuli, including parasitic infection. They affect the behaviour of other cells both locally and at a distance, by binding to specific cytokine receptors. *See* **chemokine**.

cytological band an area of the chromosome that stains differently from areas around it.

cytological map a type of chromosome map whereby genes are located on the basis of cytological findings obtained with the aid of chromosome mutations.

cytology the study of living cells, their origin, structure, function and pathology.

cytolytic granules present in cytotoxic T-cells and natural killer cells contain an array of proteins with varied functions from forming pores in the cell membrane and entering the cytoplasm of a target cell to initiating a cascade of events that ultimately leads to the destruction of the target cell.

cytopharynx the chamber behind the mouth (cytostome) in protozoa into which food may be taken then passed into the cytoplasm via food vacuoles formed at its base.

cytophilic binds to cells.

cytoplasm the protoplasm of a cell surrounded by the cell membrane exclusive of the nucleus.

cytoplasmic granules granular-appearing materials within the cytoplasm, especially in amoebae.

cytoplasmic incompatibility (CI) a phenomenon that results in sperm and eggs being unable to form viable offspring. The effect arises from changes in the gamete cells caused by intracellular parasites, such as *Wolbachia*, which infect a wide range of insect species.

cytoplasmic trait a genetic characteristic in which the genes are found outside the nucleus, in chloroplasts or mitochondria. Results in offspring inheriting genetic material from only one parent.

cytopyge in *Balantidium* and some other ciliates, the permanent opening in the ectoplasm through which the residue of digestion of food materials is discharged (anal opening). In other organisms, the wastes may be discharged through a number of points in the body surface.

cytosine (C) is one of four chemical nitrogenous bases in DNA, the other three being adenine (A), guanine (G), and thymine (T). Within the DNA molecule, cytosine bases located on one strand form chemical bonds with guanine bases on the opposite strand. The sequence of four DNA bases encodes the cell's genetic instructions. *See* **bp**, **nucleotide**.

cytosol the intracellular fluid found within cells, excluding that within membrane-bound organelles.

cytosomal tube actin dependent tubular elongation that arises from the cytostome into the parasite cytoplasm. These tubes were suggested to either bud-off to small haemoglobin vacuoles or be directly connected to the food vacuole.

cytostatic chloroquine resistance (CQRCS) CQR phenotype as quantified by the shift in IC50 or cytostatic potency.

cytostomal flagellum a flagellum that lies in a grove behind the cytostome (usually within the cytopharynx) and directs solid food particles to the posterior end where food vacuoles are formed (e.g. in *Chilomastix*).

cytostome 'cell mouth,' a feature used in some protozoan parasite for nutrient ingestion. In *Plasmodium falciparum*, the cytostome is involved in haemoglobin uptake from the cytoplasm of the infected RBCs. The cytostome is formed by endocytic invagination of the parasitophorous vacuole membrane and parasite plasma membrane into the cytoplasm of the parasite.

cytotoxic (cytolytic) a T-cell that kills target cells bearing appropriate antigen within the groove of an MHC class I molecule that is identical to that of the T-cell.

cytotoxic T-cell T-cell subsets characterized by expression of the CD8 co-receptor, restricted to association with antigen/ MHC class I complexes, and critical for protective immunity against intracellular pathogens via induction of apoptosis in pathogen-infected cell targets.

cytotoxic T lymphocyte (CTL, Tc) T-cells (usually $CD8^+$) which kill target cells following recognition of foreign peptide–MHC molecules on the target cell membrane.

cytotoxic T-lymphocyte antigen 4 (CTLA4) a co-inhibitory molecule expressed by activated and regulatory T-cells and is essential to immune homeostasis maintenance, but can also down-regulate T cell response and proliferation.

D gene a small segment of immunoglobulin heavy-chain and T-cell receptor DNA, coding for the third hypervariable region of most receptors.

DAF-12 (dauer formation-12) a nuclear hormone receptor that is downstream of the insulin/IGF-1 and transforming growth factor (TGF)-β pathways and that regulates dauer diapause in many nematode species.

dahomey gap the portion of the Guinean forest–savanna mosaic that extends all the way to the coast in Benin, Togo, and Ghana, thus separating the forest zone that covers much of the south of the region into two separate parts. The forest region west of the gap is called the Upper Guinean forests or Guinean forest zone, and the portion east of the gap is called the Lower Guinean forests, Lower Guinean–Congolian forests, or Congolian forest zone.

damage- (or danger-) associated molecular patterns (DAMPs) molecules that are generally present due to a non-infectious threat to cellular integrity and are capable of initiating signalling cascades, through PRRs such as NLRs and TLRs. Some examples are proteins released during DNA damage such as cytosolic DNA and RNA fragments, or nucleotides such as ATP. Fragments of damaged tissue such as hyaluronan, uric acid, and heparin sulfate have also been described as potent DAMPs.

damage-associated molecular patterns (DAMPs) host-derived conserved molecular patterns associated with damage to cells and tissues, which, via interactions with PRRs, trigger contingent immune responses.

data integration combining data residing in different sources to provide a unified view. For example, admission date, vital signs (temperature), laboratory tests (white blood cells in urine), and microbiology (urine culture) to assess for symptomatic urinary tract infection.

data mining/text mining using computer programs to assess a set of data for associations that are not obvious in clinical practice or everyday life.

data warehouse a relational database (subsequently defined) designed for query and analysis rather than for transaction processing. It usually contains historical data derived from transaction data, but it can include data from other sources. It separates the analysis workload from the transaction workload and enables an organization to consolidate data from several sources. In simple terms, the data warehouse is a separate place from the electronic medical record (where transactions, or data entry, are performed) that receives the data from all of the records and stores in a way that can be queried and analysed. In general, the electronic medical record is good at storing chronological data by individual patient, and clinicians can get the history of that patient from the electronic medical record.

dauer diapause in response to environmental stimuli indicative of poor growth conditions, *Caenorhabditis elegans* can enter an alternative developmental state. This arrested dauer larva is extremely stress-resistant and long-lived, and when conditions improve animals can exit dauer diapause and resume development.

DC-SIGN (dendritic cell-specific intercellular adhesion molecule-3-grabbing non-integrin) a receptor for mannosylated glycoconjugates that is found on macrophages and mature dendritic cells. DC-SIGN has been implicated in pathogen recognition as well as leucocyte adhesion via attachment to intercellular adhesion molecules (ICAM)-2 and -3.

de novo novel, not inherited from parents. A genetic alteration that arises either in the gamete or in the fertilized egg during its early development but is not observed in the somatic cells of the parent.

DE-52 weak anion exchanger chromatographic resin based on the diethylaminoethyl (DEAE) tertiary amine functional group.

dead-box RNA helicases member of Superfamily 2 (SF2) of helicases. Characterized by presence of 12 conserved motifs: Q-motif, motif I, motif Ia, motif Ib, motif Ic, motif II, motif III, motif IV, motif IVa, motif V, motif Va, and motif VI. Motif II contains the amino acid sequence DEAD (Asp-Glu-Ala-Asp), from which the family name was derived. The motifs are involved in ATP binding, hydrolysis, ATP-dependent RNA binding, and helicase activities. DEAD-box helicases occur in all cells: central to metabolism of RNA, from transcription to decay.

dead-end host one in which the parasite reaches an end point and is unable to continue its life cycle. *See* **aberrant host**.

death domain (DD) defines a large superfamily of proteins primarily involved in apoptosis and inflammation. There are four main subfamilies based on the homotypic interaction domains present including the death domain (DD), death-effector domain (DED), caspase recruitment domain (CARD), and the pyrin domain (PYD). These domains mediate interactions that allow formation of large signalling complexes.

death-effector domain (DED) a homotypic interaction motif that belongs to the DD superfamily and is present in proteins found to regulate a variety of cell signalling pathways including apoptosis.

death-inducing signalling complex (DISC) a ternary assembly comprising the FAS receptor, FADD, and caspase-8 or caspase-10, which is formed upon FasL binding. The DISC activates downstream caspases.

deazaflavin (F_{420}) an unusual 8-hydroxy-5-deazaflavin cofactor closely related to FMN.

deazaflavin-dependent nitroreductase (Ddn) a nitroreductase from *Mycobacterium tuberculosis* which catalyses the deazaflavin-dependent reduction of nitroimidazoles such as PA-824.

decay rate the rate at which a variable in a model exponentially declines.

deep sequencing WGS of a pure sample (e.g. DNA extracted from liquid culture that was inoculated from a single bacterial colony) is usually sequenced to at least a 10- to 15-fold depth (i.e. each position in a genome is sequenced 10–15 times). However, when sequencing samples that might be heteroresistant, or might contain multiple strains of the same species, greater sequencing depth is required to identify minority variants or to separate the strains.

defensins a family of small basic antimicrobial peptides, produced by both animals and plants.

definitive host a host in a life cycle in which the parasite attains sexual maturity and produces propagules.

degradation a gradual wearing down or away. Also, with regard to soil, a lowering of the nutrient content and associated ability to support continuing crop growth.

degradome the complete repertoire of peptidases in an organism or the subset of peptidases that is involved in a defined biological event.

degranulation the process in which cytoplasmic storage granules fuse with fagosomes, discharge their contents and disappears from the cytoplasm.

delayed death effect a phenomenon observed in *Toxoplasma* and *Plasmodium* in which treatment with antiparasitic drugs does not kill the treated parasites directly but cause them to produce progeny that are unable to complete a replicative life cycle. The death of the progeny occurs even if the drugs are removed before the completion of the parents' life cycle.

delayed-type hypersensitivity (DTH) a cell-mediated inflammatory reaction in the skin, so-called because it takes 24 to 48 hours to reach maximum intensity. Typical example: tuberculin reaction.

deletion a type of mutation involving the loss of genetic material. It can be small, involving a single missing DNA bp, or large, involving a piece of a chromosome. *See* **chromosome**, **mutation**.

deletion analyses a method used to narrow down regulatory elements in 3′-UTRs. A series of 3′-UTR deletion fragments are fused to a reporter gene; changes in the expression level of the reporter gene suggest a regulatory region of interest.

deletion map a description of a specific chromosome that uses defined mutations – specific deleted areas in the genome – as 'biochemical signposts,' or markers for specific areas.

delta pH difference in pH across a membrane.

delta psi difference in potential across a membrane.

demodectic mange (or follicular mange) mites of *Demodex* spp. infest hair follicles of all species of domestic animals. Mites infesting the different host species are considered to be specific and are designated as *Demodex bovis* for cattle, *D. ovis* for sheep, *D. caprae* for goats, *D. equi* for horses, and *D. phylloides* for pigs. The entire life cycle is spent on the host. as well as reducing the value dramatically.

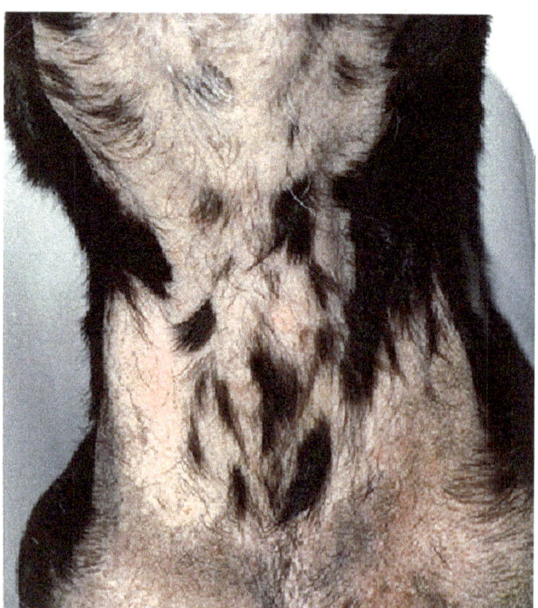

dendritic cell (DC) a key antigen-presenting phagocytic cells responsible for activating naive T cells and initiating cellular adaptive immunity.

dengue haemorrhagic fever (DHF) dengue viral infection with high fever, haemorrhagic phenomena, and hepatomegaly. There are common haemorrhagic diatheses such as positive tourniquet test (TT), thrombocytopenia, petechiae, easy bruising, and/or gastrointestinal haemorrhage with evidence of plasma leakage, such as increased haematocrit in severe cases.

dengue shock syndrome dengue infection presenting all of the criteria of DHF plus circulatory disturbance and shock as evidenced by rapid and weak pulse or age-specific hypotension and cold, clammy skin, and restlessness.

dense granules specialized secretory organelles found in apicomplexan parasites that contain a variety of proteins predominantly involved in modifying the host cell immediately following invasion.

density estimation tool for enzyme classification (DETECT) a probabilistic method for enzyme prediction based on the distribution of global alignment scores for an entire enzyme family. High quality predictions are based on a probability score greater 0.2 with at least five positive matches.

density gradient the varying sequence of densities of materials suspended in a liquid column with the most dense material at the bottom and the least dense material at the top. A phenomenon that results from gravity sedimentation.

density regulation a reduction in the number (the density) of individuals of a population.

density-dependence populations tend to increase in number until they are limited by a number of potential forces. The slower growth at higher numbers represents density dependence. Competition for food or other resources, predation, infectious disease and parasitism are common mechanisms of density dependence. A consequence is that most control methods will lead to suppression, but not elimination of the target population, possibly with additional effects on population dynamics. It is also possible in some circumstances to produce fewer but healthier individuals; such individuals might even be more capable of disease transmission.

deoxyribonucleic acid (DNA) a molecule of DNA consists of a long chain of nucleotides that are composed of deoxyribose, a five-carbon sugar, a phosphate group linked to the bases (nucleotides) adenine (A), guanine (G), cytosine (C), and thymine (T). A pairs with T and C pairs with G. DNA contains the genetic code that controls the production of proteins in living organisms.

deoxyribonucleotides compounds that are the fundamental units of DNA. Each nucleotide contains a nitrogenous base, a pentose sugar and phosphoric acid. The four bases characteristic of deoxyribonucleotides are adenine, guanine, cytosine and thymine. *See* **nucleotide**.

deoxyribose a type of pentose sugar that is one component of DNA.

Dermacentor reticulatus a hard tick that bites humans and transmits several *Rickettsia* spp., Omsk haemorrhagic fever virus or Tick-borne encephalitis virus. It can transmit *Babesia canis* to dogs. *D. reticulatus* ticks have a long lifespan and can survive for up to 4 years without taking a blood meal. They are able to tolerate −10°C for 150 days under laboratory conditions and are active during the winter in many climatic zones at temperatures at which *Ixodes ricinus* is not active.

dermal DCs (DDCs) a subset of DCs which patrol the dermis (lower layer) of the skin for invading pathogens.

dermatophytosis fungal infection caused by *Trichophyton verrucosum* or *Trichophyton mentagrophytes* that may cause multiple, round crusted skin lesions that may coalesce on the head, neck, tail, and perineum.

dermatotropic a higher affinity for the skin.

desorption removal of a substance in the reverse of absorption or adsorption.

detection of resistance this usually involves the treatment of infected hosts with a recommended dose of drug followed by a faecal egg count reduction test that compares pre-treatment egg counts with those of untreated controls. Reductions of less than 90–95% (based on group arithmetic means) score as clinical resistance.

detergent-resistant membrane (DRM) liquid ordered cholesterol- and sphingolipid-rich membrane which is insoluble in non-ionic detergents such as Triton X-100.

determinant any definable factor that effects a change in a health condition or other characteristic.

determinant part of the antigen molecule which binds to an antibody-combining site or to a receptor on T-cells. *See* **epitope**, **hapten**.

deterministic model a model where no randomness or chance is present and every state always leads predictably to the next.

detritivore an organism that uses particulate dead organic matter for food.

detritus particulate matter originating from the disintegration of biological materials such as microbiota and tissues.

diagnostic biomarker reports the concurrent presence or absence of injury.

diapause an anticipated, typically long-term, cyclical interruption in growth or development of an organism due to one or more environmental factors that occur well before adverse environmental conditions are manifest.

diapause incidence proportion of the insect population entering diapause under unambiguous short daylengths.

diapedesis the emigration of blood cells through intact capillary walls, that occurs mainly during inflammation.

diatom type of microscopic, photosynthetic algae known for its intricately designed, silica-containing shell. Thousands of diatom species are known; most are unicellular, but some form colonies. Diatoms are responsible for a large portion of photosynthetic carbon assimilation in marine and freshwater environments.

DIC disseminated intravascular coagulation.

dicer complex initiates RNAi by processing long double-stranded RNA (dsRNA) into small interfering RNA (siRNA).

dichlorophene a narrow-spectrum cestocide used to treat *Taenia* spp. and *Dipylidium* spp. infections in dogs and cats, with variable efficacy against *Echinococcus* spp. It causes uncoupling of oxidative phosphorylation, to deplete ATP from tapeworms, and disrupts the pH difference across the external tegumental membranes. Vomiting and diarrhoea may be seen after dichlorophen administration. Animals that are hypersensitive to dichlorophen may exhibit salivation, vomiting and anorexia.

diclazuril a benzeneacetonitrile, a class with potent anticoccidial activity. It is effective against various stages of the life cycle by inhibiting nuclear division. It is used for the treatment and prevention of coccidial infections in lambs caused in particular, by the more pathogenic *Eimeria* species *E. crandallis* and *E. ovinoidalis*. Also, it is used to aid in the control of coccidiosis in calves caused by *Eimeria bovis* and *E. zuernii.*

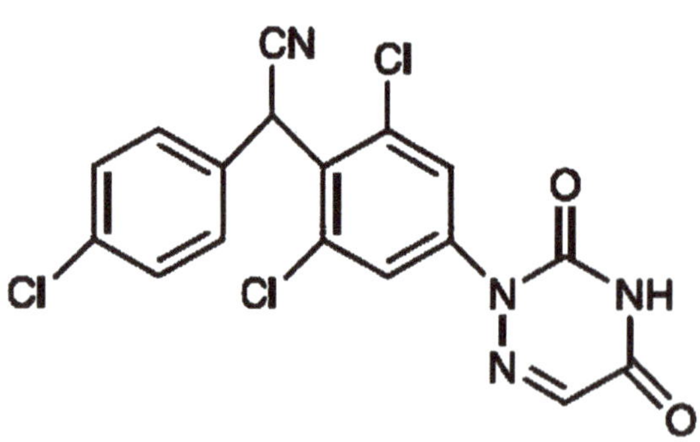

***Dicrocoelium dendriticum* (small lanceolate fluke)** *Dicrocoelium dendriticum* can infect sheep, goats, cattle, deer and rabbits, and occasionally horses and pigs. The life-cycle differs from other fluke in that there are two intermediate hosts; namely, land snails and brown ants. Miracidia are already embryonated within the eggs when shed in faeces, these are ingested by suitable snails and hatch in the snail. The miracidium becomes a mother sporocyst which becomes a daughter sporocyst releasing cercariae. Cercariae are released in slime balls, which are subsequently eaten by ants where the cercariae encyst to become metacerariae. These infected ants are, in turn, ingested by grazing animals. The chances of this happening are greatly enhanced by the parasite within the ant's brain altering host behaviour, such that infected ants migrate to the tips of blades of grass. The transmission of *D. dendriticum* is, thus, more terrestrial and not dependent on access to water, unlike *F. hepatica*. Because there is no migratory phase within the liver, infected livers can be relatively normal. However, in heavy infestations, there can be extensive focal liver damage accompanied by distension of the bile ducts.

Dictyocaulus filaria is similar to *Dictyocaulus viviparus* and is found in bronchi and bronchioles. It has a direct life cycle with eggs being released in either sputum or faeces and they infect the next host without an intermediate host. Young animals are more susceptible. Affected animals exhibit respiratory signs, such as cough, dyspnoea, lethargy, weight loss, and catarrhal bronchitis.

didelphic having two ovaries and hence a uterus with two horns; usually applied to nematodes.

diel over a 24-hour period of night and day.

diel vertical migration a pattern of diurnal movement between surface and deeper waters used by some aquatic organisms, mitigating the conflicting demands of finding food and avoiding predators.

Dientamoeba fragilis a unicellular intestinal non-flagellated flagellates parasite common in humans, also found in some non-human primates and pigs. Two genotypes are known, one of which appears to be rare.

difference equation a type of recurrence equation, which describes the dynamic behaviour of a system in terms of discrete time-steps where the present value of a state variable is calculated from its value at a previous time.

differentiation of Th2 lymphocytes occurs preferentially due to infection by extracellular pathogens (parasitic worms). The Th2 pathway inhibits Th1 activity and activates the humoral (antibody) immune system via the production of cytokines, which promote B lymphocyte growth and differentiation (such as IL-4) and production of IgG1 (IL-4), IgA (IL-5) and IgE (IL-4 and IL-13). The Th2 response also stimulates effector cells that use these antibody isotypes or are activated by Th2 cytokines. For example, mast cells and basophils will degranulate if IgE, bound to their surface, subsequently binds parasite antigen, while IL-5 (a Th2 cytokine) induces degranulation of eosinophils.

differential splicing the utilization and splicing of different exons from a primary RNA transcript in order to generate different mRNA sequences.

differentiation the process by which cells and tissues undergo a series of changes resulting in their specialization to a specific form or function. A differentiated cell such as a muscle cell or a skin cell contains a full set of genes of that organism, but only expresses the genes necessary for its specific function. In animals, stem cells (both embryonic and adult) are the only cells capable of undergoing differentiation to form more specialized cell types found in the body.

differentiation antigen a cell surface molecule expressed at a particular stage of development or on cells of a given lineage.

digenetic a subclass of parasitic flatworms with complex life cycles, with a sexually reproducing adult in a definitive host, and asexually reproducing larvae in a first intermediate host, which is almost always a mollusc.

digestive vacuole (DV) lysosomal-like organelle within mature RBC stages of *P. falciparum* involved in digestion of red cell haemoglobin., and the free haeme is converted into hemozoin.

dihydrofolate reductase inhibitors such as trimethoprim, ormetoprim, are always combined with sulfonamides to inhibit sequential steps in folate metabolism and increase their antiprotozoal action (this is an example of synergy). These potentiated sulfonamides are effective against *Toxoplasma* and *Neospora* spp. Pyrimethamine is another dihydrofolate reductase inhibitor used in the treatment of toxoplasmosis in small animals and for equine protozoal myelitis. It is often combined with other drugs that share some antiprotozoal action such as clindamycin or sulfonamides. The drug is associated with a higher incidence of gastrointestinal disturbances than trimethoprim, particularly in cats. Bone marrow suppression may occur. Concurrent administration of folinic acid prevents the development of mammalian folate deficiency. Very high doses of pyrimethamine are teratogenic in laboratory animals.

dilatation (dilation) an abnormal distension of the tubule lumen.

dilute environment spacious habitat with few apparent boundaries where organisms occur in extremely low densities.

dilution effect the disruption of parasite transmission in a natural ecosystem because the low suitability as hosts of the species added as diversity increases. It is not a universal effect because new species in a community may be very suitable hosts for existing parasites, having positive impacts on their dynamics, creating a 'spillback' effect on the original hosts, which suffer from higher infection rates as a consequence.

dinoflagellate a diverse clade of biflagellates that share a novel flagellar apparatus and novel nuclear features including permanently condensed chromosomes that apparently lack canonical histones. Some dinoflagellates are predators, some are parasites, and roughly half are photosynthetic. Some of the photosynthetic species rank among the most important players in the primary fixation of carbon in the oceans, whereas others are responsible for toxic algal blooms.

Dioctophyma renale a nematode found worldwide with a wide range of mammalian host species, such as dog, wolf, cheetah, mink, horse, swine, and humans. Adult worms are found in the host's kidney and eggs are excreted in the urine. Thereafter, they are ingested by an aquatic annelid, which is an intermediate host. Fish-eating mammals ingest the infected annelid and serve as transport hosts. Ingestion of the annelid by the definitive hosts results in migration of the larva from the intestine to the kidney, which is the maturation site for adult worms. Infected patients develop haematuria and pain, probably as a result of migration of worms.

dioecious refers to species in which the sexes are distinctly separate. Male reproductive organs are present in one individual, the female organs in another. Example in flukes – the schistosomes.

dioecy *See* **dioecious**.

diploid a full set of genetic material consisting of paired chromosomes, one from each parental set. *See* **haploid**.

diplomixis a process by which the two nuclei of *G. intestinalis* (and probably other diplomonads) fuse in the cyst stage, leading to exchange of genetic material between the nuclei.

diplostomids referring to a superfamily of digenetic trematodes, Diplostomoidea (includes the families Cyathocotylidae, Diplostomatidae, and Strigeidae), that are commonly found in birds and mammals.

diprotic able to bind two protons.

dipterans insects of the order Diptera (meaning two wings) (e.g. flies, midges, mosquitoes) that includes several families with vectors of African trypanosomes, malaria and other haemosporidians.

direct enzyme detection native enzymes from a biological specimen may be localized by coupling the enzyme's activity to production of an electron-dense deposit that is detectable by electron microscopy. The restricted set of enzymes appropriate to this application greatly limits the use of this method.

direct feeding assay (DFA) the numbers of oocysts are counted in the midguts of mosquitoes fed on the skin of a treated host are compared with those of a replicate group fed on the same host before treatment.

direct life cycle a life cycle of a parasite in which the pre-parasitic larvae develop in the environment as free-living stages in which the eggs will either hatch and the pre-parasitic larvae are entirely free-living (e.g. *Ostertagia ostertgia*) and the infective stage is usually an L3 and infection occurs via ingestion of infective stage or skin penetration or the eggs do not hatch and pre-parasitic larvae develop inside their eggs (e.g. *Ascaris suum*) and infective stage is usually the egg containing the infective larva and infection of the definitive host occurs via ingestion of the infective stage.

direct membrane feeding assay (DMFA) as for the DFA (*see* **direct feeding assay**) except that blood is withdrawn before treatment, divided into replicate aliquots and experimental reagents, and relevant control reagents are added to separate aliquots. These are then fed to replicate groups of mosquitoes through artificial membranes.

direct transmission a process in which a parasite is transmitted directly from an infected host to an uninfected host without the involvement of an intermediate host or vector. Sexually transmitted diseases are spread by direct transmission.

direct variant repeat (DVR) a repeated sequence, composed of a direct repeat (DR), plus an adjacent non-repetitive spacer sequence.

directed evolution a laboratory process used on isolated molecules or microbes to cause mutations and identify subsequent adaptations to novel environments.

directed mutagenesis alteration of DNA at a specific site and its reinsertion into an organism to study any effects of the change.

directed sequencing successively sequencing DNA from adjacent stretches of chromosome.

directional selection selection (negative or positive) that would cause the loss or fixation of an allele in a population.

Dirofilaria immitis the aetiological agent of heart worm. Adult female worms measure up to 30 cm long and live in the pulmonary arteries of dogs, leading to endothelial damage and myointimal proliferation. Pulmonary artery damage results in lobar arterial enlargement, tortuosity, and obstruction, causing pulmonary hypertension and thrombosis. *D. immitis* has a complex and indirect life cycle, with dog as definitive host and mosquito plays a crucial role as intermediate host. Female adult worms release L1 (microfilaria) into circulation (can live for 2 years). L1 in circulation may infect mosquitoes during feeding; develop to infective larval stage (L3) in mosquito and injected back into dog to migrate to heart. Microfilaria cannot develop in the dog of origin. They must pass through a mosquito to develop. Prepatent period ranges from 6 to 7 months.

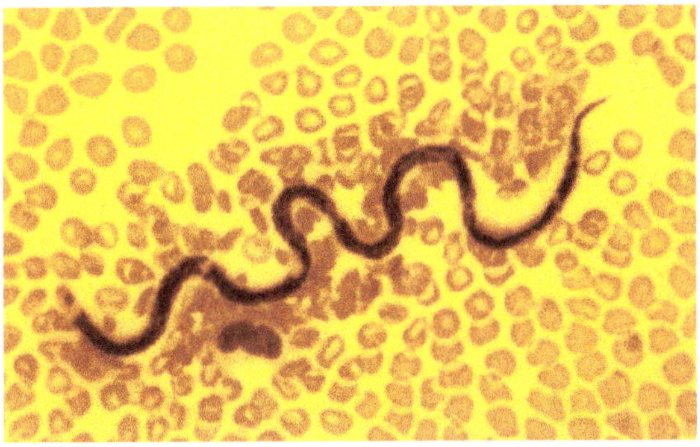

disability-adjusted life years (DALYs) the number of healthy years lost to illness, morbidity, or death; an assessment of the overall disease burden.

disassortative mating non-random mating system in which unlike individuals pair.

discordance between phylogenies comparison between two phylogenetic trees gives differing topologies (structures), suggesting that the genes or species in the tree have different evolutionary histories. A frequent use is to identify instances where parasite speciation does not mirror that of their hosts, a probable consequence of host switching.

discrete typing units (DTUs) subdivision used to classify *Trypanosoma cruzi* strains. Strains within a DTU are genetically more similar to each other than to strains from other DTUs.

discrimination or fractionation factor the difference in isotopic composition between an animal and its diet.

disease abnormal performance of physiological functions as a result of injury to cells performing those functions.

disease cost estimate estimates of the value of losses in the expected output and/or resources due to a specific disease, together with the expenditures associated with treatment or preventative measures.

disease liability an underlying or latent continuous variable such that those with a liability above a threshold are considered diseased. The quantitative trait of liability reflects both genetic and environmental factors.

disease-inducing threshold parasite infection level at which pathogenic signs related to infection begin to emerge (i.e. skin ulceration, lethargy and morbidity). Infection levels below this threshold are considered to be subclinical.

disgust a system based in neural tissue that evolved to detect reliable signals co-occurring with disease-causing infectious agents and which stimulates behaviours tending to reduce the risk of disease. Disgust and immunity are part of a defensive continuum: disgust acts before or immediately following contact, whereas immunity deals with infectious agents that evade evasion and penetrate the body boundary.

disinfection killing of infectious agents outside the body by direct exposure to chemical or physical agents. High-level disinfection refers to the inactivation of all microorganisms except some bacterial spores.

disomy the presence of two copies of a given chromosome. In normal, diploid organisms, all chromosomes are disomic.

dispersal the movement of individuals that results in gene flow (i.e. transfer of genetic material) between populations.

disruptive selection natural selection within a single population towards two different (usually extreme) phenotypes, for example, when large or small individuals have an advantage over those of intermediate size.

disseminated bronchiectasis permanent dilatation of the airways (bronchi) throughout the lungs.

disseminated intravascular coagulation (DIC) pathological activation of blood coagulation in peripheral tissues, resulting in consumption of clotting factors and increased risk of bleeding.

distribution phase the phase following drug administration during which the drug distributes to and exchanges with the tissues. During this phase, blood concentrations fall faster than during the elimination phase.

disturbance any abrupt event that drastically changes ecosystem characteristics such as population diversity, behaviour, or climate response.

divergence the magnitude of DNA sequence difference between different species or populations.

divergence rate the rate at which DNA sequences in a species pair accumulate differences after splitting from a common ancestor. The divergence rate is twice the substitution rate along one species/lineage.

divergent selection natural selection in different directions within each of several populations, for example, when large size is favoured in one population, whereas small size is favoured in another. Intermediates are never favoured.

diversification rate the net rate of increase in species numbers within a group, equal to the speciation rate minus the extinction rate over a given period.

diversity a term used to describe the effective number of taxa (that is, the richness) of a particular rank and their respective abundances (that is, the evenness). Both negative frequency-dependent selection and diversifying selection for microbial resistance have been suggested to result in the diversity observed in natural systems.

diversity (D) gene segments found in the immunoglobulin heavy chain gene and T-cell receptor β and δ gene loci between the V and J gene segments. Encode part of the third hypervariable region in these antigen receptor chains.

diverticulum a sac-like structure extending out from a tubular organ such as a blood vessel or intestine.

dixenous parasite a parasite with a life cycle split between two host species; for example, as during insect-mediated transmission among vertebrates.

dLN for draining lymph node; is the lymph node draining the site of vaccination, infection or tumour.

d*N*/d*S* the ratio of non-synonymous (amino acid-altering) changes per non-synonymous site to synonymous changes per synonymous site. This ratio is used to infer the type of evolutionary pressure acting on a protein: when $dN/dS > 1$ positive Darwinian selection is inferred; when $dN/dS = 1.0$, strict neutrality is inferred; and when $dN/dS < 1$, purifying selection is inferred.

DNA barcodes a taxonomic method that uses a short genetic marker (e.g. the DNA sequence of the mitochondrial gene cytochrome c oxidase I) as a rapid means to identify that an organism belongs to a particular species. DNA barcodes rely on significant variation in DNA sequences between species and relatively uniform DNA sequences within species.

DNA barcoding the diversity of life can be discriminated through the analysis of a small segment of the genome, in animals a stretch of the mitochondrial cytochrome oxidase I (COI) gene. This promising approach to the diagnosis of biological diversity has gained broad acceptance in the scientific community. Currently, the major Barcoding of Life projects focuses on fish, birds and Lepidoptera (butterflies and moths).

DNA carrier substance or particle that can transfer genes into a cell. These include viruses, liposomes (fat globules) and artificial chromosomes (sequences of DNA created in a laboratory) that can transport large amounts of DNA. *See* **vectors**.

DNA damage nucleic acids are constantly being compromised by various endogenous and exogenous insults. This damage can include cross links, base modifications, base mismatches, stalled replication forks, single-strand breaks (SSBs), and particularly dangerous double-strand breaks (DSBs). To preserve genome integrity eukaryotes use pathways to detect and repair damaged DNA, and checkpoints to arrest cell cycle progression to allow for DNA repair and prevent transmission of damaged DNA to daughter cells.

DNA fingerprinting based on hybridization of non-specific probes to genomic DNA fragments obtained after digestion with restriction endonucleases. Restriction fragments are separated by polyacrylamide gel electrophoresis (often abbreviated as PAGE) or on agarose gels and blotted onto a membrane before hybridization. The resulting banding patterns are transformed into presence or absence matrices and analysed by different phylogenetic methods.

DNA fluorescence *in situ* hybridization (DNA FISH) an *in situ* hybridization technique in which probes consisting of short (<500 nucleotides) fluorescently labelled DNA fragments are hybridized to the complementary target sequences in the nucleus.

DNA laddering cleavage of genomic DNA between nucleosomes during apoptosis; can be visualized by appearance of a DNA ladder after agarose gel electrophoresis.

DNA methylation a biochemical process where a methyl group is added to a cytosine nucleotide, typically found in a CpG sequence. DNA methylation represses gene expression when found in CpG-rich (CpG island) promoter regions.

DNA methyltransferase (DNMT) an enzyme that catalyses DNA methylation. There are two functional DNMT classes in metazoa. DNMT1 carries out maintenance DNA methylation using a pre-methylated DNA sequence as a template, whereas DNMT3 is responsible for de novomethylation of DNA. DNMT2 was originally defined as a DNA methyltransferase but was later correctly recognized as a transfer RNA methyltransferase.

DNA polymerase an enzyme that helps in the replication of DNA molecules.

DNA profiling a laboratory technique used to establish a link between biological evidence and a suspect in a criminal investigation. A DNA sample taken from a crime scene is compared with a DNA sample from a suspect. If the two DNA profiles are a match, then the evidence came from that suspect. Conversely, if the two DNA profiles do not match, then the evidence cannot have come from the suspect. DNA fingerprinting is also used to establish paternity.

DNA repair genes genes encoding proteins that correct errors in DNA sequencing.

DNA replication the process by which a molecule of DNA is duplicated. When a cell divides, it must first duplicate its genome so that each daughter cell winds up with a complete set of chromosomes.

DNA sequence the relative order of bps, whether in a DNA fragment, gene, chromosome, or an entire genome. *See* **base sequence analysis**, **genome sequence**.

DNA sequencing a laboratory technique used to determine the sequence of bases (A, adenine; C, cytosine; G, guanine; T, thymine) in a DNA molecule. The DNA base sequence carries the information a cell needs to assemble RNA molecules and subsequently proteins. DNA sequence information is important to scientists investigating the functions of genes.

DNA substitution models evolutionary models that estimate the probability of nucleotide substitutions per site (a site is a position in a group of aligned DNA sequences) correcting for different factors.

DNA taxonomy nucleotide sequences are, as a first approximation, taken as a tag for the identification of the species that was the source for the extracted genetic material. These sequences are deposited in public databases, and the DNA extracts serve as molecular vouchers to be stored in natural history collections.

DNA:RNA immunoprecipitation (DRIP) a technique to isolate DNA–RNA hybrids from cells using a specific monoclonal antibody.

Dobzhansky–Muller incompatibilities epistatic incompatibilities that appear in hybrids as a result of interaction between alleles that did not lower fitness when they were substituted within the diverging lineages.

Dobzhansky–Muller model a model of evolution based on the accumulation of genetic incompatibilities between populations (i.e. post-zygotic isolation). According to this model, post-zygotic isolation is a multigene trait that evolves by the accumulation of new mutations, eventually leading to fixation of different alleles in separate populations. Once hybrids are formed, their heterozygotic composition (for such loci) significantly reduces or disrupts hybrid fitness, thus leading to post-zygotic reproductive isolation. The model also predicts that post-zygotic reproductive isolation cannot be a trait encoded by a single locus.

domain discrete portion of a protein with its own function. The combination of domains in a single protein determines its overall function.

domestication a process under which new sequences, obtained by horizontal gene transfer and endogenization, may acquire novel functions in the host genome.

dominance a genetic interaction between the two alleles at a locus, such that the phenotype of heterozygotes deviates from the average of the two homozygotes.

dominant refers to the relationship between two versions of a gene. Individuals receive two versions of each gene, known as alleles, from each parent. If the alleles of a gene are different, one allele will be expressed; it is the dominant gene. The effect of the other allele, called recessive, is masked. A dominant trait will be expressed in individuals who are either homozygous or heterozygous. *See* **gene**, **genome**.

donor fatigue a state in which donors no longer contribute to a cause because they have become tired of receiving appeals for donations.

dopamine a monoamine neurotransmitter and member of the catecholamine family that plays a fundamental role in many neurological processes including reward-driven learning. Abnormal levels of dopamine are characteristic for many neurological and psychiatric diseases, such as Parkinson disease and schizophrenia.

dormancy and reversible cell cycle arrest cell quiescence, hibernation, dormancy, or reversible cell cycle arrest are denominations of a common and important physiological response in free-living microorganisms to control cell size and growth that grants protection against environmental insults including poor nutrient and micronutrient levels.

dorsal stress fibres long parallel actomyosin bundles that are anchored to focal adhesions at one end.

dosage this information is always available on the package insert and so there is no excuse to make a mistake. Always be sure that you are using the correct dosage for the species.

dosage sensitive pertaining to genes for which the associated phenotypes are directly affected by modifications of their transcriptional output (such as changes in the number of active copies in the genome). For example, genes that encode subunits of a protein complex are dosage sensitive, as individual gene duplication leads to an imbalance among the parts of the complex.

dose–response relationship a relationship in which change in amount, intensity, or duration of exposure is associated with a change – either an increase or decrease – in frequency or intensity of a specified outcome.

dosing/treatment regimen the timing and number of doses of an antimalarial used to treat malaria. This may vary by patient weight or age.

double helix the description of the structure of a DNA molecule. A DNA molecule consists of two strands that wind around each other like a twisted ladder. Each strand has a backbone made of alternating groups of sugar (deoxyribose) and phosphate groups. Attached to each sugar is one of four bases: adenine (A), cytosine (C), guanine (G), or thymine (T). The two strands are held together by bonds between the bases, A forming a bp with T, and C forming a bp with G.

double peak signals from two alternative bases in a single position in a sequence of a PCR product.

dourine a chronic or acute contagious disease of breeding *Equidae* directly transmitted from animal to animal during mating. It is notifiable to OIE. An infected mare may also transmit the disease to its foal in the milk or through udder lesions. The causative agent of dourine is *Trypanosoma equiperdum*, a protozoan parasite of the subgenus *Trypanozoon* that also includes *Trypanosoma brucei* and *Trypanosoma evansi*. *T. brucei* is cyclically transmitted by tsetse flies and causes 'nagana' in domestic and various wild animals, and 'sleeping sickness' in humans. *T. evansi* is mainly spread by biting insects through mechanical transmission, and causes 'surra' in a very wide range of domestic and wild animals. The diagnosis of dourine is usually based on clinical signs (urticarial plaques, oedema, genital depigmentation and cachexia) and/or an epidemiological link to a confirmed case of dourine, together with serological evidence from complement fixation tests.

downwards incorporation the life cycle extends by incorporating an intermediate host lower down the food chain, i.e. at some stage before the definitive host.

DR antigens MHC class II molecules found on B-cells and antigen-presenting cells of humans.

draft sequence a DNA sequence that, while incomplete, offers a virtual road map to many of the known genes. Draft sequence data are mostly in the form of large bp fragments whose approximate chromosomal locations are known. *See* **sequencing**, **complete genome sequence**.

drainage/secretion precautions precautions used to prevent infections transmitted by direct or indirect contact with purulent material or drainage from an infected body site.

droplet transmission the transmission of infectious agents in droplets from respiratory secretions.

drug pressure constant exposure of an infectious agent (such as *P. falciparum*) to drug(s) and, thus, the generation of a selective force on the parasite to develop resistance.

drug repurposing investigating the effectiveness of an existing drug for a new indication.

drug resistance refers to the resistance to antimicrobials as the ability of a parasite strain to survive and/or to multiply despite the administration and absorption of a medicine given in doses equal to or higher than those usually recommended but within the tolerance of the subject, provided drug exposure at the site of action is adequate.

drug toxicity the extent to which a drug causes adverse effects.

drug-like a small molecule that falls within a pre-defined range of physicochemical parameters. The parameters differ slightly depending upon the indication and the desired mode of administration. The most commonly applied parameters for an orally available drug are: molecular weight <500 Da, partition coefficient log $P<5$, ≤ 5 hydrogen bond donors and ≤ 10 hydrogen bond acceptors.

drug-susceptibility testing (DST) the process of classifying an isolate or a mix of isolates from the same species as susceptible (high likelihood of clinical response) or resistant (low likelihood of clinical response) to an antimicrobial.

DTH delayed-type hypersensitivity.

DTP diphtheria, tetanus and pertussis vaccine.

dual acylation addition of both palmitoyl and myristoyl groups to a protein; often this modification is required for lipid raft association.

dual labelling method a technique that detects two different epitopes within the same cells, thereby allowing testing of whether the two epitopes colocalize. In fluorescence microscopy this may be achieved through the conjugation of fluorophores of different colours to alternate antibodies.

DUB (deubiquitinating peptidase) a peptidase that can catalytically remove the ubiquitin covalently attached to specific lysine side chains. These enzymes are also referred to as isopeptidases.

Duffy binding-like protein domains (DBL) an adhesive cysteine-rich protein domains present on PfEMP1 and initially identified as part of the Duffy binding protein of *Plasmodium vivax*. It is involved in parasite : receptor adhesion to both protein (e.g. ICAM-1) and carbohydrate (e.g. CSA). Six different DBL domains have been identified.

Duffy receptors receptors on the reticulocyte that play an important role in *Plasmodium vivax* and *Plasmodium knowlesi* invasion.

duplication a type of mutation that involves the production of one or more copies of a gene or region of a chromosome. This evolutionary mechanism occurs in all organisms.

duration of epidemic events the time elapsed between the onset and the end of the epidemic.

duration of infection *See* **infectious period**.

dysbiosis an imbalance in the relative abundance of microbial community within an ecosystem that is associated with a disease (e.g. inflammatory bowel disease). Dysbiosis can be either the cause or the consequence of disease.

dysgammaglobulinaemia the abnormal production of gammaglobulins in blood.

E3 ligase an enzyme catalysing the final step of a three-step process that serves to activate and transfer ubiquitin (or ubiquitin-like molecules) to a protein that will be targeted to the proteasome.

EC number Enzyme Commission number for the classification of enzymatic function.

EC50 the concentration of a nutrient or essential solute that supports half-maximal parasite growth under *in vitro* conditions.

ecdysis periodic shedding of the exoskeleton or cuticle.

ecdysozoa a group of animals that possess a cuticle that is periodically shed during development, includes arthropods and nematodes.

echinococcosis (or hydatidosis) a zoonotic disease that causes considerable economic losses and public health problems in many countries. CE, which is considered as an emerging disease in various regions, is economically important and constitutes a threat to public health in many countries. Human infection with *Echinococcus granulosus* typically results in a slowly growing parasitic cystic disease most frequently seen in the liver. The cysts may be asymptomatic for many years, and occasionally spontaneous regression has been noted. More commonly the disease is slowly progressive, and symptoms and complications eventually arise. The symptoms include pain from expansion or rupture, fever from pyogenic infection due to intrabiliary rupture and jaundice, or anaphylaxis from intrabiliary or extrahepatic rupture. There are three treatment options for hydatid disease of the liver. These treatment strategies include: surgery, percutaneous aspiration and medical treatment from which surgery has been known as the most efficient treatment option in humans.

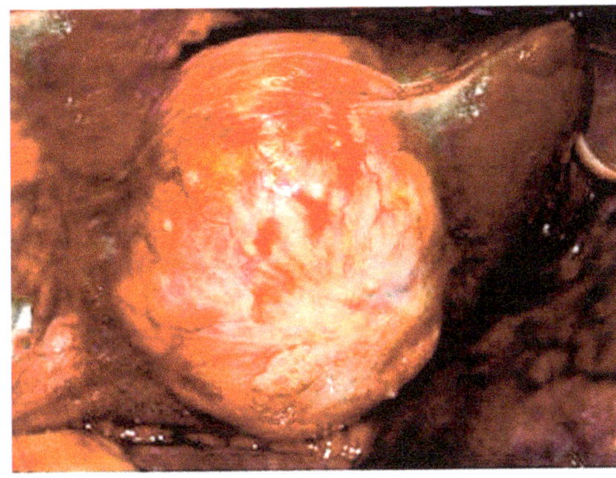

Echinococcus there are 10 identified *Echinococcus* sp. distinct genetic types (G1–10) and it has been reported the presence of fertile and infertile cysts produced by the same *E. granulosus* strain.

eclosion the moulting and emergence of an adult insect from its pupal case.

ecocentrism the view that considers the whole environment or ecosphere as important and deserving of consideration, without giving preference to organisms such as animals and humans. It states that all elements of the environment have worth and should be valued and cared for.

ecological genetics a branch of evolutionary biology focused on understanding the process of adaptive evolution through the study of natural selection and inheritance in wild populations. Ecological genetic studies employ direct observation or experimentation in present-day populations and focus on ecologically important traits that could influence lifetime fitness.

ecological mechanism of reduced hybrid fitness the reduced hybrid fitness resulting from direct natural selection against intermediate phenotypes, and whose presence depends on niche characteristics of environment in which hybrids are tested.

ecological niche modelling the process of using computer algorithms to capture the niche of a species and project such preferences into the prediction of its distribution in the geographic space on the basis of a mathematical representation of its known distribution in the environmental space. The environment is in most cases represented by climate data (such as temperature and precipitation), but other variables such as soil type, water depth, and land cover can also be used. The concept may be alternatively known as environmental niche modelling, species distribution modelling, ecological niche modelling, predictive habitat distribution modelling, and climate envelope modelling.

ecological speciation the evolution of reproductive isolation caused ultimately by divergent natural selection on traits between populations (or disruptive selection between phenotypes of a single population) in different environments (including use of different resources).

ecology the science of environmental interrelationships, studying all factors that affect the ability of organisms to live within their environment and the effects of the organisms on their environment.

ecophenotypic change non-evolutionary change in the morphology, or other characters, of an organism, usually induced by environmental pressures.

ecophysiology study of the physiological functions of organisms as they pertain to their ecology or interactions with each other and their environment.

ecosystem natural unit that contains living and non-living components (e.g. a community and its environment) interacting and exchanging materials, and generally balanced as a stable system, e.g. grassland or rainforest.

ecotope a small, environmentally uniform area, which serves as a living habitat for a group of interacting organisms (= biocenosis).

ect(o)- a prefix denoting external, outside.

ectoparasites live on the outer surface of the host. They either suck the blood and lymph or feed upon feather, hair and/or skin and its secretions, e.g. lice, ticks, fleas, true bugs, and flies including mosquitoes.

ectoparasiticides compounds that kill parasites living on the outside of an animal, including fleas, ticks, mites and lice. These compounds can be classified according to their chemical constituent or their mode of action. Based on chemical structures they include botanical compounds, synthetic pyrethroids, chlorinated hydrocarbons, organophosphates, carbamates, formamidines, macrocyclic lactone endectocides, fipronil, imidacloprid, insect growth regulators and development inhibitors, and spinosad.

ectoparasitism the state in which the ectoparasite is living on the surface of the host's body.

ectoperitrophic space (ES) the space between the peritrophic matrix and gut epithelium in the insect midgut.

ectopic literally this means displaced. In the context of parasitology it refers to a parasite that is in an atypical location within its host.

ectopic recombination recombination between genes that are not alleles, often between members of gene families positioned in different chromosomal locations.

ectoplasm the more hyaline, outer peripheral cytoplasm of a protozoan. The portion of the cytoplasm from which cilia usually originate. The surface of the ectoplasm apparently gives rise to the cell membrane, the periplast.

ectotherm an animal that is dependent on the external environment to maintain its body temperature.

ectothermic organisms having a body temperature determined primarily by the temperature of the environment. Those organisms that maintain a body temperature independent of the temperature of the environment are endothermic.

effective concentration 50 (EC_{50}) the concentration of drug required to kill 50% of parasites in an *in vitro* assay.

effective population size (N_e) the effective number of individuals in a population contributing to the offspring of the next generation. In natural populations, because not every potential parent actually mates, N_e equals the size of an idealized random-mating population that would have the same evolutionary behaviour of the population under study. Its value depends on mode of inheritance, inbreeding level, sex ratio, variance of reproductive success, population size changes, and any genetic or geographic structuring of the effective population. Natural populations with low values of N_e are more affected by random genetic drift.

effectiveness a measure of the benefit resulting from an intervention administered under usual conditions of clinical care for a particular group of patients.

effector cells immune cells that respond directly to parasite stimuli through pathogen recognition receptors (innate), or indirectly through cytokines produced by lymphocytes that recognize parasite antigens (adaptive).

effector memory T-cell an antigen-experienced T-cell that maintains effector differentiation, resides in, or expresses cell surface receptors that are required for homing to, extra-lymphoid effector sites and that has limited expansion potential.

effector-triggered immunity (ETI) an immune response induced by the recognition of pathogen effector activity by a host nucleotide-binding domain and leucine-rich repeat-containing protein (NLR). The ETI response includes the fortification of the plant cell wall, the generation of reactive oxygen species, the induction of pathogenesis-related proteins and a hypersensitive response.

efficacy 1 evaluative measure of the extent to which a defined intervention produces a beneficial result under ideal conditions, typically in the context of a randomized placebo-controlled trial. **2** the ability of an intervention to produce the desired beneficial effect under ideal circumstances (assessed by randomized controlled trials). **3** a measure of the benefit resulting from an intervention for a given health problem administered to patients under ideal conditions (i.e. perfect compliance). **4** a statistical test or measure is efficient when its sampling distribution has smaller variance (i.e. is more reliable) than other similar tests or measures. For example, in estimating the mean of a normal population, the sample mean is more efficient than the sample median. There is often a tradeoff, however, between efficiency and robustness. The sample mean is less robust than the sample median.

efflux movement out of a cell.

Ef-Tu a GTP/GDP-binding protein required for correct placement of the amino acid-transferRNA into the acceptor site of the ribosome during elongation of the peptide strand.

egg (cestode) life cycle stage emitted on the soil with faeces after rupture of the proglottids; they are often ingested with contaminated food. The six hooks can be seen in the eggs (used for species determination). Taenid eggs are activated by the acid pH in the stomach and hatch releasing microscopic larvae, which actively cross the intestinal barrier.

egg capsule a membranous structure containing eggs of a tapeworm, in the absence of a uterus.

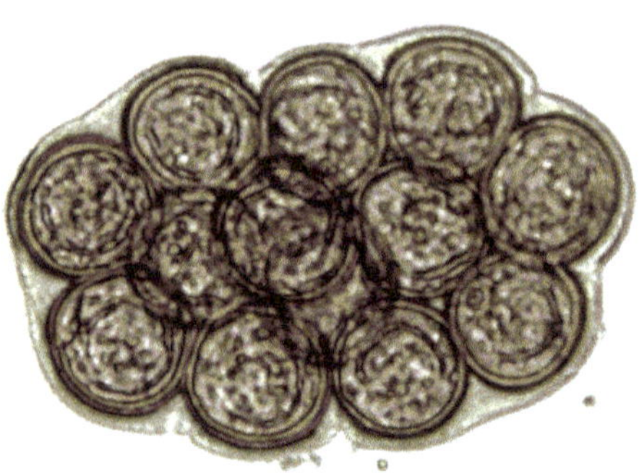

egg count counting of helminth eggs as an estimate of the parasite status in the animal or group. Flotation techniques and special counting chambers are used. The results are expressed as eggs per gram (e.p.g.) of faeces.

egg membrane thin protective membrane that surrounds the fertilized ovum.

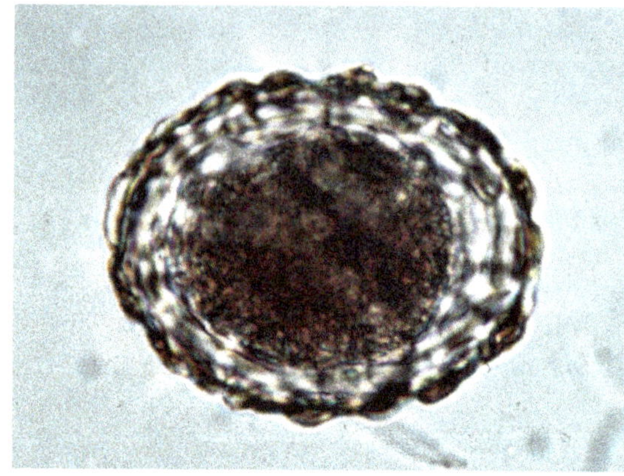

egg wall the external surface of eggs shed by parasites; often characterized by features useful for a specific diagnosis upon microscopic analysis.

EhMLBP (*Entamoeba histolytica* methylated LINE-binding protein) an essential constituent of *E. histolytica* epigenetic machinery and a potential drug target.

ehrlichiosis a tick-borne disease caused by rickettsiae of the genus *Ehrlichia*.

eicosanoids derivatives of polyunsaturated fatty acids that are important in the oviposition and reproduction of insects. There is also some evidence that eicosanoids mediate the immune response of insects to bacteria in the haemolymph.

electroimmunodiffusion an immunodiffusion technique run over a slide in which antigens and antibodies are driven towards each other in an electric field resulting in precipitation lines.

electroimmunotransfer blot detection of antigens by antibodies on a nitrocellulose membrane, after separation of the antigenic preparation by electrophoresis on an acrylamide gel and electrotransfer of the protein to the nitrocellulose membrane. *See* **western blot**.

electron acceptor substance that gains electrons from another substance in an oxidation-reduction reaction.

electron donor substance that loses electrons to another substance in an oxidation-reduction reaction.

electron microscope a microscope using an electron beam of very short wavelength as the source of illumination. It has a resolving power of 2 nm, which is 100 times greater than with the light microscope. Types include the transmission electron microscope.

electrophoresis a method of separating large molecules (such as DNA fragments or proteins) from a mixture of similar molecules. An electric current is passed through a medium containing the mixture, and each kind of molecule travels through the medium at a different rate, depending on its electrical charge and size. Agarose and acrylamide gels are the media commonly used for electrophoresis of proteins and nucleic acids. The conditions used during electrophoresis can be adjusted to separate molecules in a desired size range.

electroporation a process using high-voltage current to make cell membranes permeable to allow the introduction of new DNA. It is commonly used in recombinant DNA technology. *See* **transfection**.

electrospray ionization (ESI) and nano-ESI technique used in mass spectroscopy to produce ions, generating ions in a gas phase without fragmentation of the sprayed sample. ESI is adapted to the ionization of polar to very polar molecules, with a molecular weight ranging from around 100 to 100,000 kDa. ESI is widely used in lipidomics.

elicitor a protein or other compound from the nematode that is recognized by the host R-gene.

Eliminate Dengue Project (EDP) a global effort to eliminate dengue virus through the spread of *Wolbachia*-infected mosquitoes.

elongation initiation factor 1α (EIF1α) the gene encoding EIF1α is expressed at all stages of the parasite life-cycle and is highly up-regulated in the female but not the male gametocytes.

eltonian noise hypothesis an ecological hypothesis in which the biotic factors become much less relevant than the abiotic ones in the definition of the niche of organisms.

embryo the stage in development following fertilization and subsequent cleavage of the zygote up to, but not including, the first larval or first juvenile stage.

embryonated egg or larvated egg a nematode egg with a developed larva inside it. Most nematode eggs leave the host in the morula stage and develop in the environment to the embryonated stage (the stage just before hatching). A few nematode eggs are embryonated at the time they leave the host.

embryonic stem (ES) cells undifferentiated stem cells derived from the inner cell mass of an early-stage embryo, known as a blastocyst. ES cells are pluripotent and can differentiate into any type of cell in the body. These cells can be genetically modified and are still germline competent.

embryophore the envelope immediately surrounding the onchosphere (embryo) in tapeworm eggs.

emerging parasite a parasite population responsible for a marked increase in disease incidence, usually as result of changed societal, environmental, or population factors.

emodepside an anthelmintic that belongs to the class of drugs known as cyclooctadepsipeptides and is a semisynthetic derivative of the fermentation product PF1022A from the fungus *Mycelia sterilia*. It is effective against a number of gastrointestinal nematodes. Emodepside acts by binding to presynaptic latrophilin-like receptors (HC110-R) and by pre- and post-synaptic interactions with a Ca^{2+}-activated K^{+} ion channel (SLO-1), leading to inhibition of pharyngeal pumping, paralysis, and death of the worm.

empirical Bayes procedure a statistical approach whereby codons under positive selection are identified on the basis of maximum likelihood estimates of the parameters of interest.

encapsulation a cellular immune response of insects in which moderate-sized foreign material, microorganisms, parasites or parasitoids are surrounded or enveloped by haemocytes. Encapsulation occurs around objects too large for haemocyte phagocytosis. It involves co-operation between haemocytes and is mediated by cytokines and adhesion molecules.

encephal(o)- a prefix denoting brain.

encephalitis an acute inflammation of the brain. Encephalitis presenting with meningitis (inflammation of the protective membranes covering the brain and spinal cord, the meninges) is known as meningoencephalitis. The latter may be due to bacterial, viral, or amoebic causative agents. Amoebic meningoencephalitis is caused by the species *Naegleria fowleri*, and it is also known as PAM.

enchancing antibodies antibodies which enhance the survival of a graft or of a tumour.

encopula pairing of an adult female schistosome with an adult male within the gynaecophoric canal of the male.

encounter filter the probability of contact between a pathogen and a potential host.

encystation a term used in relation to some organisms, such as amoebae, which have the ability to encyst (i.e. form a cystic stage) from the trophozoite stage.

encysted enclosed in a sac, bladder or cyst.

endangered a species with such a low population number that it is in danger of extinction.

endectocides antiparasitic compounds that combine the activity against internal and external parasites, offering greater convenience and broader-spectrum metazoan parasite control.

endemic a disease or disease agent that present continually in a region or among a certain group of animals.

endemic area geographically defined areas within which the disease has been reported consistently.

endemic normal individuals who have lived in endemic areas for prolonged periods of time, but do not manifest any signs or symptoms of the disease. If tested, these individuals exhibit humoral and cellular immune reactivity towards parasite antigens, which indicates that they have been exposed.

endemic stability a situation in which all factors influencing disease occurrence are relatively stable, resulting in little or no fluctuation in disease incidence over time. Also termed enzootic stability.

endergonic a reaction that consumes energy (the change in the free energy is positive) and thus can only take place with external energy input.

endocrine pancreas portion of the pancreas that produces hormones (insulin and glucagon) that are essential for glucose homeostasis.

endocytic/phagocytic processes the cellular based uptake of extracellular molecules by engulfing them through processes involving invagination of the outer plasma membrane; these processes have been implicated in the movement of double-stranded RNA in some species.

endocytosis cellular ingestion of macromolecules by invagination of plasma membrane to produce an intracellular vesicle which encloses the ingested material.

endodyogeny a process of asexual development by which two daughter cells develop internally within the mother parasite cell. It occurs in certain members of the phylum Apicomplexa such as *Toxoplasma gondii*.

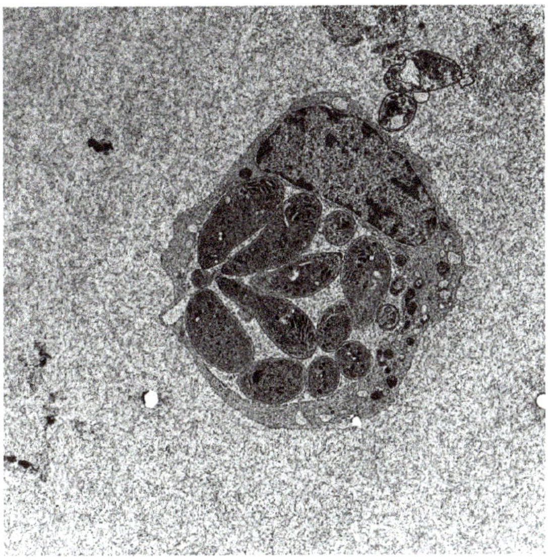

endogenization the process of inclusion of genetic material from a donor organism into a recipient genome through horizontal gene transfer. Several cases of endogenized viral sequences (mostly from retroviruses) are known in many eukaryotic genomes. Endogenized sequences may either become non-functional through pseudogenization, or be a source of new genes and functions in the recipient genome.

endogenous developing or coming from within an organism.

endogenous antigen an antigen synthesized within body cells (e.g. a virus protein).

endomastigote a morphotype characterized by round to oval cells with a very short flagellum convoluted around the nucleus and not extending outside of a flagellar pocket.

endomycorrhizae a type of mycorrhizal fungus that surrounds a plant root tip and uses its hyphae to penetrate the cell walls of the root. Endomycorrhizal fungi form vesicle-like structures at the root cell surface that enhance the transport of substances between a plant and fungus.

endonuclease *See* **restriction enzyme**.

endoparasite a parasite that lives within the body or internal body passages of its host.

endoparasiticides (internal antiparasitics) kill parasites living inside the animal, such as gastrointestinal parasites, respiratory parasites and heartworms. Endoparasiticides include anthelmintics (agents lethal to worms) and antiprotozoal (agents lethal or suppressive to protozoa).

endophenotype heritable traits that are genetically correlated with disease traits. They are often traits (such as the level of a metabolite or transcript) that can be measured in all individuals (both diseased and healthy) and that can potentially provide a predictor of disease status.

endophyte any organism (usually a fungus or microbe) that lives inside another organism and establishes a parasitic or mutualistic relationship with its host.

endoplasm that portion on the cytoplasm in which vacuoles form and in which internal organelles and inclusions are suspended.

endoplasmic reticulum a network of membranes inside a cell through which proteins and other molecules move. Proteins are assembled at organelles called ribosomes. When proteins are destined to be part of the cell membrane or exported from the cell, the ribosomes assembling them attach to the endoplasmic reticulum, giving it a rough appearance.

endosome a membrane bound compartment in eukaryotic cells involved in membrane and cargo trafficking between the cell surface and the lysosome. Like the lysosome, endosomes also have a low internal pH. In *Trypanosoma brucei* all endo- and exocytosis occurs in the flagellar pocket.

endosphere the microbial habitat inside plant organs.

endosymbiont an organism that lives within the body or cells of another organism; it can include facultative or obligate symbionts.

endothelial cells (ECs) a thin layer of cells lining the inner wall of the blood vessels and lymphatic vessels which form the endothelium.

endotherm an animal that generates heat to maintain its body temperature independent of, and typically above, its environment.

endothermic an organism that self generates and maintains a constant body temperature.

endotoxin a structural component of the Gram-negative bacterial outer cell membrane that is recognized by the innate immune system and has been associated with complex effects on human diseases.

end-replication problem a problem caused by the conventional replicative machinery that synthesizes DNA only in the 5′–3′direction and is therefore unable to complete DNA replication at the end of the lagging strand. In the absence of telomerase, organisms with linear chromosomes lose telomeric DNA with each cell division.

engorgement distension of a feeding tick with blood; cannot occur in male hard ticks, because their back is completely covered by a hard scutum.

enhanced pause dimensionless respiratory parameter derived from unrestrained whole body plethysmography.

ensheathed a larval nematode that is within the retained cuticle of the previous stage.

enteritis inflammation of the intestine, usually the small intestine.

Enterobius vermicularis a human intestinal nematode, also known as pinworm, that is common in children and, to a lesser degree, in their caregivers. Infection is a common cause of anal itching, which stems from the depositing of eggs by the adult female worm in the perianal area. Scratching facilitates transmission of the worm by eggs thus trapped under fingernails, in clothes, etc.

enteromonads a group of single-celled eukaryotes having a single nucleus as most other eukaryotes. However, molecular phylogenies have shown that they branch among 'true' diplokaryotic diplomonads.

enteropathy the presence of any pathology or inflammation in the intestine.

enterotype the classification of humans or of other living organisms with an intestine based on similarities in their gut microbial communities as viewed through multivariate statistical methods.

entisols undifferentiated soils of recent origin found in river valleys and deltas.

entomological end points outputs measured by entomological studies, for instance, mosquito density, feeding inhibition; they may be predictors of epidemiological impact.

entomological inoculation rate (EIR) the number of infectious bites an individual is exposed to in a given time period.

entomology that branch of biology concerned with the study of insects.

entomopathogenic an organism that infects insects and causes a disease.

entomopathogenic fungus a fungus that can act as a parasite of insects, killing or seriously disabling them.

entomophagous an entomophagous organism is one that eats insects (also called an insectivore). Many different animals in a range of taxonomic groups eat insects including mammals, birds, reptiles, amphibians and invertebrates. In some areas of the world, entomophagy (the practice of eating insects) is common within human communities.

entropy-based methods entropy is a key measure of uncertainty associated with a random variable in information theory. Entropy-based methods examine the information entropy difference between different models with and without interactions to detect epistasis.

env (envelope) the retroviral env gene encodes protein precursors for the virion envelope proteins. These proteins reside in the lipid layer and serve to form the viral envelope. In the HIV-1 genome, env genes encode two envelope proteins: gp120 and gp41.

environmental change refers to any variation in the environment, which includes changes in climate, in the characteristics of the landscape, and communities of animals, plants, vectors, or reservoirs of a pathogen. The simple trends or changes of climate are not the only factor in an environmental change, which is a richer and broader concept.

environmental DNA (eDNA) DNA extracted from environmental samples (e.g. water), without first isolating the target organism.

environmental stewardship the view that humans have a duty to manage and care for the whole natural environment; that we are responsible for the continued health of the whole ecosystem, not just the parts that benefit the human race. It involves integrating and applying environmental values into a process.

enzootic a disease or disease agent that occurs in an animal population at all times.

enzyme a biological catalyst and is almost always a protein. It speeds up the rate of a specific chemical reaction in the cell. The enzyme is not destroyed during the reaction and is used over and over. A cell contains thousands of different types of enzyme molecules, each specific to a particular chemical reaction.

enzyme-coupled immunodetection an immunochemical method of protein localization that uses an antibody (primary or secondary) covalently linked to an enzyme which releases an electron-dense or coloured deposit when supplied with appropriate substrate. This can be used to localize the activity of that enzyme (and hence the antibody's target) through electron or light microscopy. Horseradish peroxidase and alkaline or acid phosphatases are commonly used systems. These systems are now less frequently used than colloidal gold based detection.

enzyme-linked immunosorbent assay (ELISA) an assay in which an enzyme is linked to an antibody and a coloured substrate is used to measure the activity of bound enzyme and, hence, the amount of bound antibody.

eomesodermin (Eomes) also known as Tbr2, is a T box transcription factor related to T-bet. Eomes, like T-bet, is involved in the regulation of the CD8 T cell response. Eomes expressing cells preferentially develop in long-lived memory cells, whereas T-bet expressing cells become preferentially terminal effector cells.

eosinophil a leucocyte with large eosinophilic (i.e. red) cytoplasmic granules, containing toxic cationic protein, with a two lobed nucleus. An increase in the number of peripheral blood eosinophils is usually associated with allergic reaction or parasitic infection.

eosinophil-derived neurotoxin (EDN) a ribonuclease protein with immunomodulatory properties expressed in eosinophils.

eosinophilia increased numbers of eosinophils in the blood.

eosinophilic hyaline membranes deposition of an eosinophilic and amorphous layer of blood-derived material (including fibrin, other plasma proteins, and cellular debris) in the alveoli. This deposition reduces or completely inhibits gas exchange in the affected lung segments.

EP1 procyclin 26-mer element (EP1 procyclin ARE) a U-rich motif found in the 3′-UTR of EP1 procyclin that affects its stability and translation in bloodstream forms.

epicytic folds longitudinally arranged folds on the surface of gregarine trophozoites formed by the plasma membrane and underlying inner membrane complex. The folds increase surface area for surface-mediated nutrition and facilitate actinomyosin-based gliding motility in eugregarines (e.g. *Lecudina* and *Gregarina*). The number and density of epicyctic folds around the cell periphery varies considerably in different species of gregarines (range = 4 to > 300).

epidemic the occurrence of a number of cases of a disease (or condition) in excess of a number expected in a given time and place. In some instances a single case will constitute such an unusual occurrence.

epidemic burden the sum of the public health consequences of the epidemic within the population affected by it over its entire duration. This includes a minimum morbidity and mortality in excess of that expected, in addition to indirect consequences such as long-term sequelae of infection, increased rates of anaemia and malnutrition, economic losses and decreased educational achievement.

epidemic periodicity the average frequency with which epidemics occur in a given region or population.

epidemic stratum the subregion or subpopulation within the entire population that is characterized by specific aetiology, demography and pattern of disease transmission.

epidemiological cycle a diagrammatic illustration of the interaction between disease determinant factors.

epidemiology the study of patterns, causes, and effects of health and disease conditions in populations. It is usually used for diseases in humans; epizootiology is used tor diseases in animals.

epidermal Langerhans cells (E-LCs) langerin (a Ca^{2+}-dependent lectin receptor, CD207) positive dendritic cells that reside within the epidermis and act as major antigen-presenting cells during skin infections.

epigenetic an emerging field of science that studies heritable changes caused by the activation and deactivation of genes without any change in the underlying DNA sequence of the organism. The word epigenetics is of Greek origin and literally means over and above (epi) the genome.

epigenetic modifier proteins proteins that 'write', 'erase' or 'read' epigenetic marks on chromatin.

epigenome derived from the Greek word 'epi' which literally means 'above' the genome. The epigenome consists of chemical compounds that modify, or mark, the genome in a way that tells it what to do, where to do it, and when to do it. Different cells have different epigenetic marks. These epigenetic marks, which are not part of the DNA itself, can be passed on from cell to cell as cells divide, and from one generation to the next.

epigenomics study of the complete set of epigenetic modifications on the genetic material of a cell, known as the epigenome. Epigenetic modifications are reversible modifications on a cell's DNA or histones that affect gene expression without altering the DNA sequence.

epimastigotes a non-infective and replicative form of *Trypanosoma cruzi* present in the intestine of the vector and in axenic *in vitro* cultures. In this form, the kinetoplast and the flagellar pocket are located between the nucleus and the anterior end. The flagellum emerges at the side of the trypanosome and runs along the body, building a short undulating membrane.

epimerase an enzyme that catalyses the conversion of one epimer to another.

epimerite anterior part of gregarine cell containing the apical complex, usually involved in attachment to host.

epimers two isomers of a monosaccharide that differ only in the configuration around a single chiral carbon, for example the C5 carboxylic acid group in glucuronate versus iduronate.

epimutation a change in epigenetic organization at a locus that is causative of an adverse phenotypic outcome.

epiphyte any organism that grows upon or attaches to a living plant for physical support but not for nutrients.

episome DNA within a cell that is not integrated into any chromosome and is normally found in a closed circle.

epistasis phenomenon where the effects of one gene are modified by one or several other genes. The gene whose phenotype is expressed is called epistatic, while the phenotype altered or suppressed is called hypostatic.

epistatic interactions epistatic interactions are genetic interactions between different loci in which the combined phenotypic effect of two or more loci supersedes the sum of the effects of the individual loci. Thus, any gene that alters the expression of another non-allelic gene is epistatic. Epistasis often has a role in complex traits.

epistatic interactions non-additive interactions between two or more mutations at different loci, such that their combined effect on a phenotype deviates from the sum of their individual effects.

epithelioid cells macrophages that accumulate around a tubercle and resemble epithelial cells in histological sections.

epitope the portion of an antigenic determinant that contacts antibody.

epizootic an increase in number of disease cases in an animal population (equivalent to an epidemic in humans).

epsiprantel a praziquantel analogue with activity against intestinal tapeworms, and is approved for use in dogs and cats. Epsiprantel is administered orally. Unlike praziquantel, epsiprantel is absorbed poorly after oral administration and therefore has no activity against extraintestinal cestodes.

Epstein–Barr virus (EBV) the causal agent of infectious mononucleosis and Burkitt's lymphoma. Can be used to immortalize human B-cells *in vitro*.

equilibrium population a panmictic population of continuously constant size.

equine lungworms *Dictyocaulus arnfieldi*; a nematode with a direct life cycle, no intermediate host. Horses become infected by ingesting larvae, which migrate to the lungs. Horses will have chronic cough which is non-responsive to antibiotics. Often seen in horse grazing pasture with donkeys. The disease should be differentiated from coughing caused by *Parascaris equorum* larval migration.

equivalence the ratio of antibody to antigen at which immunoprecipitation of the reactants is virtually complete.

equivalence zone the region in which the concentration of antigen and antibody leads to maximal precipitation.

eradication a situation in which a disease have disappeared from the globe, with the possible exception of specimens held in a secure laboratory. It is permanent reduction to zero of the worldwide incidence of infection caused by a specific agent as a result of deliberate efforts.

eraser enzyme a protein that removes a specific PTM from a histone tail.

ergasilid copepods serve as intermediate hosts for other fish parasites, act as vectors of diseases or be fish parasites themselves.

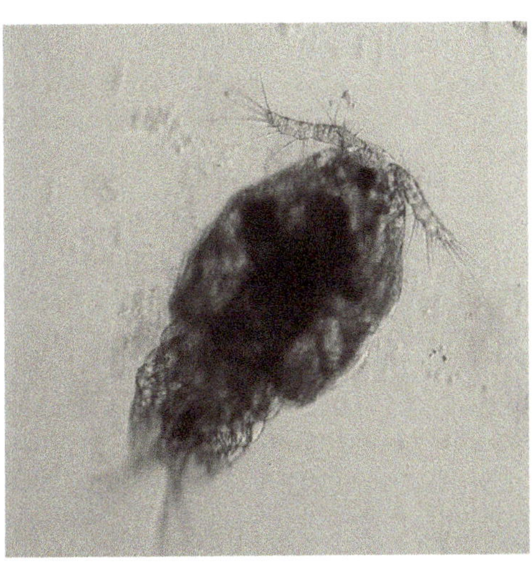

erratic parasite a parasite found in its normal hosts but in unusual organs or tissues in which they are not adapted to live (e.g. *Angiostrongulus vasorum* in the eye).

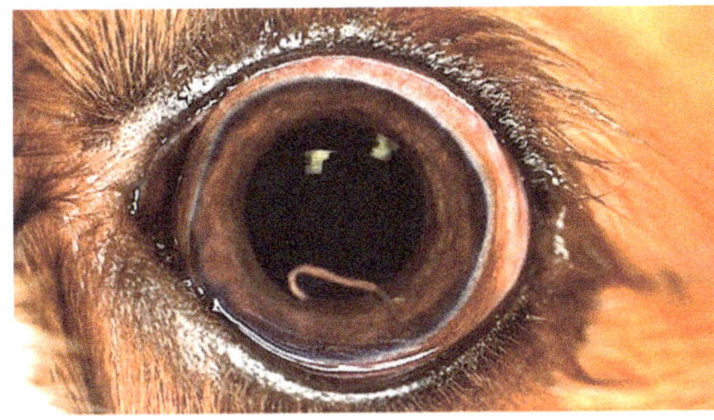

error threshold a process that limits the amount of information that a genome can store at a given mutation rate.

erythema redness produced during inflammation due to erythrocytes entering tissue spaces.

erythema migrans a characteristic 'bull's eye' rash on the skin, typically 1–2 weeks after the tick bite, which is often the first clinical manifestation of Lyme borreliosis.

erythrocytic asexual-stage schizonts a developmental form of the malaria parasite in the blood in which merozoites (the stages that invade red blood cells) are formed.

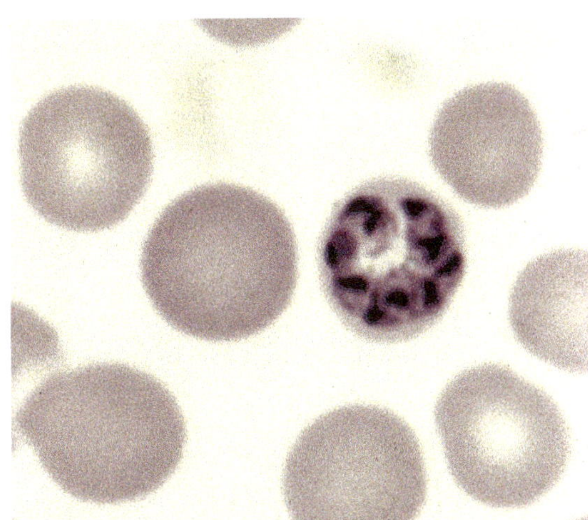

erythropoiesis erythrocyte production.

erythropoietin a hormone released from the kidneys and the liver in response to low oxygen concentrations in the blood.

It controls the rate of red blood cell production.

Escherichia coli a common Gram-negative, enteric bacterium that has been studied intensively by geneticists because of its small genome size, normal lack of pathogenicity, and ease of growth in the laboratory.

ethics a branch of philosophy that deals with morality. It is concerned with distinguishing between right and wrong human actions, both at an individual and societal level. Ethics may also apply to the rules or standards that specify how particular members of an organization should conduct themselves.

ethyl methanesulfonate (EMS) mutagenesis chemical mutagenesis acting particularly by alkylating guanines. The method usually produces point mutations. Typically, thymidine is incorporated opposed to O-6-ethylguanine during replication of EMS treated DNA.

euchromatin an 'open', transcriptional permissive chromatin state harbouring classical 'active' histone marks such as H3K4me3, H3K9ac, and H4ac.

eugregarines gregarine parasites of the intestine and other organs of marine and terrestrial invertebrates exhibiting only sporogony and gametogony; considered to be derived from the archigregarines.

eukaryote (synonym **eucaryote**) single-celled or multicellular organism (e.g. plant, animal, or fungus) with a cellular structure that includes a membrane-bound, structurally discrete nucleus and other well-developed subcellular compartments. Eukaryotes include all organisms except viruses, bacteria, and blue-green algae.

eukaryotic translation initiation factor 4F (eIF4F) a translation initiation factor complex that consists of three subunits: the cap-binding protein eIF4E; the RNA helicase eIF4A; and eIF4G, which recruits the small ribosomal subunit.

eukaryotic-like proteins bacterial proteins containing domains preferentially found in eukaryotes, or proteins having greater similarity to eukaryotic proteins than to bacterial proteins.

EuPathDB integrated database for eukaryotic pathogens providing access to genomics and functional genomics information. EuPathDB is frequently updated and support of the database is largely guided by input from the research community users.

euphotic zone the layer of a body of water that receives sufficient sunlight for photosynthesis. The depth of this layer, which is about 80 m, is determined by the water's extinction coefficient, its cloudiness, and the sunlight's angle of incidence.

euploid having the normal number of chromosomes, in other words two copies of each chromosome in a diploid organism.

European Scientific Counsel Companion Animal Parasites (ESCCAP) an independent, non-profit-making organization consisting of experts in the field of parasitology and public health from across Europe.

euryxenous having a broad host range.

eutectic phase a chemical mixture that has a lower freezing point than a composition of pure ingredients.

eutrophication death of organisms in a lake or pond due to an overabundance of algae that consume all of the dissolved oxygen in the water. This usually happens when the water becomes rich in mineral and organic nutrients, often due to fertilizer run off from farms.

E-value cutoff (expect-value) BLAST statistical significance threshold for reporting matches against database sequences. Lower E-value thresholds are more stringent.

evidence code a two- or three-letter code showing the evidence that supports a specific GO annotation (http://www.geneontology.org/GO.evidence.shtml).

evolution is the process by which organisms change over time. Mutations produce genetic variation in populations, and the environment interacts with this variation to select those individuals best adapted to their surroundings. The best-adapted individuals leave behind more offspring than less well-adapted individuals.

evolutionarily conserved *See* **conserved sequence**.

evolutionary adaptation this term is used specifically to refer to genetically based changes in phenotypes that are a response to selection arising from a changing climate.

evolutionary arms race a series of adaptations and counteradaptations that occurs in two classes of individual that are in evolutionary conflict, such as between predator and prey, cheater and co-operators, or parasite and host.

evolutionary forces five major forces that can cause evolutionary change: natural selection, genetic drift (or population size), mutation, non-random mating, migration (in the genetic sense of permanent movement of genes from one location to another).

evolve and resequence studies studies that combine experimental evolution with next-generation sequencing. They make use of controlled environmental, demographic and selective variables to facilitate genotype–phenotype mapping.

***ex vivo* ELIspots** peripheral blood mononuclear cells are stimulated with antigen for 18 hours and cytokine production is measured directly by a standard ELIspot assay.

excavata a supergroup of unicellular eukaryotes that include many human parasites.

exchangeable correlation correlation among observations measured from a single unit (e.g. multiple parasite counts measured from a single host) that is assumed constant among units (i.e. among hosts).

exchange-only a process in which movement of a substance across a membrane only occurs in exchange for the movement of a second substance in the opposite direction. A special case of counter-transport (q.v.). In the latter, the movement of the two substances can be linked, but might not necessarily be linked.

exclusion criteria stated conditions which preclude entrance of candidates into an investigation even if they meet the inclusion criteria.

excretory/secretory products (ES) biomolecules in the secretions or excretions of parasitic organisms.

excretory/secretory system a term that refers to nematode excretory systems that encompass ducts, pores and gland cells; a key role includes excess water removal.

excystation escape from a cyst or envelope, as in that stage in the life cycle of parasites occurring after the cystic form has been swallowed by the host. *See* **cyst**, **oocyst**.

excyzoite a short-lived stage in the *Giardia* life cycle immediately after the release of the parasite from the cyst prior to cytokinesis and before the initiation of infection of the host.

exergonic reaction a reaction that releases energy (the change in the free energy is negative) and thus takes place without external energy input.

exflagellation a process by which *Plasmodium* male gametes are released from the red blood cell. It is identified under the microscope as male gametes move their flagella vigorously.

exhaustion a state of T-cell dysfunction characterized by attrition of effector functions, prolonged expression of inhibitory receptors, and inability to control infections or tumours.

exhaustive search a search of all possible pairwise combinations of loci for evidence of epistatic interactions.

exo- a word element; prefix denoting outside of, outward.

exoantigen a toxin, protein or other antigen released by an abnormal cell or pathogen which recognized by a specific immune response.

exocytic pathway the exocytic pathway consists of a series of membrane-bound organelles that transport proteins and lipids in vesicles from the ER to the cell surface.

exocytosis the export of material from a cell by the fusion of cytoplasmic vesicles with the outer cell membrane.

exoerythrocytic forms (EEFs) intrahepatic forms of the *Plasmodium* parasite, which develop inside a parasitophorous vacuole to generate first-generation merozoites.

exogenous developed or originating outside the organism.

exogenous antigen a foreign antigen that originates at a source outside the body (e.g. bacterial antigens).

exogenous DNA DNA originating outside an organism that has been introduced into the organism.

exogenous infection infection caused by organisms not normally present in the body but which have gained entrance from the environment.

exome sequencing selectively capturing the exome (coding regions) and other content in a whole-genome library before sequencing. This enables deeper coverage of the genomic region that is enriched in disease-causing variants in a megabase-sized DNA library instead of sequencing a lower coverage gigabase-sized whole-genome library.

exomembrane system exported trafficking system of a membrane network that is established by *P. falciparum* in the red blood cell cytoplasm. Two of the main features of the exomembrane system are the tubulovesicular network and Maurer's clefts.

exon most gene sequences are broken up by one or more DNA sequences called introns. The parts of the gene sequence that are expressed in the protein are called exons, because they are expressed, while the parts of the gene sequence that are not expressed in the protein are called introns, because they come in between – or interfere with – the exons.

exon skipping a form of alternative splicing in which internal exons are alternatively included or excluded (skipped) in the mature mRNA.

exonuclease an enzyme that cleaves nucleotides sequentially from free ends of a linear nucleic acid substrate.

exophagous preferring to feed outdoors; contrast with endophagous (preferring to feed indoors).

exophilic preferring to rest outdoors; contrast with endophilic (preferring to rest indoors).

exotic disease a disease that does not occur in the host country.

exotoxin soluble protein toxins, usually produced by Gram-positive pathogenic bacteria.

Expanded Program on Immunization (EPI) World Health Organization initiative aiming to improve access to standardized vaccination schedules.

experimental event rate (EER) the percentage of intervention/exposed group who experience outcome in question.

experimental huts the main semi-field means of evaluating vector control tools used in indoor interventions against mosquitoes in the field. Experimental huts are standardized structures representative of local housing that have been modified to make the measurement of entomological endpoints more standardized.

exploitation extraction or diversion of host resources to enhance parasite fitness. Extraction of host resources for production of transmission stage parasites. Diversion of host resources (e.g. host immune cells) for replication and dissemination of transmission stage parasites in the host. Empirical studies often use within-host replication or growth rate as a measure of exploitation.

exportome the compliment of parasite-encoded proteins that are exported from the parasite into the host RBC.

expressed gene *See* **gene expression**.

expressed sequence tag (EST) short complementary DNA sequences that are used to identify genes expressed in a cell at a particular time. Because each transcript can be represented by multiple ESTs covering different parts of the sequence, a single gene can be represented by many ESTs.

expression noise stochastic fluctuations in the protein level of a gene within a clonal cell population maintained in a homogeneous environment. Noise is generally defined as the coefficient of variation (CV); that is, the ratio between standard deviation and mean value of protein abundance for a given gene in a population of cells. In a more general sense, expression noise may also refer to variation in the abundance of biomolecules, such as mRNA.

expression quantitative trait locus (eQTL) a locus that controls variation in expression of a particular gene. An eQTL may lie adjacent to the gene being controlled (*cis*-acting control) or some distance away (*trans*-acting control).

expression site-associated genes (ESAGs) the set of genes co-expressed from the polycistronic transcript of a particular VSG expression site (ES). The expression site associated genes (ESAGs) are thought to help 'optimize' a *Trypanosoma brucei* cell for growth in a particular mammalian host.

expressivity measures the extent to which a given genotype is expressed at the phenotypic level, due to variation in the genomic background or the effect of environmental factors.

exsheathed the removal, escape from or shedding of the residual outer-surface/cuticle retained from a previous moult by some species of nematode.

exsheathment the final step in moulting of the infective stage when the old cuticle (sheath) is lost and the new nematode stage emerges inside the definitive host. It is a requirement for the establishment for infection in the host.

extensible markup language (XML) a language commonly used for web page development. *See* **hypertext markup language**.

external parasites ectoparasites of medical and veterinary importance include insects (fleas, lice, flies) and acarines (mites and ticks).

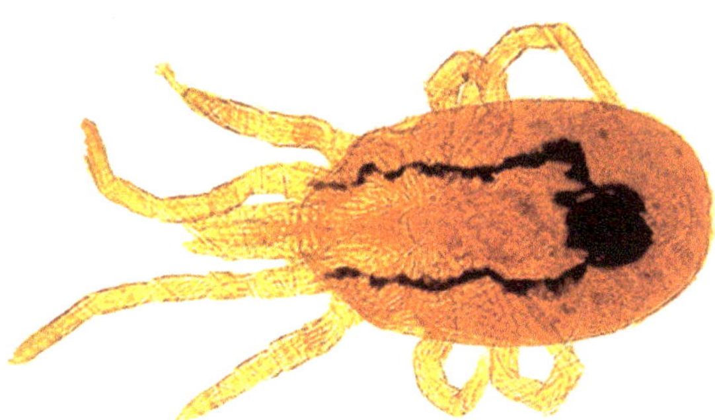

extinction the specific organism no longer exists in nature or the laboratory.

extracellular-signal-regulated kinase (ERK) a member of the mitogen-activated protein kinase (MAPK) family, ERK is a serine/threonine protein kinase that is activated by a range of extracellular signals. Once activated through phosphorylation (by a kinase), ERK can phosphorylate (activate) several cytoplasmic and nuclear targets, including transcription factors.

extrachromosomal susceptibility factor a heritable factor determining the susceptibility of tsetse to trypanosome infection that is not present on the chromosomal DNA of the tsetse, instead showing cytoplasmic inheritance. Classical examples of extrachromosomal heritable factors include organelles such as mitochondria and maternally inherited bacteria.

extra-label usage use at a different dosage or use in a different species than covered by the approved label. Extra-label usage in feeds is prohibited in food animals and horses. In the UK, 'Cascade' is a route through choices of drugs not licensed for a species.

extraneous factor a characteristic of an experiment that is not of interest but that might influence the outcome (also called a 'nuisance parameter').

extravillous trophoblast a specialized trophoblast that migrates from villous tissue and invades through the maternal decidua and transforms maternal spiral arteries.

extreme climatic events these are events at the boundaries of the range of weather experienced in the past and can include severe episodes such as heat waves, droughts, intense rainfall, or cold snaps.

extremophile an organism that can survive in physically or chemically extreme conditions that are not liveable to most other organisms.

extrinsic apoptosis apoptosis initiated by binding of extracellular ligands (e.g. TNF-α, FasL) to cell-surface death receptors.

extrinsic incompatibilities forms of reproductive isolation (usually sterility or inviability of hybrids) whose presence or intensity depends on the particular environment in which the hybrids are tested.

extrinsic incubation period (EIP) the development time of parasites within the vector.

exudate the extracellular fluid, containing proteins and cellular debris, which accumulates during inflammation.

eyespots a concentrated area of light sensitive pigment cells.

A B C D E **F** G H I J K L M N O P Q R S T U V W X Y Z

F(ab′)2 a bivalent antigen-binding fragment obtained following pepsin digestion of immunoglobulin. Consists of both light chains and the N-terminal part of both heavy chains linked by disulfide bonds.

F1 the first generation of a monohybrid cross. For example, F1 could refer to the first generation of mice from a C57BL/6 and BALB/c parental cross.

F2 analysis analysis of mapping populations generated by the F_2 design. The F_1 progeny from crossing two phenotypically different parental strains are themselves crossed to produce an F_2 population that is segregating for the phenotype of interest. The F_2 mapping population may carry up to three genotypes at every marker and therefore allows the detection of additive and dominance effects, as well as interactions between loci.

Fab a fragment of antibody containing the antigen-binding site, generated by cleavage of the antibody with the enzyme papain, which cuts at the hinge region N-terminally to the inter-H-chain disulfide bond and generates two Fab fragments from one antibody molecule.

facilitation a density-dependent process thought to regulate filarial larval infection in anopheline vectors, which can be thought of in ecological terms as an example of depensatory population dynamics (positive density dependence), arising when factors operating at low population densities cause a rate parameter to decrease such that, if depressed below some critical threshold population, growth or production might decline to zero. Its importance to filariasis control is that it might lead to a critical non-zero threshold of infection in the host, below which the number of parasites recruited into the mosquito population will be too low to sustain transmission, and infection dies out.

FACS (fluorescence-activated cell sorting) a laser-based technology that is used for cell sorting and cell counting, in which fluorescently tagged suspended cells pass through an electronic detection apparatus.

factor H a serum protein component of the vertebrate complement system that inhibits complement activity by binding complement protein C3b.

facultative anaerobe an organism that produces energy and ATP by aerobic respiration in environments where oxygen is present but is capable of switching to anaerobic respiration in environments where oxygen is absent.

facultative intracellular organism an organism that can, if necessary, grow within cells.

facultative parasite a parasite that can complete its life cycle in a free-living environment or, alternatively, that lives during part (or all) of its cycle within a host, e.g. *Strongyloides* species.

faecal microbiota transplantation (FMT) a transplantation process in which the faecal material (including the faecal microbiota) from a healthy donor is transferred into a recipient. Patients are often treated by enema infusion or the consumption of capsules containing donor faeces.

faeces the excrement (waste matter) eliminated from the intestinal tract which includes unabsorbed food materials, sloughed cells, mucus, bacteria, etc. (The word ‘faeces’ is plural but can be used with either a singular or plural verb.)

FAIRE–seq formaldehyde-assisted isolation of regulatory elements coupled with high-throughput sequencing; a method for identifying DNA from nucleosome-free regions.

fallowing the process of leaving cages on a farm site empty of fish so as to interrupt parasite and disease cycles and/or to enable the seabed to recover from waste sedimentation.

false discovery rate a multiple comparison adjustment approach to control the expected proportion of incorrectly rejected null hypotheses in a list of rejected hypotheses.

FAMACHA© score a score that evaluates the colour of the ocular mucosa, which is used as an indication of the degree of anaemia induced by *Haemonchus contortus* in sheep.

family the taxonomic category above genus.

family selection breeding for desired phenotypes based on families. Correlation between phenotypes of parents and offspring allows evaluation of the additive genetic variance controlling the trait or how 'true breeding' the trait is in a family. This is in contrast to mass selection.

farnesylation addition of an isoprenoid farnesyl group to proteins with a CaaX motif.

FAS receptor a death receptor, belonging to the tumour necrosis factor receptor superfamily, which contains a cytoplasmic death domain. Binding of its ligand (FasL) induces the formation of the death-inducing signalling complex (DISC) and triggers the extrinsic apoptosis pathway.

Fas/Fas ligand also known as APO-1 or CD95, a Type-I transmembrane protein, it is a widely expressed cell death receptor that has a critical role in the regulation of the immune system and tissue homeostasis. Fas ligand (FasL) is a Type-II Transmembrane protein of TNF. FasL is expressed predominantly in activated T lymphocytes and natural killer cells. In the immune system, engagement of the cell death surface receptor Fas by FasL results in apoptotic cell death, mediated by caspase activation. Fas and FasL are involved in down-regulation of immune reactions. Malfunction of the Fas-FasL pathway causes lymphoproliferative disorders. Whereas its exacerbation may cause tissue destruction.

***Fasciola hepatica* (the common or sheep liver fluke)** a trematode parasite with a worldwide distribution and has a significant economic impact on pasture-based ruminant livestock production. The life cycle of *F. hepatica* involves lymnaeid freshwater snails as the intermediate host and depends on the development and survival of larval stages both in the snail intermediate host and in the environment.

fascioliasis–ostertagiasis complex a disease of cattle in which the diarrhoea of ostertagiasis is accompanied by the anaemia of fascioliasis.

fasciolosis (fascioliasis; or liver fluke disease) is caused by the parasitic flatworm, *Fasciola hepatica*. The parasite itself is leaf-shaped, approximately 2–5 cm in length as an adult and causes severe liver damage with associated production losses, in sheep and cattle. Whilst primarily a parasite of livestock, *F. hepatica* has a broad host range and can also infect goats, deer, rabbits and man. Tropical fasciolosis is caused by the closely related species, *F. gigantica*, and is a significant constraint on livestock production in large parts of Asia and Africa. Liver fluke disease is classified into three clinical forms, with varying degrees of severity. These are dependent on the timing, level and duration of ingestion of metacercarial cysts. *Acute fluke disease* – is more often seen in sheep than in cattle and occurs when animals ingest massive numbers of infective metacercariae over a short period of time from herbage. The simultaneous migration of large numbers of immature fluke through the liver causes severe pathological damage and haemorrhage and frequently results in sudden deaths. Deaths typically occur before the immature fluke have developed into egg-laying adults so there will be no fluke eggs in the faeces. Other sheep in the flock will often be anaemic with the inside mucosa of eyelids pale and the abdomen swollen. Acute fluke infection can predispose animals to clostridial necrotic hepatitis or 'black disease', but this is less commonly seen with the advent of effective clostridial vaccines. *Subacute fluke disease* – infection is acquired over a more prolonged period. Infected animals tend to show rapid weight loss and poor body condition. *See* **chronic fluke disease**.

fast pace-of-life organisms that mature rapidly and reproduce prolifically.

fat body a large organ and major immune-responsive tissue that is located in the insect open circulatory body cavity and is functionally equivalent to the mammalian liver.

fatty acid carboxylic acids (-COOH) with a long tail composed of chains of carbon atoms. Depending on the chain length, fatty acids can be classified as short (<6 carbon atoms), medium (6–12), long (12–21) or very long (>22).

fatty acylation post-translational covalent addition of a fatty acid to a protein; usually refers to addition of a myristoyl or palmitoyl group to a protein.

Fc gamma receptor (FcγRs) receptor on the surface of phagocytes, including macrophages and neutrophils, that bind to the Fc fragment of immunoglobulin G and stimulate the phagocytosis of particles attached to the antibodies.

Fc receptor a cell-surface receptor that specifically binds antibody (Ab) molecules through their Fc region.

Fc region crystallizable, non-antigen-binding fragment of an immunoglobulin molecule partially digested. It consists of the C-terminal halves of both heavy chains and it is responsible for the biological activities of the molecule, through binding to Fc receptors.

FcεRI this serves as a high affinity immunoglobulin E receptor. It is expressed constitutively on mast cells and basophils, is inducible in eosinophils and is necessary for the allergic hypersensitivity reaction.

FcRs cell surface receptors binding to the fragment crystallizable (Fc) region of immunoglobulin molecules.

Fe speciation the redox state of iron (Fe) and the identity of its ligands. The two most common environmental Fe redox species are Fe^{2+} and Fe^{3+}.

fecundity the capacity of an organism to produce offspring.

feedback an interaction mechanism between processes in the Earth system that occurs when the result of an initial process triggers changes in a second process that in turn influences the initial one. A positive feedback intensifies the original process, and a negative feedback reduces it.

female killing (FK) a genetic strategy that initially used autosomal insecticide-resistance genes pseudo-linked to the Y-chromosome and in genetic repulsion with multiple deleterious recessive mutations. Insecticides thereby eliminated females in the laboratory and the recessive mutations, once they were established in the field in a high enough frequency, killed homozygous females. With RIDL, females are killed by a conditional lethal gene under the control of a female-specific promoter that is only active in the absence of tetracycline.

feral a feral cycle of a parasitic agent is one that takes place in the wild.

fermentative glycolysis catabolism (enzymatic degradation) of glucose to pyruvate as common intermediate and ultimately to different possible end products The better-known products are lactate in mammalian cells and ethanol in yeast. Replenishment of NAD^{+} is a crucial consequence of fermentation.

fertilization the union of male and female reproductive cells (gametes) during the process of sexual reproduction to form a cell called a zygote.

festoon rectangular raised areas separated by grooves. Festoon occurs on the posterior edge of hard ticks of certain species.

fetal growth restriction when a fetus has reduced growth, during intrauterine development, due to a variety of pathological factors.

fibrinogen a glycoprotein present in blood plasma that causes blood coagulation.

fibroblast connective tissue cell which produces collagen and plays an important part in wound healing.

fibronectin receptors integrins on the surface of a variety of cell types, such as phagocytes and fibroblasts, which bind to fibronectin, an abundant glycoprotein in the extracellular matrix that mediates tissue connectivity.

fibrosis the formation of fibrous tissue or scar tissue usually in repair or replacement of cellular elements destroyed by injury, infection or deficient blood supply.

fibrotic edge scar margin between normal liver tissue and fibrosis (scar) where collagen is being produced.

ficolin-2 secreted pattern recognition lectins that contribute to the innate immune recognition of pathogens and activate the complement pathway to clear pathogens.

ficolins bind to the lipotheichoic acid and related molecules present in various 'microbial aggressors'. They participate in the lectin pathway of complement activation.

field isolate strain of parasites originally obtained directly from the blood of a malaria patient.

field population *See* **natural population**.

fifty per cent inhibitory concentration (IC50) a dose of a drug required to slow the rate of growth of a cell population by 50%.

fifty per cent lethal dose (LD_{50}) the dose of a drug required to kill 50% of a cell population.

filament a thread-like structure arising from a granule-like body (the blepharoplast) and covered by a thin sheath of cytoplasm that usually projects from the body of an organism and functions as an organelle of locomotion or when lying in a groove in the cytostome (cytostomal flagellum) causes movement of the fluid medium in a certain direction. In sessile flageolets, the flagella create currents in the medium to bring food particles to the mouth of the organism and to move wastes away from it.

filarial nematodes thread-like roundworms that cause the tropical infectious disease filariasis. They have a life cycle, which involves a sexual stage giving birth to microfilarial larvae in the mammalian host, and an insect vector in which the larval stages develop.

filariform (juvenile) a post-feeding-stage of a nematode characterized by its delicate, elongate structure and its slim, capillary oesophagus. Also, the infective stage of hookworm, filarial worms, and some other nematodes.

filiform resembling a filament or thread.

filipin a cyclic compound derived from *Streptomyces filipinensis* that binds cholesterol and disrupts lipid raft function.

filopodia fine protrusions of the plasma membrane that are driven by actin polymerization.

final host *See* **definitive host**.

finite element analysis a numerical method of approximation.

firewall a software- or hardware-based network security system that controls the incoming and outgoing network traffic based on an applied rule set. A firewall establishes a barrier between a trusted, secure internal network and another network (e.g. the internet) that is not assumed to be secure and trusted.

first intermediate host the host in which asexual reproduction occurs in the digenetic trematode life cycle. It is almost always a mollusc (usually a snail), and rarely an annelid.

first-line treatment the government-recommended treatment for uncomplicated malaria.

first-order kinetics a reaction rate in which the rate is proportional to the concentration. In the case of drug (or malaria parasite) elimination, the rate of reduction in blood concentration at any time is proportional to the concentration at that time. The result is that a fixed fraction of the drug (or parasites) is cleared per unit time. When plotted on a semi-log scale, the plot is linear and, if the vertical axis is in natural logarithm ($\log_e$), the slope gives the first-order rate constant (k).

first-set reaction the initial rejection of a foreign tissue graft.

first-stage larva the earlier stage in the larval development of parasitic nematodes, occurring after mating of adult stages.

FIS (inbreeding coefficient) the proportion of the variance in the subpopulation contained in an individual. High FIS implies a considerable degree of inbreeding.

fishborne zoonotic trematodes (FZT) are well-known causes of fluke diseases in humans. These trematodes include the liver flukes *Opisthorchis viverrini* and *Clonorchis sinensis*, and intestinal flukes of the families Heterophyidae, Leicithodendriidae and Echinostomatidae. Among these trematodes, *C. sinensis* is one of the most harmful to humans, which can cause cholangiocarcinoma. *See* **clonorchiasis**.

FITC abbreviation for fluorescein isothiocyanate.

fitness 1 an evolutionary concept, applicable to individuals, comprising (i) the ability to survive, and (ii) the number of offspring produced (fecundity). It is ideally measured as lifetime reproductive success. 2 the relative reproductive success of a genotype as measured by survival, fecundity or other life-history parameters.

fitness cost describes the disadvantage in growth of an organism with a mutation (e.g. in metabolically important genes) in competition to wild-type organisms under no selection (such as drug pressure).

fixation elimination of polymorphism. This occurs when an allele or mutation reaches the frequency of 100% at a particular locus in one population.

fixation index (F_{ST}) a measure of genetic differentiation due to genetic structure, a special case of Wright's F-statistics, and a common statistics in population genetics. It is estimated from genetic polymorphism data including microsatellite loci and SNPs. The values range from 0 to 1. A zero value implies complete panmixis (i.e. random mating among individuals), implying that the two populations are interbreeding. A value of 1.0 suggests that the two populations do not share any genetic diversity.

fixed pertaining to the point at which an allele (such as a mutation) has completely displaced other alleles; that is, the allele is present in a homozygous state in every individual in the population.

fixed and random effects definitions of fixed and random effects vary with the specific context. Here, a covariate coefficients (parameters) specified as fixed exert a constant effect among individuals while coefficients specified as random effects exert a variable effect among individuals. Parameters exerting random effects include a fixed component which represents the hypothetical effect on the 'average individual' but not necessarily the average effect among individuals.

fixed differences nucleotide differences between two species (or defined populations) that are not polymorphic among sequences sampled from either species (or population).

flagellar calcium-binding protein (FCaBP) dually acylated EF-hand calcium binding proteins found in the flagellar membrane of *T. cruzi*.

flagellar matrix a luminal compartment of the flagellum. Although the matrix is contiguous with the cytoplasm, protein entry is restricted by a diffusion barrier at the base of the flagellum.

flagellar pocket the large invagination at the base of a *T. brucei* cell's single flagellum proximal to the Golgi apparatus. It is the only region of the cell where the plasma membrane is not supported by a dense corset of microtubules, the subpellicular array, and hence is the only site at which vesicle invagination can occur for endocytosis. It is thought the evolutionary pressure for the confinement of endocytosis and exocytosis to this region is for immune evasion; membrane proteins in the flagellar pocket appear not to be immunogenic.

flagellum a defining feature of eukaryotes that is composed of basal body and axoneme; in addition to locomotion, the flagellum has many other conserved and specific functions across eukaryotic phyla. The structure consists of nine pairs of tubulin rods with a centre pair.

flame cell a structure in a primitive excretory system in Platyhelminthes. Cilia in flame cells apparently move fluid wastes from the body into the protonephritic tubules to be excreted. Sometimes flame cells can be seen in living schistosome eggs. The excretory cell in cestodes and trematodes, the number and arrangement of which is used as a basis for identification. The cell has a tuft of cilia, whose beating resembles the flickering of a flame. The flame cells open into a collecting tubule. Also known as solenocyte.

flame photometry a method for measuring metal ion concentrations based on the colour and intensity of flames in a controlled vaporization.

flanking region the flanking regions are the stretches of DNA outside the simple sequence tandem repeat. These sequences are used as primer pairs. The flanking regions are usually invariant across a population or species, but mutations in the flanking region can be a cause of null alleles as well as a potentially serious source of homoplasy.

flavin a cofactor for redox reactions based on pteridine. Depending on the attached moiety, flavins are found as flavin adenine dinucleotide (FAD) or flavin adenine mononucleotide (FMN).

flavin mononucleotide (FMN) a cofactor in many oxidoreductase reactions.

flip-flop when opposite alleles are associated with risk for a disease or a phenotype in different groups (e.g. ethnic groups or groups stratified by environmental exposure).

flow cytometry analysis of biological material by detection of the light-absorbing or fluorescing properties of cells or subcellular fractions (i.e. chromosomes) passing in a narrow stream through a laser beam. An absorbance or fluorescence profile of the sample is produced. Automated sorting devices, used to fractionate samples, sort successive droplets of the analysed stream into different fractions depending on the fluorescence emitted by each droplet. Also known as flow sorting.

flow karyotyping use of flow cytometry to analyse and separate chromosomes according to their DNA content.

fluctuating environment an environment whose physical, chemical and nutritional components are subjected to frequent variations.

fluke a helminth organism of the class Trematoda in the phylum Platyhelminthes. Characterized by a body that is usually flat and often leaf-like. Trematodes can infect the blood, liver, intestines and lungs. Some, such as *Chlonorchis sinensis*, the liver fluke, parasitize humans and cause disease. All of the species that infect domestic animals are in the subclass Digenea, that is they are digenetic flukes. Members of the subclass Monogenea occur on the gills or scales of fish.

fluorescein isothiocyanate (FITC) a green fluorescent dye that can be conjugated to proteins (antigens or antibodies) for use in immunofluorescence reactions.

fluorescence a property of molecules to absorb light at one wavelength and emit light at a longer wavelength.

fluorescence *in situ* hybridization (FISH) a laboratory technique for detecting and locating a specific DNA sequence on a chromosome. The technique relies on exposing chromosomes to a small DNA sequence called a probe that has a fluorescent molecule attached to it. The probe sequence binds to its corresponding sequence on the chromosome.

fluorescence microscope a microscope used for the examination of specimens stained with fluorochromes or fluorochrome complexes, e.g. a fluorescein-labelled antibody, which fluoresces in ultraviolet light.

fluorescence recovery after photobleaching (FRAP) a technique to study protein dynamics in live cells. A region of interest (e.g. a membrane-bound organelle) within a cell is intentionally photobleached with a laser, and the flow of fluorophores migrating into that region (and depletion from surrounding areas) is subsequently monitored. Biological information on the mobility of the fluorescently labelled component and the cellular barriers to such mobility may be deduced from the rate of recovery of fluorescence.

fluorescence-activated cell sorting (FACS) a technique to rapidly separate cells in a suspension on the basis of size and the wavelength of their fluorescence.

fluorescent antibody an antibody conjugated to a fluorescent dye such as FITC.

flux balance analysis (FBA) a type of constraint-based analysis that imposes mass balance and flux capacity constraints on the network and uses an optimization procedure to find a steady-state flux distribution that maximizes the biomass yield of the cell without the need for measurements of kinetic parameters. FBA can be used to identify essential reactions and to simulate growth in different environmental conditions.

foam cell macrophages which have engulfed low-density lipoproteins. They are characteristically present in atherosclerotic plaques.

focal adhesions large patches of integrins that typically develop from focal complexes. They grow and are stabilized by the application of actomyosin-driven force.

focal complexes clusters of integrin-mediated adhesions. These structures are linked to F-actin. They can form and turnover rapidly, or may mature into focal adhesions.

focal species the species that is in focus, the species of interest.

focused testing and treatment (FTAT) a focused testing for infection followed by treating all infected persons in a localized area such as a neighbourhood or village irrespective of whether they have clinical symptoms. FTAT may be conducted broadly over non-contiguous areas on high-risk populations and hotspots.

follicular dendritic cell MHC class II-negative Fc receptor-positive dendritic cells which bear immune complexes on their surface and are involved in the generation of antibody-secreting cells and maintenance of B-cell memory in germinal centres. (N.B. a different cell type to interdigitating dendritic cells).

folliculitis inflammation of hair follicles.

follow-up observation over a period of time of an individual, group, or initially defined population whose relevant characteristics have been assessed in order to observe changes in health status or health-related variables.

fomites inanimate objects that may be contaminated with microorganisms and become vehicles for the transmission of infectious agents.

foodborne a term used when referring to an infection or other damaging agent that is transmitted via the animal or human food chain.

foraging behaviour the sequence of behaviours employed by organisms to locate, subjugate, and consume food resources.

force of infection (FOI) the rate at which humans become infected.

forced oscillation measurements invasive method for the accurate measurement of several respiratory parameters, including tissue impedance, elasticity, and airway resistance.

forced separation an event of forced mate-switching that is not necessarily initiated by either partner but is instead caused by the intrusion of a third individual who displaces one member of the pair.

foreign host occasionally become parasitized.

forensic relating to courts or legal matters. Molecular markers are increasingly common in the context of forensics, both in wildlife and human cases involving identity or relatedness.

forkhead box P3 (FoxP3) is also known as scurfin, and it is the master regulator as well as transcription factor of T regulatory cells. It inhibits the transcription of IFN-γ and activates transcription of IL-10.

formamidines a group of acaricides-insecticides that have a direct effect on the parasite's central nervous system (CNS). The formamidine amitraz is used to treat local and generalized demodectic mange and sarcoptic mange in dogs. It is also marketed for use in cattle and swine to control ticks, mange mites, and lice. Although their exact mechanism of action is unknown, they may inhibit the monoamine oxidase activity, which is responsible for metabolism of neurotransmitter amines, and/or activate the inhibitory neurotransmitter octopamine. Amitraz should not be used in horses or cats.

Förster resonance energy transfer (FRET) a technique used (among other applications) to determine whether two proteins are in close proximity. Excitation wavelengths are absorbed by one fluorophore (the donor fluorophore). The emitted energy from that fluorophore can be secondarily absorbed by acceptor fluorophores provided they are sufficiently close (within several nanometres). The acceptor fluorophores re-emits light at a longer wavelength, which is detectable by fluorescence microscopy. This technique is sometimes also referred to as fluorescence resonance energy transfer.

forward genetic screens genetic screens in which mutants are isolated on the basis of their phenotypes. The mutations responsible are identified by positional cloning or by a candidate gene approach.

forward genetics an approach in which mutations induced by random mutagenesis that lead to the disruption of gene function are identified based on their phenotypes. The causative mutation is traditionally identified by positional cloning or by a candidate-gene approach.

founder effect the reduction in genetic variation that results when a small subset of a large population is used to establish a new colony. The new population may be very different from the original population, both in terms of its genotypes and phenotypes. In some cases, the founder effect plays a role in the emergence of new species.

foxp3 a transcription factor present in the nucleus of most regulatory T-cells and is essential for the differentiation and maintenance of natural and subsets of peripherally induced T regulatory cells.

frameshift mutation a type of mutation involving the insertion or deletion of a nucleotide in which the number of deleted bps is not divisible by three. 'Divisible by three' is important because the cell reads a gene in groups of three bases. Each group of three bases corresponds to one of 20 different amino acids used to build a protein. If a mutation disrupts this reading frame, then the entire DNA sequence following the mutation will be read incorrectly.

framework regions (FRs) the relatively conserved amino acid sequences which flank the hypervariable regions in Ig and T-cell receptor variable regions and maintain a common overall structure for all V-region domains.

free text information that is entered in narrative form, without constraint.

free-living living in a free and unrestrained manner in the environment, free of a host.

frequency-dependent selection selection where fitness is dependent on allele frequency. It is a form of balancing selection when the rarer the genotype, the fitter the individual.

Freund's adjuvant an additive to an antigen injected to enhance an immune response to the antigen. It is a water-in-oil emulsion that contains an immunogen, an emulsifying agent, and killed mycobacteria which enhance the immune response to the immunogen; termed 'incomplete' Freund's adjuvant if mycobacteria are not included.

frontal filament an organ attaching chalimi and moulting mobiles of *Lepeophtheirus salmonis* to the host.

***F*-statistics** a measure of genetic structure developed by Sewall Wright. Related to statistical analysis of variance (ANOVA), F_{ST} is the proportion of the total genetic variance contained in a subpopulation (the S subscript) relative to the total genetic variance (the T subscript). Values can range from 0 to 1. High F_{ST} implies a considerable degree of differentiation among populations.

full gene sequence the complete order of bases in a gene. This order determines which protein a gene will produce.

fully edited mRNA transcripts in which all the needed editing steps have been completed.

functional annotation the process of attaching biological information (e.g. biochemical function, biological function, involved regulation and interactions, and expression) to genomic elements. *See* **genome annotation**.

functional genomics the study of sequencing data to describe gene (and protein) functions and interactions. Unlike genomics, functional genomics focuses on dynamic aspects such as gene transcription, translation, and protein–protein interactions, as opposed to the static aspects of genomic information such as DNA sequence or structures.

functional interactions a general term that incorporates protein–protein interactions, classical genetic interactions, regulatory interactions (such as kinase–substrate interactions) and phenotypic interactions.

functional orthologues two genes from different organisms that share the same function, having both arisen from the same ancestral gene.

fungicide a substance or chemical that kills fungi.

furazolidone a nitrofuran derivative that interferes with parasitic enzyme systems, although its exact mechanism of action is unknown. It is effective against *Giardia* sp., coccidia, *Trichomonas*, and some enteric bacteria. It is available as a liquid suspension, which facilitates administration to puppies and kittens. Side effects include anorexia and vomiting. The nitrofurans are suspected mutagens and carcinogens and their use in food animals is prohibited.

furca(e) refers to one of the two forks in the tail of a schistosome cercaria.

Fv the variable region fragment of an antibody heavy or light chain.

G protein-coupled receptor seven-transmembrane receptor that binds small molecules such as neurotransmitters, neuropeptides, and odorants. Upon ligand binding, the receptor activates a guanine nucleotide-binding protein (G protein), which can in turn produce second messengers that activate neuronal activity.

G test a correlation between genetic distances calculated from different genetic markers.

G0 a phase of the cell cycle designating cells that are quiescent and have not yet entered the growth cycle. Normal cells in this phase have exactly one set of chromosome pairs.

G1 a phase of the cell cycle in which cells commit to division. These cells have the same number of chromosomes and the same amount of DNA as G0 phase cells. They are about to enter S phase where they synthesize DNA.

G2 a phase of the cell cycle in which proliferating cells have duplicated their DNA and formed two sets of chromosome pairs, in preparation for division. G2 follows the S phase and precedes the M (mitosis) phase.

G6PD (glucose-6-phosphate dehydrogenase) deficiencies an X-chromosome-encoded human gene. Deficient individuals are protected against overwhelming parasitaemia and severe falciparum malaria. This is mainly owing to the limited tolerance to oxidative stress of deficient erythrocytes.

GALM of growth arrest at larval maturity; the cessation of larval growth, before which ability to establishment in the next host is typically low or negligible.

GALT for gut-associated lymphoreticular tissue.

gamergates a unique reproductive caste comprising mated, fertile workers. In some ponerine species (for example, *Harpegnathos saltator*), gamergates emerge from the existing cohort of workers when a queen dies or is artificially removed from a colony.

gamete a sexually differentiated cell that fuses with a gamete of different sex to form a fertilized zygote during sexual reproduction. *Plasmodium* gametes are formed in the mosquito midgut lumen and are derived from gametocytes ingested by a mosquito in a blood meal.

gametocyte **1** the sexual reproductive stage of the malaria parasite. Female (macro-) and male (micro-) gametocytes circulate in the bloodstream, are picked up by the *Anopheles* mosquito, and undergo gamete formation and fertilization in the midgut of the mosquito. **2** the stage in the life cycle of some protozoa (Apicomplexa) which is destined to become a gamete (macro- or micro-) zygote, that can develop into a new host.

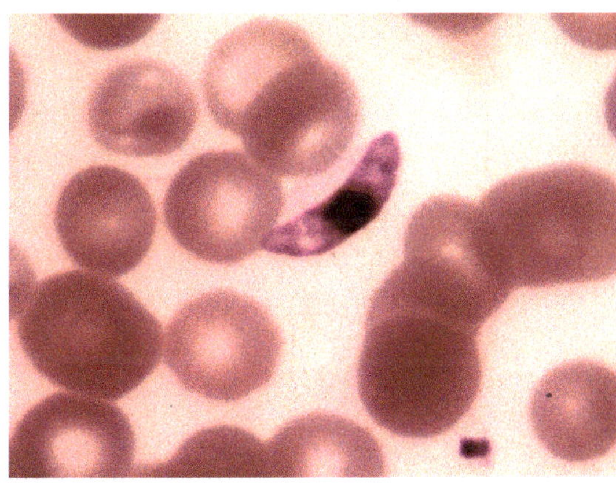

gametocyte density the number of gametocytes in gametocyte carriers.

gametocyte prevalence the proportion of a population that harbours gametocytes, also most commonly assessed by microscopy.

gametocytogenesis the developmental process that leads to the formation of male and female sexual gametocytes, which later fertilize after ingestion by the mosquito vector.

gametogenesis the intracellular development of gametocytes (macrogametocytes, female and microgametocytes, male) that by meiosis (reduction division) yield either female or male gametes as in malarial parasites.

gammaglobulins the serum proteins, mostly immunoglobulins, which have the greatest mobility towards the cathode during electrophoresis.

gammopathies abnormalitiy in gamma-globulin levels.

gamogony (=gametogony) the formation of gametes.

gamont the sexual stage in the sporozoan life cycle, produced by gametogony from a trophozoite or a merozoite (synonym gametocyte); a pre-gamete. *See* **gametocyte**.

gapeworm a common name for *Syngamus trachea*; a nematode parasite of the respiratory tract of gallinaceous birds. The parasite parasitizes the trachea of a variety of domestic and wild birds causing a condition of 'gapes'. Adults are red in colour, and measure 2 cm long. The two sexes are present in permanent copulation. The related *Cyathostoma bronchialis* causes a similar condition in geese. Disease is due to physical blockage of the airway, leading to dyspnoea, the affected bird exhibits the appearance of outstretched neck with open mouth. Other manifestations include lack of condition and mortality.

GAS for interferon gamma-activated sequence: promoter element.

gas chromatography (GC) and mass spectrometry(MS) a method separating (GC) and identifying (MS) compounds that can be vaporized without decomposition, based on their selective retention in the matrix of specific columns. GC coupled to flame ionization detection (GC-FID) allows the detection of micrograms to lower nanograms of vaporizable lipids. GC is very widely used for the determination of fatty acid profiles in glycerolipids. GC can be coupled to a mass spectrometer (GC-MS) for chemical identification of compounds based on their mass to charge ratio following ionization.

Gasterophilus a genus of flies, called bot flies, which lay eggs on the hairs of horses (and a few other animals). The adults are free living, but the larvae, the maggots, live and develop only in the stomach of the horse.

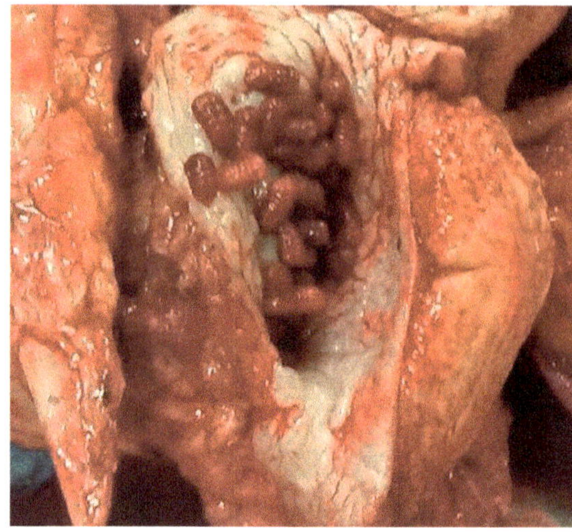

gastroallergic anisakiasis an acute allergic reaction (urticaria/angioedema or anaphylaxis) which is produced in the context of an acute gastric parasitism by *Anisakis simplex*, when the parasite attempts to penetrate the gastric mucosa.

gastropod a class in the phylum Mollusca that includes snails and slugs.

gating the kinetic transitions between open and closed states of a functioning channel.

Gaussian mixture models (GMMs) the parametric probability density functions that fit the Gaussian distribution to the data set to model the presence of sub-populations within an overall population; they are represented as weighted sums of Gaussian component densities.

GC-biased gene conversion (gBGC) a non-adaptive evolutionary process that favours the fixation of GC alleles over AT alleles in highly recombining regions; it is likely to result from biased mismatch repair in heterozygous individuals during meiotic recombination.

GC-rich area many DNA sequences carry long stretches of repeated guanine (G) and cytosine (C), which often indicate a gene-rich region.

geitonogamy fertilization between gametes originated in separate, but genetically identical individuals.

gel diffusion an immunoprecipitation technique that involves letting antigen and antibody meet and precipitate in a clear gel such as agar.

GenBank an online database of publicly available DNA sequence data maintained by the National Center for Biotechnology Information (NCBI), a part of the National Institutes of Health (NIH).

gene a sequence of DNA that either codes for the synthesis of a specific protein or has a specific regulatory function. It is a fundamental physical and functional unit of heredity. *See* **gene expression**.

gene amplification an increase in the number of copies of a gene sequence. Cancer cells sometimes produce multiple copies of genes in response to signals from other cells or their environment. The term also can refer to PCR, a laboratory technique that is used by scientists to amplify gene sequences in a test tube. *See* **gene**.

gene chip technology the development of cDNA microarrays from a large number of genes. Used to monitor and measure changes in gene expression for each gene represented on the chip.

gene circuit a small number of genes influencing the expression of one another via regulatory interactions.

gene cluster a group of closely linked genes that are collectively responsible for a multistep process such as the biosynthesis of a metabolite.

gene complex a cluster of related genes occupying a restricted area of a chromosome.

gene conversion the type of recombination in which a gene in a heterozygote of genotype Aa is converted from one allele to the other, resulting in a homozygote AA or aa.

gene diversity (expected heterozygosity) a measure of genetic variation in a population. It is calculated from the squared gene (=allele) frequencies.

gene duplications one or more duplications of a gene or gene regions in a given genome. The extent of a duplication is largely dependent on the process that produces it: unequal crossing over, replication slippage, or retrotransposition may involve one or a few genes, whereas aneuploidy and polyploidy involve single chromosomes or whole genomes, respectively. Duplicated sequences may become either non-functional through pseudogenization, or a source of new genes. Compared with endogenization, gene duplications are commonly recognized as a major source of new gene functions.

gene–environment interaction an influence on the expression of a trait that results from the interplay between genes and the environment. Some traits are strongly influenced by genes, while other traits are strongly influenced by the environment. Most traits, however, are influenced by one or more genes interacting in complex ways with the environment.

gene expression the process by which a gene's coded information is converted into structures present and operating in the cell. Expressed genes include those transcribed into mRNA and then translated into proteins, as well as those transcribed into RNA but not translated into proteins [e.g. transfer ribonucleic acid (tRNA) and rRNA].

gene family a group of closely related genes that make similar products.

gene flow 1 the exchange of alleles between populations or subpopulations. The lower the gene flow, the higher the genetic differentiation between populations or subpopulations. It is the mechanism by which migration is assessed in most population genetic models. 2 the transfer of alleles or genes from one population to another.

gene frequencies refers to allele frequencies.

gene function the biochemical reaction, protein–protein interaction, metabolic or signalling pathway association, cellular localization, phenotype, and changes in protein function that are mediated by shifts in protein structure.

gene knockout specific targeting of a gene by homologous recombination (pairing and rearrangement of 2 copies or alleles of a gene) commonly performed in mice. Frequently, this is designed to produce a complete and specific loss of function of the gene.

gene library *See* **genomic library**.

gene mapping the process of determining where genes are located on individual chromosomes, their position in relation to other genes and the distance between them.

gene ontology (GO) an ontology consisting of three separate ontologies (cellular component, biological process and molecular function) that is used to describe genes and, more often, gene products in a species-independent manner.

gene pool the total genetic diversity found within a population or a species. A large gene pool has extensive genetic diversity and is better able to withstand the challenges posed by environmental stresses. Inbreeding contributes to the creation of a small gene pool and makes populations or species more likely to go extinct when faced with some type of stress. *See* **allele**, **gene**, **polymorphism**.

gene prediction the predictions of possible genes made by a computer program based on how well a stretch of DNA sequence matches known gene sequences.

gene product the biochemical material, either RNA or protein, resulting from expression of a gene. The amount of gene product is used to measure how active a gene is; abnormal amounts can be correlated with disease-causing alleles.

gene regulation the process of turning genes on and off. During early development, cells begin to take on specific functions. Gene regulation ensures that the appropriate genes are expressed at the proper times. Gene regulation can also help an organism respond to its environment. Gene regulation is accomplished by a variety of mechanisms including chemically modifying genes and using regulatory proteins to turn genes on or off.

gene regulatory network (GRN) an intracellular network of regulatory proteins that control the expression of gene subsets involved in particular cellular functions. A simple network would consist of one or more input signalling pathways, regulatory proteins that integrate the input signals, several target genes (in bacteria a target operon), and the RNA and proteins produced from those target genes.

gene related to anergy in lymphocytes (GRAIL) a type I transmembrane E3 ligase identified as an early gene that promotes T-cell anergy.

gene splicing a technique used to join segments of DNA to form a new genetic combination.

gene technology *See* **genetic engineering**.

gene testing methods that identify the presence, absence or mutation of a particular gene in an individual.

gene therapy the addition of a functional gene or groups of genes to a cell using recombinant DNA techniques to correct a hereditary disease. *See* **gene splicing**.

gene transfer the incorporation of new DNA into an organism's cells, usually by a vector such as a modified virus. *See* **mutation**, **vector**.

gene trees the evolutionary history of a gene represented as a branching diagram. A gene tree may or may not be identical to the species tree.

genealogy the history of the descent of genes, often within a single species.

gene-drive efficacy the ability of a gene-drive system to bias the frequency of inheritance of a genetic element, which is measured as the departure from Mendelian inheritance in the gametes of a heterozygous individual and ranges from 0 (when half of the gametes carry the genetic element) to 1 (when all the gametes carry the genetic element).

gene-for-gene interaction the allele-specific interaction between a host R-gene and a pathogen Avr gene that leads to a host resistance response.

gene-for-gene model a model of pathogen–host or parasite–host interactions, in which each specific pathogen avirulence factor (that is, effector) interacts with a specific cognate host resistance factor [i.e. nucleotide-binding domain and leucine-rich repeat-containing protein (NLR)]. The specific combination of avirulence and resistance factors carried by the pathogen and the host, respectively, will determine whether the interaction is compatible (virulent) or incompatible (avirulent).

genera taxonomic categories of organisms that rank between family and species. Genera for higher organisms generally consist of species with similar characteristics.

general circulation model (GCM) a class of computer-driven models (sometimes called global circulation models) that provide weather forecasts and climate projections. GCMs integrate a variety of fluid dynamical, chemical, and biological equations that represent processes in Earth system components (e.g. land, ocean, atmosphere, and biosphere).

generalist parasite a parasitic organism (or its developmental stage) with a broad host spectrum.

generalized estimating equation (GEE) a technique for estimating the parameters of a marginal model fitted to correlated repeated measures (observations). The GEE approach is semi-parametric because it relies on the first two moments of the observed data, but not on the full likelihood.

generalized linear mixed model (GLMM) an extended GLM that includes a linear predictor comprising covariate coefficients that exert both fixed and random effects.

generalized linear model (GLM) an extension of the simple linear regression model that is compatible with error distributions from any of the exponential family of probability distributions, including the normal, Poisson, binomial, and gamma distributions. The simple linear regression model is a GLM with normally distributed errors.

genetic anticipation a mode of disease inheritance characterized by progressively earlier ages of disease onset as generations progress. Generally caused by the gradual expansion of STRs.

genetic architecture the complete description of the genetic factors influencing trait variation, such as the number of genetic loci, their effects, allele frequencies, actions and interactions.

genetic assignment test a test to identify immigrant individuals that exhibit a strong signal of membership in a population other than the one in which they are sampled. Immigrants in this sense could be true immigrants or recent descendants of immigrants.

genetic background the entire genetic and genomic context of an organism; the complete genotype of an organism across all loci.

genetic bottleneck population genetics term describing a temporary reduction in population size that causes the loss of genetic diversity.

genetic code the instructions in a gene that tell the cell how to make a specific protein. A (adenine), C (cytosine), G (guanine), and T (thymine) are the 'letters' of the DNA code that make up the nucleotide bases of DNA. Each gene's code combines the four chemicals in various ways to spell out three-letter 'words' that specify which amino acid is needed at every step in making a protein.

genetic counselling counselling individuals and prospective parents who are at risk of a particular genetic disease (either themselves or their potential child). Provide them and their families with education and information about genetic-related conditions such as the probabilities, dangers, diagnosis and treatment, and helps them make informed decisions.

genetic discrimination a prejudice directed against people who have or may have a genetic disease. Genetic discrimination can involve being denied employment or health insurance. In a healthcare context, it can refer to people being treated based on their genetic status rather than by some more relevant criterion.

genetic disorder a hereditary condition that results from a defective gene or chromosome.

genetic distance a measure of the genetic relatedness of populations. Typically, these measures are based on allele frequency differences between the populations that may have resulted from isolation (i.e. no immigration), random drift, founder effects, selection and so on.

genetic draft an evolutionary process through which the frequency of an allele in a population changes through time simply by a random sampling effect; that is, some alleles can increase or decrease in frequency by chance over time, even in the absence of an effect on the fitness of the individuals. The magnitude of this phenomenon is stronger for lower effective population sizes.

genetic engineering a term used to describe various forms of genetic manipulation of organisms usually based on gene cloning. May describe insertion of 'foreign' genes into organisms, that is, genes not naturally occurring in the species or strain. This term covers all laboratory or industrial techniques used to alter the genetic material of organisms. These techniques assist organisms to produce new substances or perform new functions, for example increase yields of compounds already produced by the organism, form new compounds, or allow organisms to adapt to drastically altered environments.

genetic engineering technology *See* **recombinant DNA technology**.

genetic epidemiology a relatively new medical discipline that seeks to understand how genetic factors interact with the environment in the context of disease in populations. Areas of study include the causes of inherited disease and its distribution and control.

genetic equilibrium the state at which the allele distribution and the frequencies of mutations are constant over time.

genetic fixation an increase of the frequency of an allele by selection until no other alleles are preserved in a specific finite population.

genetic heterogeneity the phenomenon whereby different loci possess alleles which cause the same trait or disease independently of one another. Allelic heterogeneity occurs when different mutations within the same gene can cause a trait or disease. Genetic heterogeneity is one complexity associated with multilocus inheritance.

genetic hitchhiking the process by which neutral mutations increase in frequency by virtue of being in the neighbourhood of (or linked to) an advantageous mutation that is undergoing a selective sweep.

genetic informatics *See* **bioinformatics**.

genetic lineages lineages defined by DNA sequence features, such as a gene or portion of a gene, rather than phenotypic features.

genetic map a type of chromosome map that shows the relative locations of genes and other important features. The map is based on the idea of linkage, which means that the closer two genes are to each other on the chromosome, the greater the probability that they will be inherited together. By following inheritance patterns, the relative locations of genes along the chromosome are established. *See* **linkage map**.

genetic marker a DNA sequence with a known physical location on a chromosome. Genetic markers can help link an inherited disease with the responsible gene. DNA segments close to each other on a chromosome tend to be inherited together. Genetic markers are used to track the inheritance of a nearby gene that has not yet been identified, but whose approximate location is known. The genetic marker itself may be a part of a gene or may have no known function. *See* **chromosome**, **deoxyribonucleic acid**, **gene**.

genetic material *See* **genome**.

genetic microadaptation selective evolutionary changes in the genetic content of closely related microorganisms.

genetic modification (GM) any process that alters the genetic material of a living organism. This includes duplicating, deleting or inserting one or more new genes or altering the activities of an existing gene. It can be performed on microbes, plants or animals (humans included). Where this is done in humans, it is gene therapy, and only human genes are used.

genetic monogamy a situation where two sexual partners only have offspring with one another, irrespective of social mate-switching, and the individuals of the pair are recognized as being genetically 'faithful'.

genetic mosaic an organism in which different cells contain different genetic sequence. This can be the result of a mutation during development or fusion of embryos at an early developmental stage.

genetic or locus heterogeneity a disease with high locus heterogeneity has several variants at the population level. Locus heterogeneity is common in syndromes resulting from failure of a complex pathway.

genetic rescue reversal of the phenotype from a genetically defined mutant by introduction of foreign DNA encompassing a wild type copy of the mutant gene that can complement (act against) the loss-of-function allele in the genome.

genetic screening the process of testing a population for a genetic disease in order to identify a subgroup of people who either have the disease or the potential to pass it on to their offspring.

genetic testing the use of a laboratory test to look for genetic variations associated with a disease. The results of a genetic test can be used to confirm or rule out a suspected genetic disease or to determine the likelihood of a person passing on a mutation to their offspring. Genetic testing may be performed prenatally or after birth. Ideally, a person who undergoes a genetic test will discuss the meaning of the test and its results with a genetic counsellor.

genetic variance the variance of trait values that can be ascribed to genetic differences among individuals. The total genetic variance of a trait can be dissected into additive, dominance and other components.

genetic variants DNA sequence differences found across human individuals. These include SNPs and structural variants such as insertion-deletions, block substitutions, inversions, and copy-number variants. The variants that are statistically associated with human diseases or traits are called genetic risk variants.

genetic variation differences between individuals or to differences between populations. Mutation is the ultimate source of genetic variation, but mechanisms such as sexual reproduction and genetic drift contribute to it as well.

genetically attenuated parasites (GAP) *Plasmodium* parasites lacking one or more genes that encode for essential functions, resulting in impaired parasite growth.

genetically modified organism (GMO) organisms whose genetics has been altered by humans. Could apply to virtually all commercial crops and livestock that humans have modified over generations of selection. More often refers to modification based on molecular technologies such as cloning of genes. In this case, the modification could involve inserting genes into an organism derived from the same or different species, or even a completely synthetic gene.

genetics the branch of science dealing with the principles of heredity and variation in animals and plants.

genital primordium a group of cells in juvenile nematodes that are the precursors of a reproductive system. Seen in rhabditoid juveniles of *Strongyloides* but usually not detectable in those of hookworms.

genome all the genetic material in the chromosomes of a particular organism. Most prokaryotes package their entire genome into a single chromosome, while eukaryotes have different numbers of chromosomes. Genome size generally is given as total number of bps.

genome annotation process of identifying elements in the genome and attaching biological information to these elements. Automatic annotation tools perform this process by computer analysis, as opposed to manual annotation (i.e. curation), which involves human expertise. Ideally, these approaches co-exist and complement each other in the same annotation pipeline. *See* **functional annotation**, **structural annotation**.

genome assembly process of taking a large number of short DNA sequences and putting them back together to create a representation of the original chromosomes from which the DNA originated. In a shotgun sequencing project, all the DNA from a source is first fractured into millions of small pieces. These pieces are then 'read' by automated sequencing machines, which can read up to 1000 nucleotides or bases at a time. A genome assembly algorithm works by taking all the pieces and aligning them to one another, and detecting all places where two of the short sequences, or reads, overlap. These overlapping reads can be merged, and the process continues.

genome engineering techniques for the targeted, specific modification of the genetic information (or genome) of living organisms.

genome instability an increase in the rate of mutations or chromosomal abnormalities experienced by a cell, such as aneuploidy.

genome integrity the ability to maintain the integrity of genetic information is fundamental for the life of a cell and the survival of a species. Cells can encounter DNA damage as a consequence of normal cellular metabolism or as a result of exposure to chemical or physical agents.

genome project research and technology-development effort aimed at mapping and sequencing the genome of organisms.

genome scanning typing of multiple polymorphic markers (usually 300–400 microsatellites) located at intervals (10–20 cM) across the genome in multicase families to allow detection of linkage between disease susceptibility and regions of the genome not yet known to carry susceptibility loci.

genome sequence order of nucleotides or bases within DNA molecules that make up an organism's entire genome. The four bases are adenine, guanine, cytosine, and thymine, represented as A, G, C, and T, respectively.

genome size the genome is the collective term for the entire complement of hereditary material found in an organism (e.g. all the DNA in the set of chromosomes in eukaryotes). Genome size ranges from approximately 10^4 bp in some viruses to approximately 10^{10} in many angiosperm plants, to $>10^{10}$ in some salamanders and fishes. Mammals have approximately $2–3 \times 10^9$ bp. Although polyploidy can increase genome size, most increase seems to be due to relatively small duplication events (because genome sizes within taxa tend to be approximately normally distributed around an intermediate modal size.

genome streamlining a selective process that drives genome reduction by the elimination of the genes that make the lowest contribution to fitness.

genome walking a PCR-based method to sequence flanking genomic regions adjacent to a known sequence by the use of primers specific to the known sequence.

genome-wide association (GWA) an examination of genetic variation across the entire human genome used to identify genetic association with specific phenotypic traits, most notably the presence or absence of a disease. Typically, these studies involve cases and controls, and the frequency of alleles is compared between the two groups. If certain alleles are more common in cases than in controls, that allele is determined to be associated with the trait. Causality cannot be inferred from this approach, but it can guide further hypothesis-driven investigations.

genomic data any data that relate to a species genome or transcriptome.

genomic imprinting the ability of a gene to be expressed depends upon the sex of the parent who passed on the gene. In some cases imprinted genes are expressed when they are inherited from the mother. In other cases they are expressed when inherited from the father. Unlike genomic mutations that can affect the ability of inherited genes to be expressed, genomic imprinting does not affect the DNA sequence itself. Genomic imprinting affects gene expression by chemically modifying DNA and/or altering the chromatin structure. Often, genomic imprinting results in a gene being expressed only in the chromosome inherited from one or the other parent.

genomic inflation the presence of excess false-positive results, measured by quantifying the ratio of the median of the empirically observed distribution of the test statistic to the expected median.

genomic library a collection of clones made from a set of randomly generated overlapping DNA fragments that represent the entire genome of an organism. *See* **arrayed library**, **library**.

genomic profile risk a predicted measure of genetic risk for individuals constructed from a set of loci, the risk alleles and corresponding effect sizes of which have been estimated in an independent sample.

genomics the study of the DNA sequence of an organism. This includes the genes that code for proteins, the regulatory sequences that control the genes and the non-coding DNA segments.

genotype the collection of genomic alleles of an organism that result in the observable characteristics (phenotype) of the organism. Microsatellite-based genotype refers to the determination of both alleles for a number of microsatellite loci in a single individual organism.

genotype-by-genotype (G × G) describes the intrinsic specificity of compatibility among host and pathogen genotypes. It can be measured as the statistical deviation from the additive combination of host and pathogen genotypes in their effects on the infection phenotype.

genotyping the process of determining the genotype of an individual for one or more loci.

genus low-level taxonomic rank used in the biological classification; genus lies above species and below family.

geometric mean a type of mean or average which quantifies the central tendency of a set of numerical observations using the product, rather than the sum, of their values. Typically, geometric means are calculated by first taking the arithmetic mean of the log-transformed values before taking the exponent of the result to transform back onto the original scale.

germ cell the gametes or reproductive cells (eggs and sperm), or precursor cells that will develop into eggs or sperm.

germ line are reproductive cells – the egg and the sperm. Any changes to the germline will be passed on to offspring.

germ line stem cells germline stem cells have the capacity to generate all cells in the body. Under normal fate the germline stem cells are unipotent, producing gametes and self-renewing to maintain the generative population.

germ-free animals animals re-derived and reared in isolators without exposure to exogenous microorganisms.

germinal centre are specialized microenvironments within secondary lymphoid organs that support the maturation, proliferation, and differentiation of B-lymphocytes during T-cell-dependent immune responses. These microanatomical structures are divided into light and dark zones; in the dark zone, B-cells expand and diversify their antigen receptors through somatic hypermutation (that potentially increases antibody affinity) and class-switching (that alters antibody effector functions), whereas in the light zone B-cells are selected for binding to their cognate antigen.

germinate seeds starting to grow by putting out shoots and roots (can also apply to fungi and bacterial spores returning to the vegetative stage).

germline (referring to immunoglobulin genes) genes encoding immunoglobulin (Ig) variable, diversity, joining and constant segments are ordered and localized in three main chromosomal loci: IgH (14q32.22), Igκ (2p11.2) and Igλ (22q11.2). Ig genes outside these loci are non-functional orphan genes.

giant cell large multinucleate cell derived from fused macrophages and often present in granulomas.

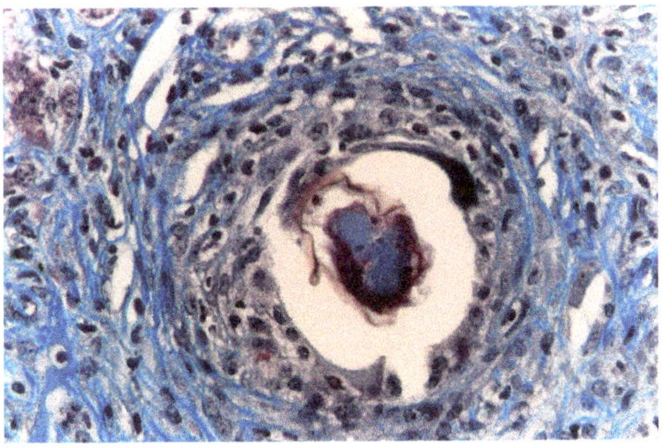

giant kidney worm *See* ***Dioctophyma renale***.

giardan a recently described, unique polysaccharide (β1,3-linked polymer of N-acetyl-D-galactosamine). This polysaccharide constitutes approximately 67% of the fibrillar cyst wall of *Giardia intestinalis* (synonym *lamblia*) and has not been detected in any other organism to date. Previously, these fibrils were thought to be chitin, a β1,4 homopolymer of N-acetyl-D-glucosamine. *See* **cyst wall synthase**.

Giemsa-stained light microscopy detection of malaria parasites in blood by drying thin blood film on a glass slide, fixing with methanol and staining with a mixture of aniline dyes, designed to highlight *Plasmodium* parasites when examined under a high-powered light microscope. Sometimes other aniline dye-based stains are used for example, Field's stain.

gigaton (GT) one billion metric tons; a metric ton is a unit of mass equal to 1000 kg (about 2200 lb).

glia limitans composed of the parenchymal basement membrane and astrocyte foot processes and covers the entire surface of the brain and spinal cord on external surfaces towards the leptomeningeal/subarachnoid space (glia limitans superficialis) and internally towards the perivascular spaces (glia limitans perivascularis).

gliding motility the gliding movement of apicomplexan cells, requiring constant contact with a substrate surface (i.e. the host cell) but functioning in the absence of cellular appendages such as cilia, flagella or lamellapodia. It occurs through the linking of an intra-cellular actin–myosin motor to the extra-cellular surface via secreted adhesins.

glomerulonephritis pathological lesions in the glomeruli of the kidney, usually as a result of immune complex depositions.

gluconeogenesis the metabolic reaction triggering the biosynthesis of glucose from pyruvate, lactate, glycerol and the amino acids alanine and glutamine.

glucose 6-phosphate dehydrogenase (G6PD) deficiency a genetically inheritable disease characterized by abnormally low levels of G6PD, a metabolic enzyme involved in red blood cell metabolism. In severe G6PD deficiency, primaquine is contraindicated for the treatment of *Plasmodium vivax* infection. Patients carrying mutations in the gene encoding for G6PD suffer from severe haemolysis when treated with the 8-aminoquinoline class of antimalarials and in particular with primaquine. The current hypothesis is that the primaquine metabolite responsible for the antihypnozoite activity would trigger massive lysis of the host erythrocytes in addition to the one due to the burst of parasites after the completion of every proliferative cycle (48 hours for *P. falciparum*) in patient blood.

glucosylation the process of transferring a glucose residue from a nucleotide sugar derivative, such as from uridine diphosphate glucose (UDP-Glc) to a target molecule.

glutamic acid/alanine-rich protein (GARP) an abundant glycoprotein expressed on the surface of epimastigote *Trypanosoma congolense.*

glutaminolysis an alternative source of biomass and electrons due to the relative abundance of glutamine in human plasma. After deamination of this amino acid, glutamate feeds part of the TCA cycle. Intermediates such as malate and oxaloacetate can transit to the cytoplasm from mitochondria and be decarboxylated to replenish glycolytic pyruvate with the production of NADPH.

glutathione (GSH) a tripeptide (γ-glutamyl-cysteinyl-glycine) and the most abundant low molecular weight thiol in cells. It plays an important role in oxidation reduction (redox) reactions because GSH with glutathione peroxidase scavenges toxic hydrogen peroxide.

glycogen vacuole a vacuole, usually in a cyst, in which glycogen (carbohydrate storage material) is accumulated until needed in the development process. Such vacuoles usually form early in the development of an amoebic cyst and are usually present in *Iodamoeba butschlii* cysts.

glycolipid a glycerolipid or a phospholipid, with a polar head containing a sugar.

glycolysis the degradation of sugars to pyruvate, which results in the generation of ATP by substrate phosphorylation reactions. The two most common glycolytic pathways are the Embden–Meyerhof–Parnas and Entner–Doudoroff pathways.

glycoprotein a protein with sugar attached.

glycoprotein 63 (gp63) a major surface glycoprotein of the genus *Leishmania* and is considered a major virulence factor of the parasite. Functionally, it is a surface Zn metalloprotease, and it has the ability to modulate macrophage signalling pathways.

glycosaminoglycans (GAGs) large unbranched polysaccharides made of repeating disaccharide units (N-acetylhexosamine and uronic acid).

glycosylphosphatidylinositol (GPI) a conserved glycolipid particularly abundant on the plasma membrane of trypanosomatid parasites. In trypanosomes, it can occur as free or anchoring surface proteins (e.g. variant surface glycoproteins, VSGs).

glycosylphosphatidylinositol (GPI) anchor a post-translational modification on proteins consisting of phosphatidylinositol, phosphoethanolamine, and a variety of sugar residues.

glycosylphosphatidylinositol linkage a membrane anchor comprising a sugar bridge linking phosphatidylinositol in the membrane to the carboxyl (C) terminus of the protein.

glycosylphosphatidylinositols (GPIs) Ionophoric coccidiostats a class of anticoccidial drugs that disrupts the flow of ions into and out of *Eimeria* parasites, thus causing an ion imbalance leading to the death of the parasite.

glycosylphosphatidylinositol-specific phospholipase C (GPI-PLC) an enzyme responsible for the cleavage of GPI-anchored proteins including variable surface glycoproteins.

glycosyltransferase an enzyme that transfers a sugar from a nucleotide-sugar donor, to a substrate.

glypicans a family of Glycosylphosphatidylinositol-linked, heparan sulfate-modified proteoglycans, currently represented by six distinct genes in vertebrates; homologues found in *Drosophila* (dally), and *Caenorhabditis elegans*.

GM-CSF (granulocyte–macrophage colony-stimulating factor) a potent cytokine adjuvant based on its activity of inducing the activation and migration of dendritic cells.

gnotobiotic a host animal containing a defined set of microorganisms, either synthetically implanted or transferred from another host; often used to refer to model organisms with microbiota humanized by faecal transfer or gavage.

gold standard (also reference standard) ideally, the criterion used to unequivocally define the presence of a condition; or practically, the method, procedure or measurement that is widely accepted as being the best available to detect the presence of a condition.

Golgi apparatus or Golgi body a cell organelle that helps process and package proteins and lipid molecules, especially proteins destined to be exported from the cell. Named after its discoverer, Camillo Golgi, the Golgi body appears as a series of stacked membranes.

gonad ridge within an embryo, the area of cells that will develop into the gonads of the foetus. This usually develops around 32 days after fertilization.

gonochorism a reproductive strategy that describes a sexually reproducing species in which there are two distinct sexes. In contrast, hermaphroditism describes a form of sexual reproduction in which individuals are not separated into distinct male and female types.

gonochoristic species species with separate male and female individuals.

good genes' hypothesis an evolutionary theory that seeks to explain mate choice preferences, whereby mates are selected on the basis of possessing genes that enhance the viability of offspring.

GP63 (glycoprotein 63) the major metalloprotease that is abundantly expressed on the surface of virulent *Leishmania* promastigotes. When promastigotes are opsonized by C3b, GP63 cleaves this to iC3b, which facilitates promastigote recognition by CR3. GP63 is also called leishmanolysin, PSP (promastigote surface protease) or MSP (major surface protease).

gp90 and gp35/50 mucins surface glycoproteins in *T. cruzi.* Their levels relative to other *T. cruzi* surface glycoproteins modulate the ability of the parasite to invade host cells.

GpC methyltransferase (M.CviPI) an enzyme from *Chlorella* virus that methylates all cytosines within the double-stranded dinucleotide recognition sequence 5′... GC...3′.

G-protein-coupled receptors (GPCRs) a class of receptors that traverse the plasma membrane seven times, sometimes termed seven transmembrane domain or serpentine receptors. They transduce extracellular signals through their interaction with heterotrimeric G proteins.

G-quadruplex (G4) an alternative, non-Watson–Crick nucleic acid structures that arise from particular G-rich sequences. Guanine residues interact with one another within the same strand by means of hydrogen bonding and stacking interactions to produce four-stranded higher-order architectures. A broader definition also includes intermolecular structures formally composed of more than one strand.

graft-versus-host reaction (or disease) a reaction occurring when grafted T-cells recognize and attack host cells.

grantham difference score a score that predicts the effect of non-synonymous mutations based on the chemical properties of the substituted amino acids.

granulocyte circulating leucocytes with a multilobed nucleus; characterized by the presence of a large number of granules in the cytoplasm. Neutrophils have phagocytic functions. Eosinophil numbers tend to increase during parasitic worm infections.

granuloma a lesion composed of collagen fibres and cells that include $CD4^+$ T-cells, eosinophils, and macrophages, that is formed as a result of a $CD4^+$ T-cell response to antigens presented by tissue-trapped eggs during chronic schistosome infection. **2** a small inflammatory lesion containing an organized collection of immune cells that is often produced in response to an infectious agent or a foreign substance.

granulomatous amoebic encephalitis (GAE) encephalitis caused by *Acanthamoeba* and *Balamuthia* amoebae. The term GAE is most commonly used for *Acanthamoeba* related infections (also recently known as AGE), whereas the term BAE is used when referring to GAE caused by *B. mandrillaris*.

granulomatous reaction a localized immune reaction in which a foreign agent is surrounded by immune cells, fibroblasts, and collagen.

granzyme serine protease present in the granules of cytotoxic T-lymphocytes and natural killer cells. They induce apoptosis in the target cell which they enter through perforin channels inserted into the target cell membrane by the cytotoxic lymphocyte.

gravid filled with eggs, as a gravid pinworm or gravid proglottid of a tapeworm.

gravidae (nulli-, primi-, secundi- or multi-) the number of a woman's current pregnancy (none, first, second, or third and higher).

green fluorescent protein (GFP) the gene encoding GFP is inserted behind the promoter of a gene of interest to identify the time and, if inserted into the protein-coding sequence, the place of expression of putative stage-/sex-specific proteins.

group I introns introns that catalyse their own excision from RNA precursors. Group I introns have an elaborate folding pattern of large stems and loops due to nine specific paired regions (P1–P9). Group I introns can also have ORFs in their loop regions.

group selection selection that acts on a group of entities as a whole (such as animals living in a social group or molecules inside a protocell) and that favours survival of the whole group, in contrast to selection acting on individual members of a group that leads to competition between the individuals.

growth factors biologically active molecules that promote cell growth.

GTPase a family of hydrolase enzymes that can bind and hydrolyse guanosine triphosphate (GTP).

Guanine (G) one of four chemical bases in DNA, with the other three being adenine (A), cytosine (C), and thymine (T). Within the DNA molecule, guanine bases located on one strand form chemical bonds with cytosine bases on the opposite strand.

The sequence of four DNA bases encodes the cell's genetic instructions. *See* **bp, nucleotide**.

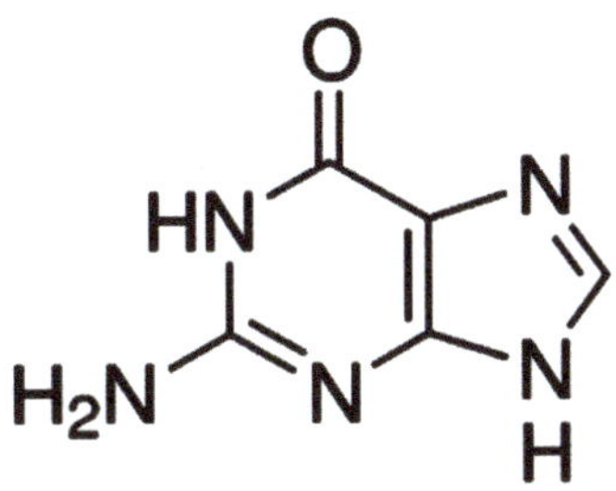

gubernaculum a sclerotized plate located on the dorsal surface of the cloaca of certain male nematodes; serves as a guide when the copulatory spicules are protruded.

guide RNA (gRNA) small RNAs that act in *trans* to supply sequence information for decrypting edited mRNAs (one gRNA contained in the 3′-UTR of the coxII mRNA acts in *cis*).

gut-associated lymphoid tissue (GALT) lymphoid tissue which includes Peyer's patches, appendix, and solitary lymph nodes in the submucosa.

gynecophoral canal the longitudinal groove on the ventral surface of male schistosomes in which the filiform female worm lies.

gyrase a topoisomerase enzyme that can introduce or relax supercoils in a DNA molecule by temporarily introducing double-strand breaks to wind or unwind the strand.

A B C D E F G **H** I J K L
M N O P Q R S T U V W
X Y Z

H-2 complex the mouse MHC. It is situated on chromosome 17 of the mouse and contains subregions K, I and D.

H&E staining (haematoxylin and eosin) routinely used staining approach to help visualize tissue features. Haematoxylin is a blue dye that stains basophilic structures such as nuclei, and eosin is a red dye that stains eosinophilic structures, such as cytoplasm and other protein-rich materials.

habitat the environment inhabited by a specific organism or animal.

habitat suitability map a geographic map that relates measurable environmental variables to the suitability of a site for a species. These maps are useful for planning disease monitoring and control, and for evaluating the impact of environmental changes.

habronemiosis (or habronemiasis) also known as 'summer sores', is caused by a reaction to larvae of the equine stomach worms *Habronema muscae*, *H. majus* (synonym *H. microstoma*), and *Draschia megastoma*. Cutaneous habronemiasis is a condition characterized by variably pruritic granulomatous lesions. The occurrence of lesions corresponds with the fly season, being caused by the aberrant intradermal migration of the larvae of stomach worms, *Habronema* species. This condition occurs worldwide especially where the climate is warm and wet. Adult worms live in the horses's stomach producing first-stage larvae (L1) that are passed in faeces. The L1 are ingested by fly maggots. The house fly (*Musca domestica*) is an intermediate host for *H. muscae* and *D. megastoma*; the stable fly (*Stomoxys calcitrans*) is the intermediate host for *H. microstoma*. After maturation, infective third-stage larvae (L3) are deposited around a horse's mouth or in horse feed; the larvae are ingested completing the life cycle. Cutaneous form of habronemiasis occurs when L3 are deposited on wounds or other moist areas of the body.

haem oxygenase an essential enzyme in haem catabolism that cleaves haem to form biliverdin and carbon monoxide. Haem oxygenase has two isoforms: haem oxygenase 1, which is inducible, whereas haem oxygenase 2 is constitutive in nature. Haem oxygenase-derived carbon monoxide and biliverdin have potent anti-inflammatory activity and can suppress macrophage function.

haemagglutinin any molecule (usually antibodies) that agglutinates red blood cells.

***Haematobia irritans* (horn fly) and *Haematobia exigua* (buffalo fly)** oligate parasites of livestock, particularly cattle and buffaloes. In addition to causing irritation and reducing livestock productivity in terms of weight gain and milk production, *Haematobia* flies could be the mechanical vectors of different pathogens (i.e. *Staphylococcus aureus*, *Staphylococcus saprophyticus*, *Morganella morganii*, *Serratia marcescens*, etc.).

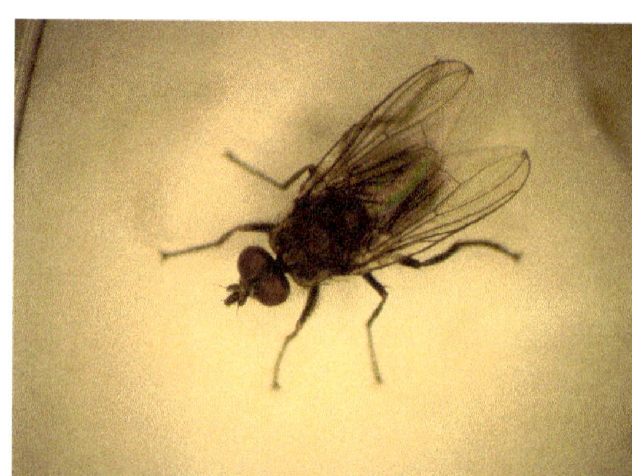

haematocrit a laboratory test that measures the volume of red blood cells in whole blood. Haematocrit is normally ordered as part of a CBC.

haematogenous dissemination when related to pathogens, it refers to their distribution through the bloodstream.

haematophagous bloodsucking; usually refers to the feeding habits of various insects (biting flies) and acarines (ticks).

haematopoiesis the production of blood cells of all types (i.e. erythrocytes, leucocytes and platelets).

haematopoietic stem cells stem cells that make all the blood cells (i.e. leucocytes, erythrocytes and platelets) in the body. They are found in the bone marrow – the tissue that fills most bone cavities.

haematopoietic system all cells and tissues responsible for the formation of blood cells.

haematotoxic poisonous to the blood and to the organs and tissues involved in the production of blood, such as the bone marrow.

haematozoa a group of intracellular parasites of vertebrates whose apical complex lacks a conoid; parasites transmitted by vectors, including *Plasmodium* spp. causing malaria and *Babesia* spp. causing tick fever.

haematuria blood in the urine; a condition seen in some individuals infected with blood parasites, e.g. *Plasmodium falciparum*.

haeme prosthetic group composed of a porphyrin ring and a central iron. It is present in haemic proteins such as haemoglobin.

haemocoel a body cavity in mollusks and arthropods through which the haemolymph (haemocoele fluid) circulates carrying nutrients, etc. to the organs but typically not oxygen.

haemocyte aggregation the process of forming adhesions between haemocytes that surround and trap invading microorganisms.

haemocytes insect cells suspended in the haemolymph. Certain haemocyte types are professional phagocytes that patrol the haemolymph and engulf invaders.

haemoglobin a protein in red blood cells that carries oxygen from the lungs to the tissues of the body. Haemoglobin also refers to the haemoglobin test, which is normally ordered as part of a complete blood count.

Haemogregarine apicomplexan intracellular adeleorinid parasites that parasitize the blood of vertebrates and are transmitted by a wide variety of invertebrate definitive hosts.

haemolymph or 'insect blood' occupies the haemocoel, in which the organs and tissues are suspended in an open circulatory system. The haemolymph contains circulating immune cells called haemocytes, of which some are functionally equivalent to macrophages and granulocytes present in vertebrates. Different types of haemocytes confer immune reactions, such as phagocytosis, encapsulation and melanization.

haemolymphatic stage initial stage of African trypanosomiasis (sleeping sickness) in which parasites are found in the blood and lymph. Symptoms include fever and headaches.

haemolysin any substance (but usually refers to an antibody) that can lyse red blood cells.

haemolysis destruction of red blood cells. Haemolysis occurs as part of the natural life cycle of red blood cells, or it may be caused by certain conditions or drugs.

haemolytic anaemia anaemia caused by increased destruction of erythrocytes.

haemolytic disease disease occurring as a result of the destruction of red blood cells by complement fixing antibodies transferred to the young individual from its mother.

Haemonchus contortus a Strongylida (bursate) nematode. Known as 'barber's pole worm' or 'wire worm' because, white ovaries are spirally coiled around the red intestine producing appearance of barber's pole. Adults are abomasal parasites of small ruminants. Female reach about 3 cm in length, making this species one of the largest of the strongylid nematodes of ruminants. Life cycle is direct without migration. Unlike many other parasites of the GI tract, *H. contortus* is not a primary cause of diarrhoea. Anaemia due to feeding worms can be very severe. If the worm burden is very high, the removal of blood may overtake erythropoeesis and this may cause death. Infected sheep can develop submandibular oedema (bottle jaw). McMaster slide is used to detect and quantify eggs in faeces. Packed cell volume may be used to assess extent of anaemia in infected animals.

haemoparasite an animal parasite, specifically a haemoflagellate, apicomplexan, trematode or filarial worm, living in the blood of a vertebrate, e.g. *Babesia* spp., *Plasmodium* spp. or *Trypanosoma* spp.

haemorrhage excessive or uncontrollable bleeding from one or more blood vessels. A haemorrhage, which may be internal or external, is often caused by injury or surgical complications or from advanced disease. Some parasites may cause internal haemorrhage.

haemosporidians parasites in the order Haemosporida (phylum Apicomplexa), including the genera *Plasmodium*, *Haemoproteus*, *Leucocytozoon*, *Fallisia*, *Garnia* and others.

haemozoin (Hz) **1** a crystal or pigment consisting of orderly arranged haem molecules, which *Plasmodium* produce to detoxify free haem. **2** metabolite from the digestion of haemoglobin by plasmodial parasites.

Haldane's law states that, when two different animal races are crossed and one of the sexes of the F1 offspring is absent, rare or sterile, it is the heterozygous sex (XY) sex.

Halicephalobus gingivalis a soil-borne, free-living nematode parasite. This saprophytic worm was first identified 1954 and since then sporadic cases have been reported as a cause of neurological disease in horses, in many countries. It is able to opportunistically colonize the host in different ways, such as oral infection via ingestion of contaminated food facilitated by lacerations of the buccal mucosa, transplacental transmission from mare to foal, percutaneous through breaks in skin/open wounds, or through the respiratory tract. Disease caused by *H. gingivalis* is difficult to treat due to the tendency of the parasite to migrate. Its high proliferative ability and its ability to reproduce parthenogenetically (i.e. without male nematodes) within the host.

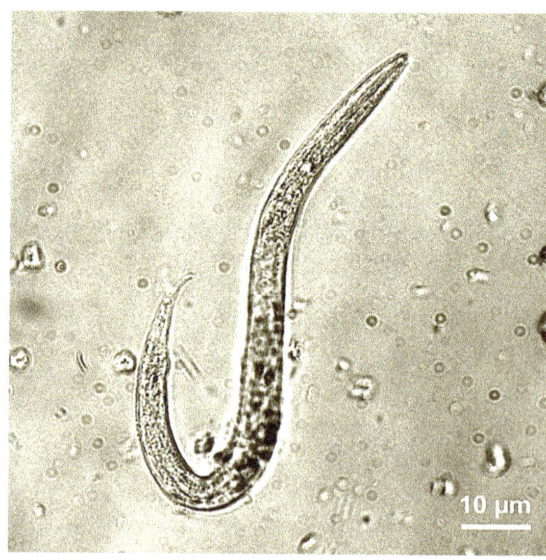

haplodiploid a genetic system of sex determination that is mainly found in the insect order Hymenoptera, including ants, bees and wasps. Hymenopteran females are diploid (that is, they have two complete sets of chromosomes), whereas males are normally haploid and have only one set of chromosomes. Some species, such as the fire ant *Solenopsis invicta*, also produce viable diploid males.

haplogroup a group of similar haplotypes that share a common ancestor.

haploid a single set of chromosomes (half the full set of genetic material) present in the egg and pollen cells of plants. Organisms that reproduce asexually are haploid. Sexually reproducing organisms are diploid (having two sets of chromosomes, one from each parent). In humans, only their egg and sperm cells are haploid. *See* **diploid**.

haplotype a set of DNA variations or polymorphisms, that tend to be inherited together (i.e. found on the same chromosome).

haplotype network a visualization of the evolutionary relationships among DNA sequences.

HapMap (haplotype map) the nickname of the International HapMap Project, an international project that aims to relate variations in human DNA sequences with genes associated with health. The HapMap describes common patterns of human genetic variation by examining the combination of alleles at multiple loci that are transmitted together on the same chromosome.

hapten a small molecule that cannot initiate an immune response unless first bound to an immunogenic carrier protein. Thus although not immunogenic, it is able to react with an antibody of the appropriate specificity.

hapten-induced contact-dependent hypersensitivity a Langerhans cell-mediated $CD8^+$ T-cell-dependent immune response elicited by epicutaneous sensitization with haptens (i.e. chemicals including metals). Such reactions in mice and humans are mediated by Th1 or T cytotoxic 1 (Tc1) effector cells and down-regulated by Th2 or T regulatory $CD4^+$ T cells. The response develops in two distinct phases: (i) a sensitization phase where the haptens penetrating the skin are captured by resident dendritic cells, which migrate to regional lymph nodes and induce activation of specific T cell precursors; (ii) the elicitation phase, which is induced by re-exposure to the same hapten at a remote skin site. Within a few hours, this leads to the rapid recruitment and activation of specific T cells and a local inflammatory response, which peaks at 24–48 hours after challenge.

haptotaxis movement directed by an immobile gradient in matrix components.

hard selection the result of regulation that is frequency and density independent, usually occurring on the whole population after genotypes have migrated from their patches and mixed.

hard ticks (family Ixodidae) are so-called because they have a dorsal scutum in all developmental stages (larva, nymph, and adults).

Hardy–Weinberg equilibrium (HWE) a model stating that allele and genotype frequencies of variants in a population will remain constant from one generation to the next in the absence of evolutionary disturbing factors such as mutation, selection and genetic drift.

heartworm disease (HWD) a vector-borne disease caused by the nematode *Dirofilaria immitis*. It primarily affects dogs, cats and other carnivores.

heavy chain (H chain) the larger of the two types of peptide chains that constitute a normal immunoglobulin or antibody molecule.

heavy-chain disease a group of immunological disorders characterized by the presence of monoclonal but incomplete H chains without L-chains in the sera and urine.

helminth a general term referring to the nematodes (roundworms), trematodes (flat worms, flukes), cestodes (tapeworms) and acanthocephalans (thorny-headed worms). Heminthes are complex, multicellular organisms that are elongated and bilaterally symmetrical. They are significantly larger than the protozoan parasites and generally are macroscopic, ranging in size from less than 1 mm to 1 m or more. The external surface of some worms is covered with a protective cuticle, which is acellular and may be smooth or possess ridges, spines, or tubercles. The protective covering of flat-worms is known as a tegument. Frequently, helminths possess elaborate attachment structures such as hooks, suckers, teeth, or plates. These structures are usually located anteriorly and may be useful in species classification and identification. Helminths typically have primitive nervous and excretory systems. Some have alimentary tracts; however, none has a circulatory system.

helminth hypersensitivity a hypersensitivity state that may occur following an invasion with helminths of, for example, the lungs.

helminthology the scientific study of parasitic worms.

helper T lymphocyte (Th) a subclass of T-cells which provide help (in the form of cytokines and/or cognate interactions) necessary for the expression of effector function by other cells in the immune system.

helper T-cells T-cell subsets characterized by expression of the CD4 coreceptor and restricted to association with antigen/MHC class II complexes. These cells are critical for the activation of macrophages, orchestration of antibody-secreting B-cell responses, and regulation of immunity via their differentiation into one of several functionally distinct subsets (e.g. Th1, Th2, Th17, Treg, Tr1, etc.).

haemagglutination the agglutination of red blood cells.

haemagglutination inhibition a technique to detect small amounts of antigen in which the agglutination of antigen bonded red cells is inhibited by homologous antigen.

hemicellulose any of several polysaccharides (e.g. xylans, mannans, and galactans) that cross link and surround cellulose fibres in plant cell walls. Where cellulose is regular in organization and consequently strong, hemicelluloses are more commonly random in structure and more easily hydrolysed.

hemimetabolous a type of development in insects in which there is a gradual change in the external structure as development proceeds the egg to nymphal stages to the adult stage; hemimetabola is the name of the taxonomic group. *See* **ametabolous**.

hemiparesis partial paralysis of one side of the body.

hemizygous a condition in which there is only one copy of a gene in an otherwise diploid cell or organism (e.g. all genes on the X chromosomes of male mammals).

haemolymph the fluid that fills the body cavities of invertebrates. It has analogous functions to blood, except that it typically does not transport oxygen. The insect blood that circulates around the body, bathing the tissues directly. It consists of fluid plasma in which haemocytes are suspended.

Henneguya myxosporidan parasites with a world-wide distribution that infect both freshwater and marine fish.

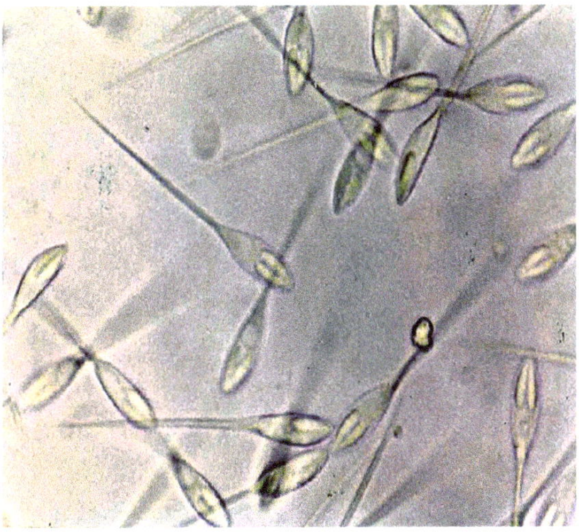

HEp2 cells epithelial cell line that originates from a human laryngeal carcinoma.

heparin a form of highly sulfated heparan sulfate made in mast cells.

hepatic pertaining to the liver.

hepatic fibrosis build-up of increased extracellular matrix, scar tissue, and collagen in the liver.

hepatic necrosis death of liver cells. Hepatic necrosis may be caused by certain antiretroviral (ARV) drugs.

hepatic stellate cells (HSCs) also known as perisinusoidal cells, HSCs are located in the space of Disse between hepatocytes and sinusoidal endothelial cells of the liver. HSCs in normal liver are in a quiescent state, become activated when the liver is injured, and are the major cell type involved in liver fibrosis.

hepatitis liver inflammation.

hepatization the transformation of the pulmonary parenchyma into a liver-like tissue, resulting in a marked lung consolidation, which makes the pulmonary alveoli no longer suitable for air exchange.

hepatomegaly abnormal enlargement of the liver.

hepatopathology pathology occurring within the liver.

hepatosplenic schistosomiasis a severe form of chronic schistosomiasis usually observed in groups of individuals who are moderately or heavily infected.

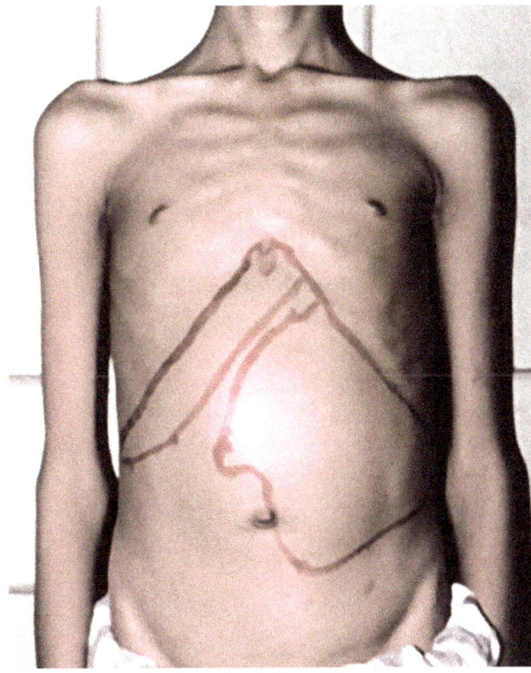

herbicide a substance that kills plants. Used in agriculture, horticulture and gardening. It can be selective (kill selected species) or non-selective (broad spectrum – kill all plants as well as insects).

herd immunity (herd effect or community immunity) a form of immunity that occurs when the vaccination of a substantial portion of a population (or herd) provides a measure of protection for individuals who have not developed immunity.

hereditary disorder a pathological condition due to changes in individual genes, or groups of genes or in sections of chromosomes or whole chromosomes. These changes may be passed from parents to offspring.

heritability refers to the extent to which a particular trait is due to inherited DNA sequence variation. A heritability of 1.0 indicates that a trait is 100% heritable and no environmental component contributes.

heritable variation proportion of the variability in a phenotype that is transmitted across generations; primarily, this will be DNA sequence variation but might sometimes represent non-genetic variation – for example, in mechanisms that regulate gene expression.

hermaphrodite self-fertilizing organisms which contain both male and female sexual organs.

hermaphroditism the presence of both male and female reproductive organs in the same organism that is capable of reproducing on its own through self-fertilization.

Heterakis gallinae **(synonym *gallinarum*)** a parasitic nematode of the caecum of galliform birds (chickens, turkeys, etc.).

heterochromatin a highly condensed form of chromatin that usually contains repetitive DNA and is generally transcriptionally repressed. Centromeres and telomeres are heterochromatic.

heterodimer a molecule consisting of two different subunits.

heteroduplex analysis heteroduplex analysis is based on the denaturation and reannealing of DNA strands to form homologous and heterologous double-stranded DNA molecules. In blood meal identification analyses, PCR products from the probe DNA (driver) are mixed with PCR products of tested DNA. During electrophoresis, the rate of migration of heteroduplex DNA through a polyacrylamide gel varies according to the three-dimensional conformations formed by mismatched base pairing between heterologous sequences and can be diagnostic for identifying DNA from different species.

heteroecious complex cycles that require several different hosts to complete.

heterogametic sex the sex whose sex chromosomes are different from each other. In mammals, most other vertebrates and most insects, males are the heterogametic sex (XY), whereas in birds, lepidopterans, and some fish it is females (WZ).

heterogenic a form of reproduction in which sexual and asexual generations alternate, as in the nematode *Strongyloides*.

heterologous prime–boost repeated immunization using different vaccines. It is used to stimulate a better immune response.

heterologous promoters non-native promoters used to drive reporter gene expression. Strong promoters may sometimes be used to maximize detection of the reporter through microscopy, but inappropriate expression may lead to artifactual protein localization.

heterologous protein expression system a system in which production of the protein of interest is carried out in an organism different from the organism source of the target protein.

heterophid having a bifurcated intestinal organ or ceca usually long and truncated.

heterophile antibodies IgM antibodies that can reach high levels in serum after febrile illnesses (for example, infectious mononucleosis), and can bind non-specifically to other proteins. They can cause false-positive reactions in immunochromatographic assays.

heterophile antigen a cross-reacting antigen that appears in widely ranging species such as humans and bacteria.

heteroresistance a mixture of susceptible and resistant populations, which may arise as a result of a mixed infection with strains with different susceptibility patterns, or the evolution of a resistant subpopulation from a susceptible parental strain during the infection. *See* **sequencing**.

heterothermy endotherms that can switch between an ectothermic or endothermic physiological state dependent on life history requirements, typically found in animals that hibernate in response to external conditions.

heterotroph organism that obtains organic carbon by consuming other organisms or the products of other organisms.

heterotrophic an organism that uses organic carbon compounds, such as dissolved organic matter and particulate organic matter, to satisfy its carbon requirement but that cannot fix carbon.

heteroxenic infecting several host species in the life cycle.

heteroxenous having more than one host required to complete a life cycle, such as in the digenetic trematodes.

heterozygosity a measure of the variability of a population, often the fraction of individuals in a population that are heterozygous for a particular locus.

heterozygote *See* **heterozygosity**.

heterozygote deficit/excess in a panmictic population of a diploid species, allele frequencies are given by the Hardy–Weinberg formula (Hardy–Weinberg equilibrium). Heterozygote excess or deficit shows that allele segregation does not occur at random.

heterozygous having two different alleles at a single gene or genetic locus. An allele can be dominant, co-dominant, or recessive. A heterozygous genotype stands in contrast to a homozygous genotype, where an individual inherits identical forms of a particular gene from each parent.

hexacanth embryo six-hooked embryo, the mature embryo within the egg of many tapeworms.

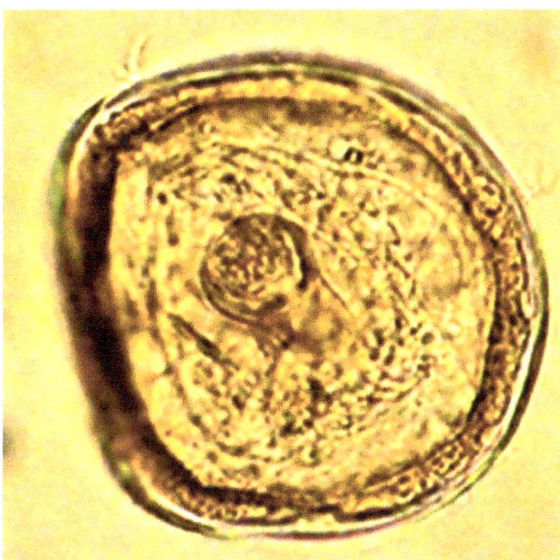

hexasomic having six copies of a given chromosome.

hexose a type of sugar molecule that contains six carbon atoms (e.g. glucose, fructose).

hidden Markov model (HMM) a statistical model in which the system being modelled is assumed to be a Markov process, that is, a random evolution of a memoryless system, for which the likelihood of a given future state depends only on its present state, its future and past are independent.

hierarchical clustering a clustering approach in which pairs of genes or clusters are sequentially connected until they form a tree-like structure.

hierarchical interaction score (HIS) a statistical method that infers functional interactions between genes if they display perturbation effects in a consistent subset of readouts, or environmental or genetic backgrounds. It also infers statistical hierarchy, in which the perturbation with a broader set of effects is placed upstream of a perturbation with a narrower subset of these effects.

hierarchical structure observations that are nested within units to define a natural hierarchy. Examples are multiple parasite counts measured within a host; multiple hosts living within a single household; multiple households within a single community. Such structure typically produces correlations among repeated measures (observations) made on the same unit. Thus, repeated measures cannot be assumed statistically independent.

high-density lipoproteins a heterogeneous population of approximately spherical blood-borne particles consisting of a lipid core (particularly triglycerides and cholesterol derivatives) with an outer hydrophilic layer (phospholipids, cholesterol and apolipoproteins), which solubilize the lipid cargo. The term 'high density' refers to high protein content relative to other lipoproteins. Their major role is the collection of cholesterol from peripheral tissues and transport to the liver.

high endothelial venule (HEV) capillary venule composed of specialized endothelial cells that mediate specific binding and migration of blood lymphocytes into lymphoid tissues.

high frequency derived alleles polymorphisms that occupy the upper bounds of the site frequency spectrum, and are of a type differing from the ancestral state.

high taxonomic ranks the taxonomic categories of genus and above.

high throughput done on a massive, automated scale.

high-content imaging a subfield of microscopy using fast and automated imaging systems that can acquire thousands of cell micrographs from different regions on a microscope slide or separate wells of multiwell plates. It is frequently applied to detect morphological cellular changes in drug and small interfering RNA (siRNA) library screens.

high-dimensional data analysis a collection of statistical methods and visualization tools used to analyse and represent data with dozens to thousands of dimensions. In spatially resolved omics, the intrinsic high dimensionality of omic data are multiplied by the number of spatial locations analysed, producing mega- or giga-dimensional data sets that need to be correlated among each other in order to identify spatially organized expression patterns and regulatory networks. Efficient handling of these mega-dimensional omic data sets requires extraordinary computational efforts and novel statistical tools that are being developed for the analysis of 'big data' in many areas of science and business.

high-performance TLC (HPTLC) automated thin-layer chromatography (TLC) system allowing the parallel analysis of a large number of samples, with higher throughput and more sensitive detection of lipids. Most recent HPTLC lines include devices that automatically couple TLC with MS (TLC-MS).

high-throughput screening (HTS) (massively parallel sequencing or next-generation sequencing) a fast method of determining the order of bases in DNA. Numerous platforms exist by which millions of short sequencing reads (35–100 bp) are performed and assembled to survey DNA or RNA abundance in a sample. *See* **sequencing**.

highly conserved sequence DNA sequence that is very similar across several different types of organisms. *See* **gene**, **mutation**.

hinge region amino acids between the Fab and Fc regions of immunoglobulin which permit flexibility of the molecule.

Hirudinea subclass of the phylum Annelida; dorso-ventrally flattened organisms ('leeches'); aquatic and terrestrial species; hermaphroditic; adults live temporarily attached to vertebrate and invertebrate hosts, or free-living.

histamine vasoactive amine present in basophil and mast cell granules which, following degranulation, causes increased vascular permeability and smooth muscle contraction.

histiocyte a tissue macrophage.

histocompatibility the ability of tissues to get along, In immunology it means identity in all transplantation antigens. These antigens are referred as histocompatibility antigens.

histocompatibility molecules cell membrane proteins that are required to present antigen to antigen sensitive cells.

Histomonas meleagridis a unicellular flagellate intestinal trichomonad parasite of birds; the cause of histomoniasis (or blackhead disease) in poultry.

histone a protein that provides structural support to a chromosome. For very long DNA molecules to fit into the cell nucleus, they wrap around complexes of histone proteins, giving the chromosome a more compact shape. Some histones variants are associated with the regulation of gene expression.

histone acetyltransferase (HAT) an enzyme that places acetyl marks. Because many epigenetic enzymes also have non-histone targets the general nomenclature of lysine acetyltransferase (KAT) might be more appropriate.

histone code collective term for the covalently bound modifications of histone tails. Histones bind DNA to form nucleosomes that are the basis of chromatin structure. Modification of histone tails plays an important part in the state of chromatin and thus transcriptional activity.

histone deacetylase (HDAC) an enzyme that removes acetyl marks. Because many epigenetic enzymes also have non-histone targets the general nomenclature of lysine deacetylase (KDAC) might be more appropriate.

histone demethylase (HDM) an enzyme that removes methyl marks. Because many epigenetic enzymes also have non-histone targets the general nomenclature of lysine demethylase (KDM), might be more appropriate.

histone methyltransferase (HMT) an enzyme placing methyl marks. Because many epigenetic enzymes also have non-histone targets the general nomenclature of lysine methyltransferase (KMT) might be more appropriate.

histone post-translational modifications (PTMs) these include acetylation (ac), mono- (me1), di- (me2) and tri-methylation (me3), phosphorylation (p), ubiquitination (ub), polyADP-ribosylation (ADP-r), and sumoylation (sumo). The histone PTM nomenclature includes the histone involved, the name and position of the modified amino acid, and the type of modification. For example, H3K9me3 refers to tri-methylation of the lysine at position 9 of histone H3.

histone variant (replacement histone) histone with a similar, but not identical, amino acid sequence compared to canonical histones. Histone variants can bear different modifications, have distinct functions and can be incorporated into the genome in a replication-independent manner.

histone-code hypothesis the idea that the combination of PTMs on histones, primarily on the histone N termini, confers epigenetic information that determines transcriptional activation or repression. Among the important modifications are: (i) methylation, acetylation, ubiquitination of lysine; (ii) methylation of arginine; and (iii) phosphorylation of serine. The combination and sequence of addition of PTMs affects the interaction of chromatin with macromolecular complexes. Many new modifications have been reported in the past two years, increasing the potential complexity of the histone code.

histones small basic proteins that make the protein core that DNA is wound around. The classic or canonical histones are H2A, H2B, H3 and H4. Histone variants include H2A.Z, H2A.X (DNA repair), H3.3 (active genes) and CenH3 (or CenPA, centromeric), as well as parasite-specific histones.

histopathology a process for visual examination of animal tissues to determine whether there are microscopic changes. Routine histopathology in pharmaceutical safety studies is conducted by fixing tissues in formalin, sectioning the tissues using a microtome, fixing these to microscope slides and staining the tissues before microscopic evaluation. *See* **H&E staining**.

hitchhiking the population genetic mechanism by which a neutral, or in some cases slightly deleterious, mutation increases in population frequency solely as a result of physical linkage with a positively selected mutation.

hitchhiking effect resulting from a selective sweep, which affects linked neutral variation in a characteristic way through the fixation of the selected haplotype.

HLA complex *See* **major histocompatibility complex**.

holo-, hyper-, hypo-, and mesoendemic classifications of malaria endemicity, with transmission occurring (respectively) year-round, intensely but with periods of no or little transmission during a dry season, very intermittently, or with regular seasonal transmission.

hologonic development in which only one sex (usually the female) is present in a colony, as in *Strongyloides stercoralis* infections.

holometabolous a type of development in insects in which there is a distinct change in the morphology of the stages; the cycle usually includes several larval stages, a pupa, and the adult; holometabola is the name of the taxonomic group. *See* **ametabolous**.

homeobox b a short stretch of nucleotides whose base sequence is virtually identical in all the genes that contain it. Homeoboxes have been found in many organisms. In the fruit fly, a homeobox appears to determine when particular groups of genes are expressed during development.

homeostasis condition of equilibrium or stability in a system, which is maintained by adjusting physiological processes to counteract external changes; a balanced relation between a host tissue and the resident microbiota that prevents destructive inflammation or disease.

homeostatic proliferation a physiological process that controls stable and constant cell number without cellular differentiation. Interleukin-7 has a crucial role in maintaining normal T-cell levels.

homeothermy animals that maintain a stable body temperature regardless of external environmental conditions.

homing receptors cell surface molecules that direct leucocytes to specific locations in the body.

homodimer a molecule consisting of two identical subunits. Example, the antibody molecule has two L-chains and two H-chains, thus 2 homodimers.

homogamy tendency to mate with partners that are genetically identical, or nearly so. It may happen even if different genotypes of the pathogen are present in the host.

homologies similarities in DNA or protein sequences between individuals of the same species or among different species.

homologous when used in reference to gene or protein sequences, those that share significant DNA or amino acid sequence similarity in contrast to non-homologous sequences, which do not share significant sequence similarity.

homologous chromosome chromosome containing the same linear gene sequences as another, each derived from one parent.

homologous protein expression system a system in which production of the protein of interest occurs in the natural host under a specific (inducible or not) promoter, often different from the native promoter.

homologous recombination is a type of genetic recombination that occurs during meiosis (the formation of egg and sperm cells). Paired chromosomes from the male and female parent align so that similar DNA sequences from the paired chromosomes cross over each other. Crossing over results in a shuffling of genetic material and is an important cause of the genetic variation seen among offspring.

homology similarity in DNA sequence or structure based on descent from a common ancestor.

homoplasy similarity of traits or genes for reasons other than co-ancestry (e.g. convergent evolution, parallelism, evolutionary reversals, horizontal gene transfer, gene duplications). Homoplasy violates a basic assumption of the analysis of genetic markers – variants of similar phenotype (e.g. bp size) are assumed to derive from a common ancestor.

homotypic interaction motif a defined amino acid sequence in a protein that mediates preferential binding to the same motif present in a second protein molecule.

homoxenic infecting only one host species in the whole life cycle.

homoxenous an adjective referring to a parasite that has a direct life cycle or one in which only a single host is required for its completion.

homozygosity mapping recessive diseases require two copies of an allele before they appear. Loci surrounding the offending disease locus will also tend to exhibit homozygosity in affected individuals. Searching for homozygous segments in affected individuals with DNA markers to identify the rough location of recessive disease loci is thus termed `homozygosity mapping.

homozygote an organism that has two identical alleles of a gene. *See* **heterozygote**.

homozygous having two forms of a particular gene that are the same, one inherited from each parent. *See* **homozygote**.

hooklet in tapeworms, the small hook-like organ of attachment present on the rostellum of the tapeworm scolex.

hookworms these are small nematode parasites of the intestinal tract of vertebrates; hookworms suck blood to various degrees, depending on the species. They are called hookworms because the anterior end is bent backwards giving the worm a 'hooked' appearance. They range in size from 55 to 90 µm by 30 to 55 µm, depending on the species.

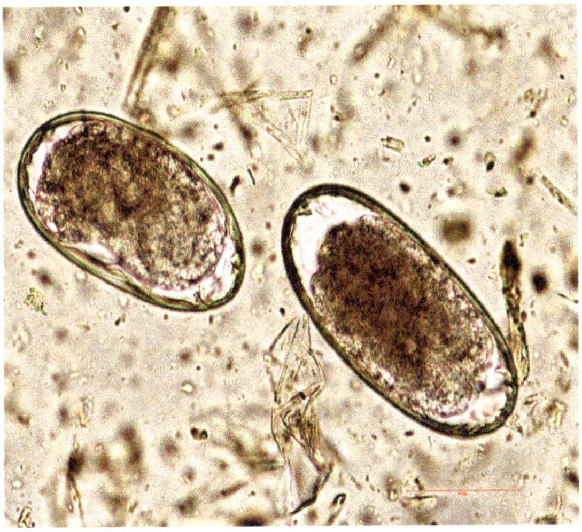

horizontal (lateral) gene transfer (HGT) the transmission of DNA encoding genes between different genomes. This phenomenon occurs, especially, frequently in single celled organisms. Horizontal gene transfer is distinct from 'genealogical or vertical transmission' of genetic material between parent and progeny.

horizontal transmission transmission of a parasitic agent among members of a group.

hormone a chemical compound released by endocrine cells that affects cells in other parts of the body. In animals, hormones are often transported in the blood.

host an animal or plant on which, or in which, a parasite lives. While the parasite receives nourishment and support from the host, the host does not benefit and is often harmed by the association.

host manipulation an evolved ability of a parasite to influence the phenotype (behaviour, morphology, physiology) of its host.

host range the number of species of hosts that can be infected by a parasitic agent.

host–parasite reaction the inflammatory reaction that sometimes occurs around a parasite in tissues, e.g. a warble fly larva in the skin of cattle.

hotpops also referred to as foci or hotspots; geographically clustered populations that are identified as having comparatively higher levels of transmission.

human African trypanosomiasis (HAT) human disease caused by infection with *Trypanosoma brucei rhodesiense* or *T. b. gambiense*.

human blood index (HBI) the proportion of bloodfed mosquitoes in a given population sample that test positive for human blood.

human cathelicidin (hCAP-18) cathelicidins are a family of antimicrobial proteins among which hCAP-18 liberates the antibacterial, cytotoxic peptide LL-37.

human serum albumin soluble blood proteins that make up about 55% of plasma proteins. They are involved in maintaining fluid balance in the blood.

human-landing catches (HLCs) the gold standard method to monitor exposure of the human population to vector populations, which involves individuals sitting with their legs exposed and collecting vectors that come to feed on them.

humoral pertaining to extracellular fluid such as plasma and lymph. The term humoral immunity is used to denote antibody-mediated immune responses.

humoral immunity the arm of the immune response that is based on the production and secretion of antibodies by plasma cells that are derived from B-cells.

hybrid something of mixed origin or composition. In the case of a plant or animal, a hybrid is produced by breeding together plants or animals of different varieties, species or race. A hybrid is the offspring of genetically dissimilar parents. *See* **heterozygote**.

hybridization the process of joining two complementary strands of DNA or one each of DNA and RNA to form a double-stranded molecule through base pairing.

hybridoma hybrid cell line obtained by fusing a lymphoid tumour cell with a lymphocyte which then has both the immortality of the tumour cell and the effector function (e.g. monoclonal antibody secretion) of the lymphocyte.

hydatid cyst a cystic larval stage of *Echinococcus* spp., which is a large fluid filled cyst containing many invaginated scolices and daughter cysts that also contain many invaginated scolices. There are two types of hydatid cysts: *E. granulosus* has uniloculular (one-chambered) cyst and *E. multilocularis* has a multilocular (many chambered) cyst. It consists of two parasite derived layers, an inner nucleated germinal layer and an outer acellular laminated layer, surrounded by a host-produced fibrous capsule called adventitial layer protoscolices, the developmental forms of the parasite infective to canine, are formed in the germinal layer of hydatid cysts.

hydatid sand hydatid protoscolices are sometimes called 'hydatid sand' because of their gritty feel when cysts are sliced open.

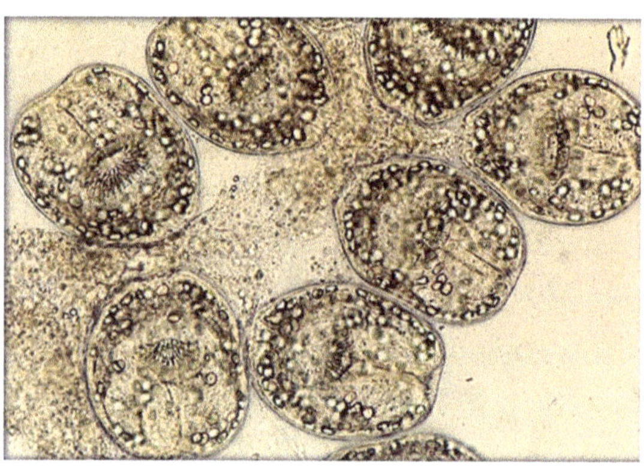

hydatidosis or cystic echinococcosis (CE) a zoonotic disease caused by the infection of mammals with the metacestode stage (hydatid cyst) of the cestode *Echinococcus granulosus*. The adult cestode inhabits the small intestine of a canid carnivore (definitive host, usually canines) and produces eggs containing infective oncospheres. After oral uptake of eggs by an intermediate host animal (herbivores as well as human), the metacestode or larval stage (hydatid cyst) developing in internal organs (mainly liver and lungs) is a fluid-filled, bladder like, unilocular structure.

hydranencephaly a condition that arises during foetal development, in which the central cavity of the brain is missing and is replaced with cerebrospinal fluid.

hydrometeoric relating to atmospheric phenomena that depend on water vapour.

hydrophilic literally means 'water-loving'. This can describe a molecule or part of a molecule that has an affinity for water, or a substance that readily absorbs or dissolves in water.

hydrophobic incapable of interacting with water; literally means 'water hating'. It means repelling, tending not to combine with, or incapable of dissolving in water.

hydrotropism ability of a plant to sense and grow towards water.

hyper- a word element; prefix denoting abnormally increased, excessive.

hypercycles co-operative replicative sets of molecules in which hyperbolic growth is possible.

hyperaemia an abnormally large amount of blood in a tissue.

hypergammaglobulinaemia in addition to the increase in specific antibodies, many parasite infections, including malaria, provoke the production of high titres of non-specific antibody, so-called hypergammaglobulinaemia. Much of this increase is probably due to antigens released from parasites acting as polyclonal mitogens (substances that cause cells, particularly lymphocytes, to undergo cell division) for B-cells.

hyperinfection infection superimposed upon an existing infection by the same parasite in which the parasite reaches high numbers. The term usually refers to internal autoinfection, e.g. strongyloidiasis, oxyuriasis (pinworm), or *Hymenolepis nana*. *See* **autoinfection**.

hyperparasite a parasite that preys on a parasite.

hyperplasia abnormal increase in the number of normal cells in a tissue or organ.

hypersensitive response a programmed cell death response of the host tissue surrounding the site of pathogen infection. It is a component of the overall effector-triggered immunity defence response and is usually induced by pathogen effector activity perceived by nucleotide-binding domain and leucine-rich repeat-containing proteins (NLRs).

hypersensitivity (= allergy) an immunologically mediated damaging inflammatory response in which a mammal is sensitized to a particular substance and has an abnormally strong reaction when the substance is encountered again.

hypertext markup language (HTML) the languages most commonly used for web page development. *See* **extensible markup language**.

hypertrophy an abnormal increase in the size of a tissue or an organ.

hypervariability high degree of variation among individuals within local populations at a given genetic marker. Examples of hypervariable markers include minisatellites and microsatellites.

hypervariable regions those amino acid sequences within the immunoglobulin and T-cell receptor variable regions which show the greatest variability and contribute most to the antigen or peptide–MHC binding site.

hypervivipary an exceptional progenetic reproduction mode that resembles 'Russian dolls': a mother contains in utero a gravid daughter with another developing embryo within. This mode of reproduction is derived from ovipary and is found in many *Gyrodactylus* species.

hyphae branching, threadlike filamentous cells of a fungus.

hypnozoite a dormant, non-dividing, form of the liver stage parasite found in several *Plasmodium* spp., including the human parasites *Plasmodium vivax* and *Plasmodium ovale*. Hypnozoites can exit dormancy anywhere between a few weeks and several months after the patient is cured of symptomatic malaria, leading to relapse in disease symptoms.

hypo-, meso-, hyper-, and holoendemic classifications of malaria endemicity based on parasite prevalence rates among children aged 2–10 years (respectively) of < 10%, 11–50%, 51–75% and > 75% where the parasite prevalence rate refers to children aged less than 1 year.

hypoalbuminaemia abnormally low concentration of albumin in the blood.

hypobiosis a temporary halt in nematode development within the host at an early stage in the parasitic phase of the life cycle in nematodes with direct life cycles only. Nematodes arrest as immature forms in the definitive hosts at a time when conditions in the external environment pose a hazard to survival of free living pre-parasitic stages (usually in winter or dry seasons). Once conditions are more favourable, the arrested development resume their development to adults the life cycle continues. This prolonged but temporarily arrested larval development, represents one of the most useful of life cycle adaptations to ensure parasite adaption enables the parasite to synchronize its life cycle to changing environmental or survival.

hypochromic erythrocytes the content of haemoglobin per erythrocyte is reduced, which is reflected by a reduced mean corpuscular haemoglobin (MCH) and/or a reduced mean corpuscular haemoglobin concentration (MCHC).

hypodermis a syncytial layer that encases the nematode body and secretes the outer cuticle.

hypogammaglobulinaemia an immunodeficiency in which low levels of all classes of immunoglobulins are found in blood.

hypothesis provisional scientific explanation that is unproven, but either thought of as probably true or used as a basis for further investigation.

hypothesis-driven an analysis that limits the combinations of loci tested for epistasis according to some prior hypothesis (for example, only loci with a marginal effect or loci involved in a particular biological pathway should be tested).

hypothesis-free an analysis in which no assumption is made about the loci involved in epistasis or their effects and so all possible pairs of single-nucleotide polymorphisms are tested (that is, an exhaustive search).

hypothetical protein a protein whose expression is not supported by any experimental evidences, and whose function is unknown.

hypoxaemia abnormally low oxygen levels in arterial blood.

IBD (identity-by-descent; also identical-by-descent) the phenomenon whereby two individuals share a genomic region as a result of inheritance from a recent common ancestor, where 'recent' can mean from an ancestor in a given pedigree, or with no intervening mutation event or with no intervening recombination event.

iberiotoxin neurotoxin found in the venom of the East Indian scorpion *Buthus tamulus* which specifically inhibits Big Potassium (BK) channels.

IBS (identical-by-state; also identity-by-state) when two homologous alleles have matching type. Some definitions of IBS exclude IBD (identity-by-descent).

IC_{50} the concentration of an antimalarial compound that inhibits parasite growth by 50%, typically estimated through *in vitro* growth inhibition dose–responses.

ICR (internal complementary repeats) related or identical sequences of DNA in inverted form occurring at opposite ends of some transposons.

idiopathic occurring without known cause.

idiopathic thrombocytopenic purpura (ITP) a condition of abnormally low platelet counts resulting from the production of Abs against platelets.

idiotope an epitope formed by the variable amino acid sequences located in or close to the antigen binding site of an immunoglobulin.

idiotype (Id)/anti-Id the antigenic characteristic of the V region of an antibody. Anti-idiotype complexes of Ab (commonly dimers) arise when Abs recognize and bind to the Ag binding site (paratope) of another Ab. The end result can be the production of a chain of autoantibodies that recognize each other and that may modulate the immune system by stimulating or suppressing it. How the immune system is regulated is of central importance for understanding, treating and eventually preventing many diseases, including allergies and autoimmune diseases, such as diabetes, multiple sclerosis, arthritis and lupus. Neils Jerne, who developed the anti-idiotypic network theory, was awarded the Nobel Prize in physiology and medicine in 1984.

idiotype network the antigen-binding site of an antibody is an idiotype. As an immune response develops and clonal expansion of B cells occurs, the prevalence of this previously rare idiotype increases and can lead to the development of an anti-idiotypic T- and B-cell response.

IFAT the immunofluorescent antibody test, which is most commonly performed on infected red blood cells (iRBC) or sporozoites that have been either air-dried or fixed on slides then incubated with serially diluted serum or plasma samples. Antibody binding is revealed with an anti-human IgG fluorescein-labelled reagent.

IgA (immunoglobulin A) the predominant immunoglobulin in secretions (e.g. saliva).

IgD (immunoglobulin D) an immunoglobulin present in the surfaces of B cells.

IgE (immunoglobulin E) is the least abundant immunoglobulin, present in mammals and is involved in the allergic response as well as in the immune response against parasites. When mast cell (or basophil) bound specific IgE recognizes the allergen, mast cell degranulation occurs. Depending on the site of action, the resulting allergic hypersensitivity reaction can elicit urticaria, angioedema, rhinoconjunctivitis, asthma or anaphylaxis.

IgG (immunoglobulin G) the predominant immunoglobulin class produced during secondary immune responses. The most conspicuous immunoglobulin in the blood.

IgM (immunoglobulin M) the predominant immunoglobulin class expressed by B lymphocytes and secreted during primary immune responses.

IgY (immunoglobulin Y) the production of polyclonal antibodies, by poultry immunization, has aroused interest in scientific community. The IgY technology is an ethical experimental procedure by replacing the bleeding for collecting eggs. The process consists in the immunization of chickens at regular intervals, with subsequent production of serum immunoglobulin and transfers these proteins to the egg yolk. The amount of IgY in the yolk is dependent upon its concentration in serum. The time between immunization and detection of antibodies in yolks is variable, depending on the biological assay performed.

IKK (IκB kinase) an enzyme complex that is part of the upstream NF-κB signalling cascade.

IL-10 an interleukin produced by monocytes which suppresses the function of macrophages, promoting anti-inflammatory pathways.

IL-12 an interleukin produced by macrophages and dendritic cells which activates natural killer cells and induces T cell differentiation, promoting proinflammatory pathways.

illumina/Solexa sequencing high-throughput sequencing based on a sequencing-by-synthesis chemistry developed by Solexa, Incorporated. Solexa was acquired by Illumina, Incorporated, and the technology currently yields ≈600 million 100-bp sequences (60 billion base reads) per sequencing run.

imageJ public domain, image processing program developed at the National Institutes of Health that can display, edit, analyse, process, save, and print several image formats.

imago the adult stage of an insect.

IMD pathway immune signalling pathway that is primarily activated by Gram-negative bacteria and responds against *Plasmodium falciparum* in *Anopheles gambiae*.

imidacloprid is used on both dogs and cats in spot-on formulations. Residual activity on the coats of dogs and cats lasts approximately 1 month. It acts through contact: fleas or lice die rapidly. To broaden its spectrum of activity, imidacloprid is now combined in formulations with moxidectin, ivermectin, permethrin, and/or pyriproxyfen.

imidazothiazoles broad spectrum nematocidal agents (e.g. levamisole). Levamisole is a depolarizing neuromuscular blocker that interferes with neuronal transmission in worms. Through mimicking acetylcholine action, it changes the permeability of the post-synaptic membranes. This leads to sustained muscle contraction (spastic paralysis) in treated worms, and their rapid removal from the host. The therapeutic margin is very narrow in dogs. Signs of toxicosis include salivation, respiratory distress, seizures, and anaphylaxis, and are related to stimulation of mammalian nicotinic cholinergic receptors.

immediate (Type I) hypersensitivity a hypersensitivity tissue reaction occurring within minutes after the interaction of antigen and antibody. *See* **allergy**.

immortalized cell line population of cells from a multicellular organism which would normally not proliferate indefinitely but, because of mutation, can keep undergoing division rather than undergoing normal cellular senescence.

immune adherence the adherence of particulate antigen coated with C3b to tissue having cells with C3b receptors.

immune complex (ICs) complexes of antigen bound to antibody and, sometimes, components of the complement system. The number of circulating ICs is increased during chronic infection with parasites (particularly worm infections) and in some autoimmune disorders (particularly SLE) in which they may be deposited in tissues causing inflammation and tissue damage.

immune deficiency (IMD) pathway a major signalling pathway of the insect immune response mainly activated by Gram-negative bacteria.

immune elimination the enhanced removal of an antigen from the bloodstream by circulating antibodies and phagocytic cells.

immune memory a property provided by specialized B and T lymphocytes (memory B and T cells) that respond rapidly on re-exposure to the parasitic infection that originally induced them.

immune modulators substances that control the expression of the immune response.

immune response the reaction of the body to substances that are foreign or treated as foreign. The response is in a variety of forms, from the recognition of antigens in the body, the production of antibodies against the foreign substance and the response of lymphocytes (white blood cells, T cells and B cells).

immune response (Ir) gene a gene controlling the immune response to a particular antigen; most genes of this type are in the MHC, and the term is rarely used to describe other types of Ir genes outside the MHC.

immune surveillance the concept that lymphocytes survey the body for cancerous or abnormal cells and then eliminate them.

immune system the cells, proteins (such as antibodies) and cellular activities that work together to fight off infection and provide resistance to subsequent infection.

immunity a specific response in vertebrates to a foreign protein in which cells respond by producing humoral and/or cellular antibodies.

immunoadsorption a method for removal of antibody or antigen by allowing it to bind to solid phase antigen or antibody.

immunoassay a biochemical assay that enables the concentration of a substance to be measured by exploiting the specific binding between an analyte and the corresponding detection antibody. The analyte can be a relatively simple chemical substance, such as a drug, or a complex entity such as a protein or a virus in biological fluids. Different variants of immunoassays exist, which can by characterized by the measurement steps (e.g. sandwich or competitive assays) and the use of non-labelled or labelled reagents (e.g. enzyme-linked immunosorbent assay).

immunocompetence a measure of the ability of an organism to minimize infections (in terms of parasite establishment or development) via innate or acquired immune defences.

immunocontraception a method (not currently used) of reducing fertility of a pest species by controlling or preventing conception and pregnancy.

immunocytological and immunohistological techniques a group of methods that visualize specific proteins or other antigens directly in cells (immunocytology) or tissues (immunohistology) using antibodies coupled to fluorophores or enzymes.

immunodeficiency disease a condition in which immune function is defective.

immunoelectrophoresis a technique involving electrophoresis to separate protein from a mixture followed by immunodiffusion; it is used to identify the proteins in a complex solution such as serum.

immunofluorescence a technique for detection of cell- or tissue-associated antigens by the use of a fluorescently tagged ligand (e.g. an anti-immunoglobulin conjugated to fluorescein isothiocyanate).

immunogen any substance that elicits an immune response. Although all immunogens are antigens, not all antigens are immunogens.

immunoglobulin in mammals, five antibody classes (or isotypes) exist. These are: IgM, which exists as a monomer on the surface of B lymphocytes (the B cell receptor) and as a pentamer secreted in circulation). IgG constitutes 75% of the circulating antibody and exists as different subclasses with IgG1 production being dependent upon Th2 cytokines and IgG2 production being dependent upon Th1 cytokine production. IgE mediates mast cell and basophil degranulation and is up-regulated by plasma cells which have induced to class switch to IgE by Th2 cytokines (IL-4 and IL-13). IgA exists in two different forms (secretory and non-secretory) with secretory IgA being an important antibody in mucosal secretions and colostrum and milk. Class switching to IgA production is also stimulated by Th2 cytokines which include IL-5, IgD is found predominantly on the surface of mature B, but non-activated, lymphocytes and may have some function in B lymphocyte maturation.

immunoglobulin superfamily a large family of proteins characterized by possession of 'immunoglobulin-type' domains of approximately 110 amino acids folded into two β-pleated sheets. Members include immunoglobulins, T-cell receptors and major histocompatibility complex molecules.

immunohistochemistry a method to localize a biomarker protein or other antigen using an antibody that recognizes that antigen. The use of labelled (e.g. chromagen, fluorochrome, enzyme) antibodies allows localization of biomarker proteins at the organ, cellular and subcellular level.

immunological memory a primary immune response requires that antigenic material is processed by APCs and presented to lymphocytes in association with the MHC. Initially this leads to a relatively weak immune response with up-regulation of IgM in serum after about 1 week (the lag or latent phase). During this period, activation of T and B lymphocytes takes place and results in the exponential phase, which marks a rapid increase in the quantity of antibodies in the circulation and antibody class switching.

immunological privilege particular tissues (e.g. brain, mucosa, placenta) are privileged in that the presence of an antigen does not elicit an inflammatory response. This privilege prevents inappropriate responses to environmental antigens (e.g. intestinal mucosa) or potentially lethal immunopathology (e.g. placenta, brain).

immunological synapse a contact point between the T-cell and antigen-presenting cell which is generated by reorganization and clustering of cell surface molecules in lipid rafts.

immunomics a field of research that bridges the disciplines of genomics and proteomics by involving the immune system, to elucidate the set of antigens that interact with the host immune system and the mechanisms involved in those interactions.

immunomodulatory cytokines cytokines that suppress immune responses in order to prevent excessive immune reactions.

immunopathology self-damage caused by non-specific immune responses.

immunoperoxidase an immunological test that uses antibodies chemically conjugated to the enzyme peroxidase.

immunoreceptor tyrosine-based activation motif (ITAM)-associated signalling networks signalling networks dependent on ITAM motifs occurring in the cytosolic domains of transmembrane adapter proteins associated with immunoreceptors including FcRs, T-cell receptors and B-cell receptors.

immunosuppression inhibition or elimination of the immune responses by drugs or other processes.

immunotoxin a biochemical conjugate, or recombinant fusion protein, consisting of an immune targeting molecule such as an antibody or antibody fragment together with a cytotoxic molecule.

impact the health and educational benefits of a defined intervention.

imprinting suppression or silencing of genes depending on which parent they were received from. When DNA is passed to daughter cells after fertilization of an egg by a sperm, certain alleles can become active only if they were received from the mother, others only if they came from the father. If a gene is suppressed through imprinting from one parent, and the allele from the other parent is not expressed because of mutation, neither allele can be expressed.

imputation statistical inference of unobserved SNP genotypes based on a reference panel of known haplotypes in a population (e.g. the 1000 Genomes Project). Imputation can greatly narrow down the distance between SNPs and causal variants and thus increase the power of detection of associations.

***in silico* approach** implies a general term to encapsulate all various computational methods that are performed strictly on a computer with the purpose to identify vaccine candidates. This term includes the computational aspects of immunoinformatics, vaccinomics and proteomics.

in situ in a natural environment.

***in situ* hybridization** a method that use DNA or RNA probes which bind selectively to specific DNA or RNA target sequences by Watson–Crick or Hoogsteen base pairing. Probes can be conjugated to fluorescent dyes (such as in fluorescence *in situ* hybridization (FISH)) or to enzymes that catalyse a chromogenic reaction (such as the horseradish peroxidase in chromogenic *in situ* hybridization (CISH)).

***in situ* methods** a vast collection of methods used for the detection of nucleic acids or proteins directly in cells or tissues while preserving spatial information.

in vitro outside of a living organism.

in vivo within a living organism.

inactive cysticercosis infection with only degenerated (calcified) cysticerci, the infection cannot be transmitted to the final host.

inactive marks histone PTMs classically associated with heterochromatin, such as H3K9me2/3, H3K27me3, and H4K20me3.

inapparent infection the presence of infection in a host without recognizable clinical signs or symptoms.

inbreeding mating between relatives. An important consequence of inbreeding in populations is the increase in homozygosity.

inbreeding coefficients the co-ancestries of the two parents of an individual.

inceptisols variable soils with horizon development in early stages.

incidence (of infection) the rate of new cases of illness commencing during a specified time period in a given population. *See* **prevalence**.

incidental parasite an accidental parasite occasionally found in other hosts than their normal ones, e.g. *Dipylidium caninum* and *Hymenolepis diminuta* in man. This is different from *Erratic parasits*, which is found in their normal hosts but in unusual organs or tissues in which they are not adapted to live, e.g. *Enterobius vermicularis* in the vagina.

incipient species two or more diverged populations that are substantially, but not completely, reproductively isolated.

inclusion bodies insoluble aggregates of denatured/unfolded protein produced in the cytoplasm or nucleus of bacteria.

inclusive fitness a measure of fitness that takes into account not only the offspring produced by a particular individual, but also the offspring whom that individual's help enabled other related individuals to produce.

incompatible insect technique (IIT) a method to reduce insect populations by sterilizing wild females with large releases of incompatible males.

incomplete antibody an antibody that can bind to a particulate antigen but is incapable of causing its agglutination.

incomplete lineage sorting the phenomenon whereby an ancestral species undergoes several speciation events in a short period of time. It is the expected consequence of natural genetic diversity from an ancestral population that sorted itself incompletely into different descendant species.

incubation period the time elapsing between infection with the organisms of a disease and the appearance of symptoms, e.g. with malaria, the interval between the entry of sporozoites and the first clinical manifestation. In any given disease it is relatively constant. People who have been exposed to an infectious disease and may be incubating it and who may become infectious, are known as contacts.

indels small insertions or deletions of 1–1000 nucleotides.

independent assortment during gamete formation segregating pairs of unit factors (e.g. genes controlling colour or shape traits) assort independently of each other. As a result, one can use multiplicative probabilities to compute multitrait or multigene phenotypes or genotypes. Linkage disequilibrium can prevent the expected probabilities from being realized. *See* **linkage**.

index case a malaria case identified by parasitological confirmation via passive case detection.

index test the test whose diagnostic accuracy is being measured against the reference or gold standard.

indirect labelling method a localization method in which an unlabelled primary antibody is added to the sample to locate the epitope of interest. The primary antibody is then detected by a secondary antibody to which a detection system (e.g. a fluorophore, a gold nanoparticle, or an enzyme) has been conjugated. A wide variety of secondary antibody reagents are commercially available.

indirect life cycle requires the parasite to pass through several hosts species (intermediate and definitive hosts) and/or a vector species, often belonging to a different phylum, to complete its development.

indirect selection the effective preservation or elimination of genomic features that do not directly affect phenotype, through causal linkage to associated phenotypic traits; also called 'second order selection'. The mutability of an SSR locus is not 'visible' to direct selection acting on phenotype, but mutability is nevertheless a characteristic property of the locus. So direct selection acting on a particular SSR allele, on the basis of its associated phenotype, necessarily but indirectly acts likewise on the mutability of that allele. If a population contains alleles that differ in mutability, then selection will favour those alleles, whether more or less mutable, that are most consistently associated with the more fit phenotypes. Whenever alleles conferring a favourable phenotype arise as a result of those alleles' high mutability, then that high mutability will itself be selected indirectly.

indirect transmission the spread of a parasite from a reservoir or host to another host involving an intermediary stage, such as airborne particles or contaminated water. Insects might also act as vectors by transmitting infections between hosts on external body parts.

individual nucleotide resolution UV cross-linking and immunoprecipitation (iCLIP) a method that provides nucleotide resolution of RNA binding sites. RNA is UV cross linked to RBPs *in vivo*, treated with RNases, and made into partial cDNAs from stalled reverse transcriptase of cross-linked RNA. Sites of cross-linking are then mapped based on the stall sites.

individual-based simulation models computer models in which the characteristic of every individual is included separately and tracked over time.

indoor residual spraying (IRS) when mosquitoes and other insects rest in houses it is possible to kill them by spraying the walls with a residual (long-lasting) insecticide. Mosquitoes resting on sprayed walls come into contact with insecticide through their feet and are killed. Some insecticides irritate mosquitoes and cause them to leave houses.

induced pluripotent stem cells adult cells that have been reprogrammed to stem cells, which can differentiate into different cell types.

infect to cause an infection, especially with pathogenic microorganisms.

infected erythrocytes (IEs) erythrocytes where one or more *Plasmodium* merozoites have invaded and are developing asexually. During this development stage in the blood, the parasite secretes several molecules to the surface of the IE, among them Variable Surface Antigens.

infected source an animal or plant (an intermediate host) in which a parasite has established an infection and can act as the source of infection for another host. Cattle infected with the cysticerci of *Taenia saginata* function as the source of infection for the adult parasite for man. Infection occurs when the infective stage is ingested along with the intermediate host or part of it is eaten.

infection invasion by and multiplication of micro-organisms in body tissue, which may or may not result in overt disease. The organisms concerned are called pathogens, and include protozoa, worms, bacteria or viruses.

infection control the utilization of procedures and techniques in the surveillance, investigation and compilation of statistical data in order to reduce the spread of infection/infestation, particularly nosocomial infections.

infection rate percentage of the population from which a specific pathogen is isolated.

infectious capable of being transmitted to and causing infection in another organism.

infectious agent an organism that is capable of producing infection or infectious disease.

infectious period interval of time during which an infected host is shedding infectious stages of a parasite or is capable of transmitting the parasite to susceptible hosts.

infectiousness a property of an infected host. This indicates the relative likelihood of transmitting the parasite to other host individuals.

infective stage the stage in the life cycle of a parasite during which it is capable of producing infection.

infertile incapable of initiating, sustaining, or supporting reproduction. Alternatively, not fertilized and therefore incapable of growing and developing.

infestation the lodgement, development and reproduction of arthropods on the surface of the body of persons or animals or in clothing.

infested source a source of the infective stage of a parasite when that stage is essentially free-living, e.g. cercariae of schistosomes that infest water (swim about freely outside of the host) are capable of initiating the infection of a suitable host. Cercariae infest water, the source from which infection is initiated.

infinite-sites model assumes that the mutation rate per site is extremely low, so the probability of multiple mutations at the same position is considered negligible. It assumes that no recombination occurs.

inflammasome a multi-protein cytoplasmic complex that promotes inflammation by converting the IL-1β precursor into active IL-1β, and additionally by stimulating the generation of IL-18. They regulate mucosal immune homeostasis and have recently emerged as a central defence mechanism against bacterial pathogens.

inflammation the complex series of responses of tissues to trauma. It is characterized by increased blood flow and entry of leucocytes into the injured tissues, resulting in the five cardinal signs of inflammation: calor, dolor, oedema, erythema and function laesa (fever, pain, swelling, redness and lessened function). These responses generally act to enhance tissue defences and initiate the repair process.

informal outlet an outlet that is not funded, licensed, or controlled by the government.

informatics *See* **bioinformatics**.

informed consent a term describing the responsibility of doctors or researchers to ensure that patients or people have an understanding of the relevant facts regarding their care or participation in research. \Consumers also have a right to practice informed consent when they buy particular foods. Informed consent relies on one having access to reliable, truthful, and complete information.

infrapopulation all the individuals of a single parasitic species from a single host at a particular time.

inherit to receive genetic material from parents through biological processes.

inherited traits or characteristics that come from one's ancestors and are transmitted from parents to offspring through genes. The traits will therefore be present at birth.

inherited trait is one that is genetically determined. Inherited traits are passed from parent to offspring according to the rules of Mendelian genetics. Most traits are not strictly determined by genes, but rather are influenced by both genes and environment.

innate immunity immunity that is not intrinsically affected by prior contact with antigen, i.e. all aspects of immunity not directly mediated by lymphocytes.

innate response the first line of defence; able to function continually in the host without prior exposure to the invading parasite. This complex system comprises, in part, cytokines, sentinel cells, complement, and natural killer cells.

inner cell mass mass of cells on one side of a blastocyst that will form the body of the embryo. This is where embryonic stem cells are taken from.

inner membrane complex (IMC) double membrane aleveolin sacs (vesicles) that play a role in parasite replication and gliding motility.

inner shell eggs of some helminths have an inner and an outer shell, e.g. *Hymenolepis* *nana*. *Ascaris* eggs have an outer mammillated coat (the cortex) that covers the thick hyaline shell.

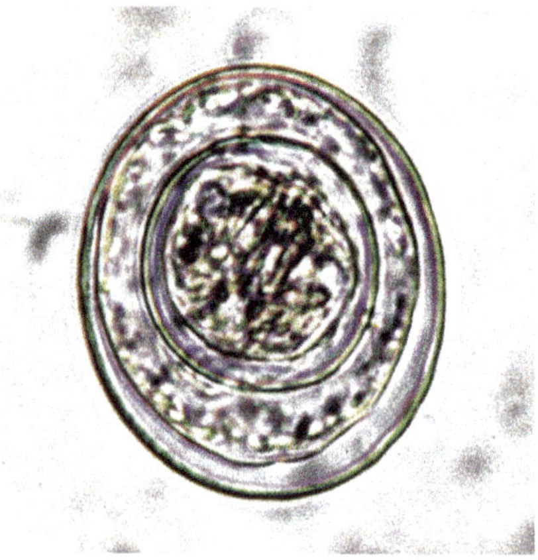

inorganic any chemical compound not based on carbon chains or rings (except oxides, sulfides of carbon and metallic carbides that are also inorganic).

iNOS (inducible nitric oxide synthase) a cytosolic enzyme found in multiple cell types and that produces nitric oxide (NO) from l-arginine, following induction by lipopolysaccharide and pro-inflammatory cytokines. It is presumed that the production of NO, in conjunction with superoxide radicals, leads to the formation of antimicrobial reactive nitrogen intermediates, such as peroxynitrite and nitrosothiols, which restrict the growth of invading pathogens.

input traits traits introduced into crop plants with the aim of lowering the cost of production and improving the performance of the crop in the field. For example: pesticide resistance, herbicide tolerance and disease resistance. This is in comparison to traits introduced into the crop to produce products with improved value. These are referred to as output traits.

insect a class of arthropods whose members are characterized by division of the body into three distinct regions: head, thorax, and abdomen, and that have three pairs of legs.

insect development inhibitor/insect growth regulator an agent that exerts its action by interfering with the development of the egg or other off-host life cycle stages of insect ectoparasites.

insect growth inhibitors (IGIs) include diflubenzuron and lufenuron. Lufenuron is marketed as an oral and parenteral flea-control product for dogs and cats. Lufenuron, a very lipophilic compound, accumulates in adipose tissue and is subsequently released slowly into the bloodstream. Adult fleas ingest lufenuron when they feed on the host and the drug is passed transovarially to the flea egg. Most flea eggs exposed to lufenuron fail to hatch and the ones that do die during the first moult. Chitin is an important constituent of exoskeleton and egg shell. Lufenuron interferes with the development of the insect's exoskeleton by inhibiting chitin synthesis or deposition in larvae and eggs. They have no effects on adult insects.

insect growth regulators (IGRs) are used in dogs and cats to control fleas and to control flies in cattle. These drugs include methoprene, fenoxycarb, and cyromazine. These preparations maintain persistently the larvae in an immature stage and interfere with reproductive organ differentiation. They are used to control faecal maggots in poultry and for flea control. IGRs are analogues and mimic the effects of an endogenous insect growth hormone, juvenile hormone. Juvenile hormone maintains the larval stage of the insect and inhibits maturation to the pupal and adult stages. In a normal insect, juvenile hormone is circulated throughout the insect's body and 'tells' the insect to stay in its current stage. After a certain amount of time, the insect stops producing juvenile hormone, and the insect mature into its next life stage. When an insect is exposed to an IGR that mimics juvenile hormone, the insect does not receive the signal to develop because, even though the insect may have stopped producing juvenile hormone, the IGR is still circulating throughout its body and sending the signal to stay in the current stage. Hence, insects poisoned with IGRs cannot moult or reproduce, and eventually die.

insect larva the second stage in the holometabolous insect life cycle, the grub, maggot or caterpillar.

insect pupa the third stage in the holometabolous insect life cycle. An inert, dormant stage from which the adult emerges.

insect vector insects may carry infection mechanically on body or mouthparts, by passage through the digestive tract but without the insect being infected, or by becoming an intermediate host with some part of the parasite's life cycle taking place in insect tissues.

insecta (antennata) the characteristics that identify an adult insect are three pairs of legs and three body parts; head, thorax and abdomen. Wings and antennae are present in most of insects. A holometabolous insect's life cycle has four stage egg, larva, pupa, and adult such as beetles, butterflies and flies. Development may be incomplete in hemimetabolous insects an immature nymph hatches from the egg; the nymph resembles the adult but is not a perfect copy, and undergoes several moults until sexual maturity, e.g. the louse. Flies are classified according to the habits of the females into: *Oviparous* females which lay eggs in early stage of development (immature) and hatching occur after certain period, e.g. house fly. *Viviparous (larviparous)* females which lay larvae, e.g. the flesh fly. *Pupiparous* females that lay well developed larvae which convert into pupa rapidly after deposition, e.g. *Glossina* and *Hippobosca*. Of veterinary medical importance are the many insects that serve as vectors for microbial diseases (mosquitoes, fleas, flies, lice, and bugs) or as venomous animals that sting (bees, wasps, and ants).

insecticide a mixture containing a toxic active ingredient used specifically for killing insects.

insecticide resistance defined as the ability of an insect to withstand the effects of an insecticide by becoming resistant to its toxic effects by means of natural selection and mutations.

insecticide-treated bed nets (ITNs) usually made of polyester and treated with pyrethroids. These nets form a protective barrier between the mosquito vector and people, and act by killing or repelling the mosquitoes. ITNs have been associated with a decrease in disease burden in malaria endemic regions.

insertion a type of mutation involving the addition of genetic material. An insertion mutation can be small, involving a single extra DNA bp, or large, involving a piece of a chromosome. *See* **chromosome**, **deoxyribonucleic acid**, **gene**, **mutation**.

insertion sequence a short DNA sequence that functions as a simple transposable element.

insertional mutation *See* **insertion**.

insertions and deletions (Indels) DNA sequences that have been inserted or deleted from a genomic region. As only phylogenetic analysis allows the distinction between insertions and deletions, indel has been used as an indifferent term.

instar a developmental stage in the life cycle of an insect, such as a larval or nymphal instar.

insulin a hormone that promotes the conversion of glucose to glycogen and typically lowers blood glucose in vertebrates.

insulin/IGF-1 (insulin-like growth factor 1) pathway a neuroendocrine pathway in nematodes that controls dauer diapause, metabolism, and lifespan. Protein hormones called insulin promote growth by binding to the insulin/IGF-1 receptor and promoting downstream signalling by the DAF-12 and other pathways.

integrated control the use of several measures to control different parasites or parasite stages present on the animal and stages present in the environment.

integrated pest management (IPM) a comprehensive strategy that incorporates chemotherapeutic and non-chemotherapeutic strategies for controlling pests.

integrin a heterodimeric molecule that binds to a wide range of extracellular matrix molecules and some cell surface molecules. The specificity of integrins is determined by the combination of α- and β-subunits. Importantly, their conformation can be regulated to modify their affinity for ligands. They frequently cluster, leading to changes in avidity for the matrix. Their cytoplasmic tails recruit a range of signalling molecules.

intellectual property (IP) the content of the human intellect, or the result of intellectual effort, which is considered to be unique and original and have value in the marketplace, and therefore requires legal protection and ownership. This includes copyrighted material such as literary or artistic works, industrial processes, and trademarks and patents.

intellectual property rights patents, copyrights, and trademarks.

intensity of infection a measure of the parasite burden of an infected host. Typically, this will vary over the course of the infection.

intensity reduction rate (IRR) the infection load after treatment expressed as a proportion of the infection load before treatment.

intention to treat analysis individual outcomes in a clinical trial are analysed according to the group to which they have been randomized, regardless of whether they dropped out, fully complied with the intervention or crossed over to the other treatment. By simulating practical experience intention to treat analysis provides a better measure of effectiveness (as opposed to efficacy).

interaction network diagram that shows numerous molecular interactions of a cell. Each point or node on the diagram represents a molecule (typically a protein), and each line connecting two nodes indicates that two molecules are capable of interacting.

interactome molecular interactions of a cell, typically used to describe all protein–protein interactions or those between proteins and other molecules.

intercellular adhesion molecule 1 (ICAM-1) a glycoprotein typically expressed on endothelial cells and cells of the immune system. It binds to integrins to stabilize cell–cell interaction.

interdigitating dendritic cell MHC class II-positive, Fc receptor-negative, antigen-presenting dendritic cell found in T-cell areas of lymph nodes and spleen. (NB: a different cell type to follicular dendritic cells.)

interface an area or system through which one machine is connected to another machine. Interfaces must be made between systems for you to access various data and information. For example, the laboratory system data are interfaced to the electronic medical record. Data travels from the laboratory system across the interface to the main electronic record.

interference one crossover event inhibits the chances of another crossover event – also known as positive interference. Negative interference increases the chance of a second crossover. *See* **crossing over**.

interferon (IFN)-γ IFN-α and IFN-β (type I interferons) can be induced in most cell types, whereas IFN-γ (type II interferon) is produced by T lymphocytes. All three types induce an antiviral state in cells and IFN-γ additionally acts in the regulation of immune responses.

interleukin (IL) any cytokine secreted by leucocytes that act as growth, differentiation and function factors for the cells of the immune system.

interleukin 1β (IL-1β) an interleukin produced by macrophages and epithelial cells that induces fever and the activation of macrophages and T cells.

interleukin 10 (IL-10) an important immunoregulatory cytokine produced by many immune cells. It promotes the development of a Th2 immune pattern by inhibiting the IFN-γ production of T lymphocytes, particularly via the suppression of IL-12 synthesis in macrophages. IL-10 is also secreted from T regulatory cells to counter balance inflammation.

interleukin 15 (IL-15) a cytokine that promotes the induction and long-term maintenance of central memory T cells.

interleukin 17 (IL-17) an inflammatory cytokine that promotes recruitment, activation, and differentiation of phagocytes and neutrophils.

interleukin-2 (IL-2) an essential mitogen and T cell growth factor.

interleukin-27 (IL-27) a regulatory cytokine that suppresses the activation and proliferation of IL-17-secreting T cells and promotes the generation of IL-10-secreting T cells.

interleukin-6 (IL-6) an inflammatory cytokine and pyrogen that promotes IL-17 expression by T cells, B cell activation and differentiation, and the antimicrobial properties of phagocytes.

interleukins (IL) a designation for some of the cytokines secreted by leucocytes.

intermediate host a host that is required by the parasite for completion of its life cycle, but one in which only the asexual or immature stages of the parasite occur. Some parasites need more than one intermediate host to complete their life-cycle. So intermediate hosts may be: *First intermediate host* – in which the early larval stage is formed, e.g. *Pirinella conica* snails as first intermediate host for *Heterophyes heterophyes*. *Second intermediate host* – which harbours the later larval stage which is usually the infective stage, e.g. fish (*Mugil* spp., *Tilapia* spp.) as second intermediate host for *Heterophyes heterophyes*.

intermittent preventative treatment (IPT) the administration of an antimalarial drug or drug combination at specified times, often corresponding to the time when a subject has access to the health services, to individuals at risk, whether or not they are infected. It is accepted that blood concentrations will fall below inhibitory concentrations between drug administrations.

intermittent preventative treatment, pregnancy (IPTp) public health intervention where pregnant women receive antimalarials at various time-points during pregnancy, regardless of infection status, with the aim of both clearing any current infections and preventing new ones.

internal image an epitope on an anti-idiotype which binds in a way that structurally and functionally mimics the antigen.

internal nodes branching points on a phylogenetic tree that reflect past speciation events leading to living species.

internal or endoparasites exist inside the host, occupying the digestive tract (enteric parasites), e.g. *Ascaris*; in blood and blood forming organs (haemoparasites), e.g. *Babesia*, *Theileria*, *Trypanosoma*; in the reproductive organs (Venereal parasites), e.g. *Tritrichomonas* in cattle and *Trypanosoma equiperdum* in equines.

internal transcribed spacer region of the ribosomal DNA (ITS) a region of non-coding nuclear ribosomal DNA used commonly to estimate phylogenetic relationships and for barcoding, especially of invertebrates. This is in large part due to the high copy number of this region and high amount of variation among closely related species.

interoperability the extent to which systems and devices can exchange data and interpret that shared data.

interphase the period in the cell cycle when DNA is replicated in the nucleus and followed by mitosis.

interspecies barrier the lack of phylogenetic similarity may be a barrier that determines the inability of a pathogen to adapt to a new host.

interspecific and intraspecific the terms are antagonists and refer to individuals of the same or different species.

interstitium the tissue in between the alveolar epithelium and the endothelium.

interviewer bias systematic error due to interviewer's subconscious or conscious gathering of selective data.

intervillous space spaces between villous tissue filled with maternal blood.

intestinal organoids epithelial 'mini-guts' grown *in vitro* from biopsies of the rectal mucosa or from stem cells from a single individual.

intracellular gene transfer (IGT) transfer of genes from an endosymbiont or organelle to the nuclear genome of an organism, or vice-versa, or between organelles. The products of these genes are often transferred back to the donor for use.

intracellular signalling a mechanism for transferring information from one part of the cell to another. For example, information received at a specific plasma membrane receptor can be relayed to the nucleus by intracellular messengers (signal transducers) such as protein kinases.

intraconoidal microtubules single pair of microtubules occurring in the centre of the apical polar ring and conoid.

intracranial pressure the pressure exerted by the cranium on brain tissue, cerebrospinal fluid and circulating blood volume of the brain.

intralesional haemorrhage the observation of bleeding in a localized lesion in the brain.

intra-locus sexual conflict opposing direction of selection between males and females for a particular locus or single trait, for instance, where a sequence variant improves the fitness of one sex but reduces fitness in the other.

intraparenchymal NCC cysticercosis of the central nervous system with cysticerci located in the parenchyma.

intravenous immunoglobulin (IVIG) a preparation of human polyclonal IgG obtained from pooled plasma samples taken from thousands of healthy blood donors. It is used as a replacement therapy for individuals with hypogammaglobulinaemia, but is also increasingly used, at much higher doses, to treat autoimmune diseases. IVIG is licensed for the treatment of idiopathic thrombocytopenic purpura, Guillain–Barré syndrome, chronic inflammatory demyelinating polyneuropathy and Kawasaki disease, and is also used in the treatment of other autoimmune diseases, including multiple sclerosis and ANCA-associated vasculitis.

intravital microscopy the direct real-time imaging of biological phenomena in exposed tissues.

intrinsic apoptosis apoptosis regulated and mediated by the Bcl-2 family of proteins in response to cellular stresses, such as UV irradiation and chemotherapeutic agents. It is also known as the mitochondrial or stress-induced pathway to apoptosis.

intrinsic incompatibilities forms of reproductive isolation that are relatively independent of the environment in which hybrids are tested. Such incompatibilities usually reflect developmental abnormalities produced by deleterious epistatic interactions between alleles from different species.

intrinsic noise gene-specific fluctuations that can be attributed to the stochastic fluctuations in the different steps of the expression process of a specific gene. Factors contributing to intrinsic noise include the promoter structure of the gene, its localization within the chromosome and the nuclear architecture.

intrinsic pathway and extrinsic pathway apoptosis can be induced from the outside of a cell via a death receptor (extrinsic pathway) or from inside the cell (intrinsic pathway). In the intrinsic pathway, procaspase 8 is activated via a so-called disc complex; in the extrinsic pathway, procaspase 9 is activated via the so-called apoptosome and includes the release of cytochrome c from the mitochondrion.

introgress introducing a genomic region from one strain or species into that of another by repeated backcrossing. By selecting for the phenotype of interest, the genomes become isogenic except for the chromosomal regions causing the selected phenotype.

introgression the transfer of a segment of the genome of one species into the genome of another species following a hybridization event and repeated backcrossings.

intron DNA sequences within the protein-coding sequences of a gene; introns are transcribed into mRNA but are cut out of the message before it is translated into protein. Introns may contain sequences involved in regulating expression of a gene.

invariant chain a polypeptide which binds MHC class II molecules in the endoplasmic reticulum, directs them to the late endosomal compartment and prevents premature association with self peptides.

invariant natural killer T (iNKT) cell a T-cell that expresses specific poorly polymorphic variable gene segments of the T-cell receptor α chain (Vα14 in mice and Vα24 in human). Typically these cells co-express cell-surface markers that are encoded by the NK locus and are activated by recognition of CD1d, particularly when α-galactosylceramide is bound in the groove of CD1d.

invasins a new term describing parasite ligands involved in establishing invasion mediated by tight junctions.

invasion assay test to observe and quantify invasion of new RBCs by the parasite.

invasome a cup generated on the membrane of a cell by a massive actin rearrangement and that mediates the phagocytosis and internalization of large bacterial aggregates.

inverse density dependence populations of some species exhibit a decrease in the per capita growth rate as their populations reach small sizes or low densities. Under such conditions, the growth rate can be ≤ 0 because of a decrease in reproduction and/or survival when conspecific individuals are few. This can be a result of inbreeding that exposes deleterious and lethal recessive genes to selection; it can also arise from demographic stochasticity and sex-ratio fluctuations. The effect is also seen in species that have co-operative interactions that occur infrequently if there are too few individuals.

inversion a chromosomal mutation involving the removal of a chromosome segment, its rotation through 180°, and its reinsertion in the same location.

ion channel a class of solute transport proteins that facilitates rapid transmembrane flux of solutes. Channels typically have an aqueous pore designed to recognize and allow passage of specific solutes; they may also be ligand gated or voltage dependent.

ionophore ionophore antibiotics are the most widely used anticoccidials. The anticoccidial action of ionophores is related to their ability to form lipophilic complexes with cations and to facilitate their transport across biological membranes. The intracellular accumulation of these ions interferes with parasite metabolism. These drugs include monensin, lasalocid and salinomycin. Ionophores, particularly monensin, can be toxic to muscle tissues in mammals and birds. Ingestion of monensin causes fatal cardiotoxicity in horses.

Ir (immune response) genes the genes, including those within the MHC, that together determine the overall level of immune response to a given antigen.

IRAK family kinases protein kinases involved in signal transduction and regulation of signalling downstream of TLRs (spell out). Recruited following interaction between MyD88 and the TLR TIR domain.

IRG (interferon-regulated guanine triphosphatases) a family of proteins that has been implicated in resistance to intracellular pathogens; sometimes called p47 GTPases to reflect their molecular weight.

iron fenton chemistry the catalysis of hydrogen peroxide breakdown to the hydroxyl radical (OH·) by ferric iron (III). OH· is a powerful oxidizing agent capable of damaging virtually all biological molecules, thus causes oxidative stress.

iron-haematoxylin stain one of several stains used to make a permanent stained slide for detecting and quantitating parasites, in particular protozoa in human faecal samples.

ischaemia–reperfusion injury a process induced by restoration of the blood flow to an area that had previously experienced deficient blood supply. Multiple cell damage and survival pathways are activated during this process.

isoacceptors tRNAs that have different anticodons but are charged with the same amino acid.

isocitrate dehydrogenase enzyme that catalyses the conversion of isocitrate to succinate in the mitochondrial matrix.

isoenzyme an enzyme performing the same function as another enzyme but having a different set of amino acids. The two enzymes may function at different speeds.

isofemale line isolated female line. A genetic strain or line of a species that had one female parent at its origin.

isofemale selection selection strategy whereby progeny are derived by mating single females.

isogeneic genetically identical or having identical (or nearly identical) genotypes.

isograft a tissue transplanted between two genetically identical individuals.

isolate refers here to clinical isolates. Parasites isolated from a patient.

isolate breaking excess heterozygosity (over Hardy–Weinberg expectation) observed when divergent populations or subpopulations establish secondary contact. The opposite of the Wahlund effect.

isolation represents separation for the period of communicability, of infected persons or animals from others in such places and under such conditions as to prevent or limit the direct or indirect transmission of the infectious agent. Categories of isolation include: strict isolation: for highly contagious infections spread by air and contact; contact isolation: for diseases spread primarily by close or direct contact; respiratory isolation: to prevent transmission over short distances through the air.

isolation by distance (IBD) a pattern of population genetic variation that derives from spatially limited gene flow. IBD is defined as a decrease in the genetic similarity among populations as the geographic distance between them increases. Statistical tests for IBD can be conducted using populations or individuals as the units of replication, although analyses at the individual level typically utilize spatial autocorrelation statistics. the increase of genetic differentiation between populations with the geographic distance reflecting the fact that dispersal preferentially occurs between neighbouring populations.

isometric tension tension within a structure that is not changing length.

isopoda (isopods) an order of the subphylum Crustacea containing free-living and parasitic species; possess dorso-ventrally flattened body with seven pairs of legs of equal size; terrestrial and aquatic species.

isoschizomer restriction enzymes that have the same recognition sequences and cleave at the same positions.

isotope atom that has the same number of protons as another atom but a different number of neutrons and hence atomic mass. For example, ^{13}C is an isotope of carbon that has one more neutron than the most common isotope of carbon, ^{12}C.

isotopic space (isotopic niche) the space occupied by a species in an *n*-dimensional isotopic niche, where each dimension corresponds to a particular isotope.

isotropic equal in all directions.

isotropic spreading spreading of cells during which their entire edge (or large parts of it) extends rapidly.

isotype an antibody constant region structure present in all normal individuals, i.e. antibody class or subclass.

isozymes enzyme variants with the same functional role, but differing in 1, 2, 3 or 4 structure. In some cases, isozymes may be multimers produced by multiple genes. Hence, they may not qualify as co-dominant allozymes for use as genetic markers.

ISRE interferon-stimulated response element: promoter element.

ITAM Immunoreceptor Tyrosine-based Activation Motifs are amino acid consensus sequences recognized by src-family tyrosine kinases. These motifs are found in the cytoplasmic domains of several signalling molecules including the signal transduction units of lymphocyte antigen receptors and of Fc receptors.

itch mites the genus *Psorobia*, formerly *Psorergates*, includes two species parasitizing domestic ruminants: *Psorobia ovis* (sheep) and *Psorobia bos* (cattle). The entire life cycle takes place on the host and comprises the egg, the larval stage, three nymphal stages (protonymph, deutonymph and tritonymph) and the adult mite. The whole life cycle of *Ps. ovis* takes about 2–3 weeks. Infestation with these mites does not lead to a skin reaction/skin lesions which are comparable to the skin reaction caused by the mange mites, and therefore they cannot be considered to be a mange mite in their true sense (= itch mite). Itch mites live in the superficial skin layers and can cause chronic irritation and skin thickening; however, the effects of this infestation may often only be very slight so that many cases are likely to remain undetectable.

ITIM (immunoreceptor tyrosine-based inhibitory motifs) presents in the cytoplasmic domains of certain cell surface molecules, e.g. Fcγ RIIB, inhibitory NK cell receptors, and which mediate inhibitory signals.

ITR (inverted terminal repeats) repeat sequences in reverse orientation that serve as the viral origins of replication. ITRs are essential for virus packaging and rescue of the integrated viral genome.

ivermectin a widely used off-label to treat sarcoptic mange and ear mange in carnivores, as well as canine demodicosis. Currently, several avermectins/milbemycins are marketed in domestic carnivores: (1) milbemycin oxime, used mainly as a broad-spectrum anthelmintic but may be indicated in dogs with demodicosis. It is administered orally. (2) Selamectin, in addition to its nematodicidal action, is used in dogs and cats as a spot-on for treatment and/or prevention of certain ectoparasitoses (pulicosis, canine sarcoptic mange, and ear mange). (3) Moxidectin, available as a spot-on treatment in combination with imidacloprid for very similar indications: nematodicidal spectrum plus insecticidal/acaricidal spectrum.

Ixodes ricinus a hard tick that infests livestock, deer, dogs and a wide variety of other species including humans. This tick has long mouthparts that can make its bites painful and annoying; the bites can also become secondarily infected by bacteria. Feeding by large numbers of ticks may result in anaemia. *I. ricinus* can also transmit a number of pathogens including *Babesia divergens* (babesiosis), louping ill virus, tick-borne encephalitis virus, *Borrelia burgdorferi* (Lyme disease) and *Anaplasma phagocytophila* (tick-borne fever of ruminants, human granulocytic anaplasmosis).

IκB a protein that inhibits the NF-κB (nuclear factor kappa-light-chain-enhancer of activated B cells) transcription factor by preventing it from entering the nucleus. It plays a role in propagating the cellular response to inflammation.

A B C D E F G H I **J** K L M N O P Q R S T U V W X Y Z

J chain (or the joining chain) a small polypeptide, expressed by mucosal and glandular plasma cells, which regulates polymer formation of immunoglobin (Ig)A and IgM. J-chain incorporation into polymeric IgA (pIgA, mainly dimers) and pentameric IgM provides these antibodies with several features.

JAK (Janus kinases) a family of intracellular non-receptor tyrosine kinases that transduce cytokine-mediated signal. This family, comprising JAK1, JAK2, JAK3 and TYK2 (tyrosine kinase 2).

J-binding proteins proteins that specifically bind to the base J (β-D-glucopyranosyloxymethyluracil), a modified form of uracil found in the DNA of a number of organisms, such as human pathogen *Trypanosoma* and the kinetoplastids.

JNK (c-Jun N-terminal kinase) signalling pathway a pathway involved in responding to oxidative stress and regulates cell growth, differentiation, survival, apoptosis, and immunity.

joining (J) gene segments found in the immunoglobulin and T-cell receptor gene loci and, upon gene rearrangement, encode part of the third hypervariable region of the antigen receptors.

junctional diversity diversity of the splice junctions in the recombined variable (V), diversity (D, for immunoglobulin heavy chains, and for TCR β and δ chains) and joining (J) gene segments of antibody and T-cell receptor (TCR) genes.

junk DNA (non-coding DNA) stretches of DNA that do not code for genes. They may have regulatory and other functions.

juvenile stage any stage in the life cycle of a helminth parasite, between the egg and the mature adult stage, which is sexually immature. *See* **larva**.

A B C D E F G H I J **K** L
M N O P Q R S T U V W
X Y Z

k cell (killer cell) a generic term for any leucocytes which mediates ADCC.

***K*0.5** the concentration of an inhibitor required to reduce enzyme function by 50%.

***Ka/Ks* (d*N*/d*S*)** the ratio of the rate of non-synonymous substitutions (that is, amino acid replacements) to the rate of synonymous substitutions (that is, changes among the synonymous codons of an amino acid).

kala-azar (or visceral leishmaniasis) the name originated from two words, kala (means black) and azar (means fever), attributed to the association of fever with hypopigmentation.

karyogamy fusion of two nuclei in a single cell. It refers to the nuclear fusion following cell fusion in the sexual cycle. However, for diplomonads it has only been observed at the cyst stage.

karyosome a structure (body) within the nucleus having a relatively constant size and location in each species and made up of two components, one achromatic and another that stains similar to chromatin and appears as a granule or bundle of granules. The size and location of the karyosome, especially the staining portion, is frequently used as an aid in differentiating species of amoebae that are similar in structure.

karyotype a photomicrograph profile of an individual's chromosomes arranged in a standard format showing the number, size, and shape of each chromosome type. It is used to look for abnormal numbers or structures of chromosomes.

Kato–Katz method (K-K) a direct parasitological technique that allows detection and quantification of *Schistosoma* eggs in faecal samples.

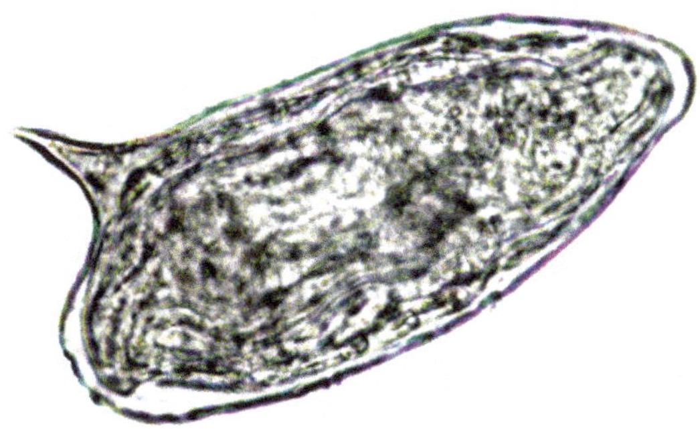

keratinization a biochemical process of keratin deposition on cells.

ketone bodies are produced by the liver during fasting and from fatty acid catabolism after prolonged exercise. They are used peripherally as an energy source when glucose is not available. The three types of ketone bodies are acetone, acetoacetic acid and beta-hydroxybutyric acid.

key a biological key is a series of questions arranged in a format designed to lead the users to an accurate identification of species within a biological group of organisms.

keystone pathogen a keystone microbial species that remodels a microbial community in ways that promote disease pathogenesis.

keystone species a species that has a disproportionately large effect on its environment relative to its abundance.

kidney worm disease a parasitic disease caused by the kidney nematode *Stephanurus dentatus*, which lives in the kidneys and perirenal tissue of pigs and pass eggs into the urinary bladder. Earthworms can serve as an intermediate host in the life cycle of this worm. Pigs are infected by ingesting eggs, larvae, or earthworms, by transmammary or transplacental transmission and by skin penetrating larvae L3. Diagnosis is usually made at necropsy; however, eggs can be identified in urine.

killed vaccine a vaccine made by taking an authentic disease-causing parasite and treating it with chemicals to reduce infectivity to non-detectable level.

killer T cell (cytotoxic T cell) a T cell with a particular immune specificity and an endogenously produced receptor for antigen, capable of specifically killing its target cell after attachment to the target cell by this receptor. *See* **k cell**.

kilobase (kb) a unit of length for DNA fragments equal to 1000 nucleotides.

kin selection a form of natural selection (an evolutionary strategy) that favours the reproductive success of relatives even at a cost to an individual's own survival and reproduction.

kin selection theory a theory of natural selection that takes into account how the relatedness of individuals within a population influences the evolution of social traits.

kinase an enzyme that transfers a phosphate group to a substrate molecule.

kinetoplast an organelle characteristic of protozoa of the order Kinetoplastida (including trypanosomes). It contains an interlinked network of circular mitochondrial DNA (kDNA).

kinetoplast DNA (kDNA) an extranuclear DNA formed by concatenated networks of mitochondrial DNA. kDNA is found within kinetoplasts, which are located at the base of flagella. The presence of a kinetoplast is a distinguishing feature of the protozoan class, Kinetoplastida.

kinetoplastid a group of single cell flagellated protozoa within the phylum Euglenozoa. The defining feature of kinetoplastids is their mitochondrial DNA, which is arranged into a tightly packed network that is known as the kinetoplast.

kinins a family of polypeptides released during inflammatory responses and which increase vascular permeability and smooth muscle contraction.

KIRs (killer cell immunoglobulin-like receptors; also known as CD158) a family of highly polymorphic activating and inhibitory receptors that serve as key regulators of human NK cell function.

KLH (keyhole limpet haemocyanin) a novel protein that stimulates an antibody response.

knockdown (Kdr)-type mechanism changes in the sensitivity of the voltage-gated sodium channels due to non-silent point mutations that confer resistance to the organochlorine DDT and pyrethroids.

knockout deactivation of one or more specific genes that is used in laboratory organisms to study gene function. Scientists create knockouts (often in mice) so that they can study the impact of the missing genes and learn something about the genes' function. The animals that are produced by this technique can be bred to homozygosity, hence allowing the generation of a null phenotype for that gene product. *See* **gene**, **locus**.

Koch's postulates a set of criteria developed by the German physician and bacteriologist Robert Koch in the late 1800s to determine if a given microorganism is the causative agent of a particular disease. These criteria include presence in disease, absence in health, isolation, induction of disease upon introduction, and subsequent reisolation.

Kolmogorov–Smirnov test a statistical non-parametric test for the comparison of continuous, one-dimensional probability distributions.

KU70–KU80 or KU70–KU86 a pair of proteins that, together with the catalytic subunit of the DNA-dependent protein kinase (DNA PKcs), form a complex called DNA-PK that is involved in non-homologous end joining, V(D)J recombination and telomere maintenance.

ku80 gene a key component of the non-homologous end joining pathway, which allows non-specific integration of DNA fragments into chromosomal DNA. Disruption of ku80 leads to higher relative levels of homologous recombination.

Kupffer cell a specialized resident macrophages lining the wall of the hepatic sinusoids that play a role in the breakdown of red blood cells.

Kyoto Encyclopedia of Genes and Genomes (KEGG) a reference pathway database based on current knowledge of biochemistry. Maps are represented as reaction networks where enzymes and compounds link to detailed entries containing information on nomenclature.

L chain light chain.

L1Tc (LINE 1 *Trypanosoma cruzi*) a non-LTR LINE1 element from *T. cruzi*.

laboratory parasite lines (e.g. Dd2, HB3, 3D7, 7G8, FVO) genetically distinct cultures of *Plasmodium falciparum*, obtained from human patients around the world. These lines are clonal and provide an important resource for examining the molecular basis of PSAC activity and other phenotypes.

lack of antiparasitic efficacy therapeutic failure can be due to several common and non-resistance-related reasons, such as (i) interactions with another drug or health condition in the patient that diminish the drug efficacy, (ii) the failure to diagnose and treat the presence of mixed infections of parasites or of other infectious agents, (iii) patient immunodeficiency, (iv) quick re-infection/infestation due to contaminated environment, (v) failure to deliver or receive the correct dosage or form of product, such as compliance failures, usage of out of date products, etc.

lactoferrin an iron-containing molecule that has antimicrobial action by binding iron needed for microbial growth.

ladder a series of known-size fragments run in a gel to allow sizing of fragments of target DNA run in other lanes. One commonly used ladder is phage lambda cut with a restriction enzyme [yields fragments of 216, 211, 200, 164 and 150 bp].

lambda phage (λΦ) DNA a useful tool in molecular biology. This *Escherichia coli* virus DNA is useful because its entire sequence is known (=50 kb double-stranded), and it is often used to create a ladder of known-size fragments for sizing bands on gels. It is also a useful cloning vector.

lamellipodia planar cell protrusions that are driven by F-actin polymerization.

lamina propria the connective tissue underlying the intestinal villous epithelium at mucosal sites is referred to as the lamina propria. It is separated from the epithelial cell layer by a basal membrane layer.

LAMP (loop-mediated isothermal amplification) a method for DNA amplification in which the entire reaction is carried out in a single tube and at a constant temperature.

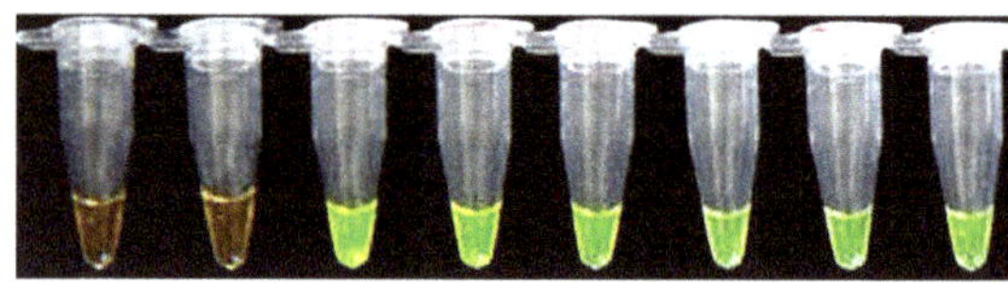

landscape genetics the study of the influence of biotic and abiotic landscape features on microevolutionary processes within and between populations. Differs from phylogeography because it is carried out in an explicit spatial framework.

Langerhans cells (LCs) a subset of dendritic cells that reside in the epidermis (upper layer) of the skin and are able to act as Fc receptor and MHC class II-positive antigen-presenting cells.

langmuir monolayers type of biomimetic membrane made by the self-assembly of lipids in a monolayer at the air–water interface.

large granular lymphocyte (LGL) large lymphocytes which contain cytoplasmic granules and function as NK and killer (K) cells. Activated CD8^{+} cytotoxic T lymphocytes (Tc) also assume LGL morphology.

large *P* small *N* problem a statistical challenge to estimate a large number of parameters based on a small number of samples.

large ribosomal subunit (28S rDNA) a region of nuclear DNA commonly used to estimate phylogenetic relationships and barcoding, especially of invertebrates.

large X effect the observation that, in genetic analyses of hybrid sterility and incompatibility, the X chromosome tends to have the largest effect among all the chromosomes, especially when examined in hybrids with heterogametic sex chromosomes.

larva (plural **larvae**) the post-embryonic stage in which internal organs are developing or are partially developed and are at least partially functioning. Any pre-adult stage (after the embryonic stage) in the life cycle of a parasite that is morphologically distinct from the adult stage. The term larva should probably not be used for immature nematodes since they are not morphologically distinct from the adults but simply smaller and less developed.

larval cysticercus stage cystic round parasite form (1–5 mm), containing the head (scolex) of the future tapeworm; they can be transported throughout the body by blood flow but are mainly located in muscles, the heart, and the brain.

laser capture microdissection (LCM) a method for isolating specific cells or specific areas from cell, tissue or organism samples using laser cutting under microscopic visualization.

latency an asymptomatic disease state in which the pathogen burden is low or undetectable.

latent class models these models estimate the test sensitivity, specificity and prevalence (i.e. unknown latent classes) using observed test results. If properly structured, they can be used to estimate the true prevalence of disease or infection in the absence of a gold standard or reference test. They can also be used to develop better reference tests.

latent infection viral infection in which the life cycle halts as a provirus integrates in the genetic material of the host and the transcription of viral genes does not occur, but the viral DNA is replicated as the bacterial genome replicates.

lateral flow immune-chromatographic assay a device that produces a visible mark to signal the presence or absence of a protein antigen, based on the accumulation of dye-labelled antibodies to form a visible line after binding directly, or through a common antigen, to a fixed line of unlabelled antibody.

lateral gene transfer (LGT) the transfer of genes from one organism to another [also known as horizontal gene transfer (HGT)]. In parasites, these genes are possible drug targets due to their potentially distant evolutionary relationship with organismal genes. This process gives prokaryotes the ability to obtain novel functionalities or cause dramatic changes in community structure over relatively short periods of time.

lateral incorporation adding an extra host species at any given point in the life cycle, without extending the length of the life cycle.

latrophilin G-protein-coupled receptors of the class B secretin family. Latrophilins are activated by latrotoxin the venom of *Latrodectus* spiders and are probably involved in cell–cell contact across synapses.

LD100 dose of parasites necessary to kill 100% of infected animals.

LD50 the dose of a drug that kills 50% of the population under test.

lead single-nucleotide polymorphism (LeadSNP) a SNP defining a risk locus identified through GWAS.

lead-time bias overestimation of survival because of earlier diagnosis – time of death does not change, just time of diagnosis.

leaky ribosomal scanning a mechanism by which internal AUG codons in an mRNA are used to initiate translation. Ribosomes occasionally skip the first AUG codon, which is often not in an optimal sequence context for initiation, and a downstream AUG is selected instead.

lectins a family of proteins which have affinity for specific sugars on glycoproteins and glycolipids. Many lectins are mitogenic (i.e. can induce lymphocytes to divide), such as Concanavalin A (Con A) and phytohaemagglutinin (PHA).

***Leishmania* RNA virus (LRV)** a cytoplasmic double-stranded RNA (dsRNA) virus residing within some strains of *Leishmania*, which may act as a virulence factor in metastatic leishmaniasis.

leishmaniasis (or leishmaniosis) a vector-borne disease caused by protozoan parasites of the genus *Leishmania*. Among the two million new human infections, 200,000–400,000 are cases of life-threatening visceral leishmaniasis (VL), culminating in approximately 20,000–30,000 deaths every year. Children and immunocompromised individuals are especially affected, but adults with active disease are frequently found in foci of recent parasite introduction. Several drugs including meglumine antimoniate and allopurinol used alone or in combination, amphotericin B, pentamidine, miltefosine, aminosidine (paramomycin), ketoconazole, metronidazole, spiramycin, and enrofloxacin have also been investigated as antileishmanial agents.

lemniscus one of a pair of lateral, fluid-filled sacs at the base of the proboscis in members of the phylum *Acanthocephala*.

leptomeningeal/subarachnoid space CSF-drained space at the surface of the CNS that protects and cushions the brain and spinal cord. This space is bordered on the outside by the arachnoid mater and towards the CNS parenchyma by the pia mater. Blood vessels at the surface of the CNS are localized in this space.

leptomixid amoeba amoebae belonging to the order Leptomyxida, class Tubulinea. These organisms are generally considered harmless, free-living soil organisms with the exception of the species *Balamuthia mandrillaris*.

let-7 The lethal-7 (let-7) gene the first discovered in the nematode as a key developmental regulator and was found to encode one of the first known miRNAs.

leucocyte white blood cells, which include neutrophils, basophils, eosinophils, lymphocytes and monocytes.

leucocytes extravasation the movement of leucocytes out of the circulatory system, towards the site of tissue damage or infection.

leukaemia an increase in the number of ineffective and immature white blood cells, causing a weakened immune system. This leaves the body susceptible to infection.

leukotrienes vasoactive metabolic products of arachidonic acid which promote inflammatory processes (e.g. chemotaxis, increased vascular permeability) and are produced by a variety of cell types including mast cells, basophils and macrophages.

liability scale an unobserved underlying risk of a binary phenotype or disease that is measured on a continuous scale and that is likely to be influenced by many genetic and environmental factors.

library an unordered collection of clones (i.e. cloned DNA from a particular organism) whose relationship to each other can be established by physical mapping. *See* **genomic library**.

lidocaine a drug that can be used to disrupt lipid rafts without affecting cholesterol content in red blood cells.

life cycle the stages an organism goes through over time containing all pre-parasitic and parasitic phases. Usually represented as a circle beginning with the formation of a individual and ending with the production of the next generation. *See* **direct life cycle**, **indirect life cycle**.

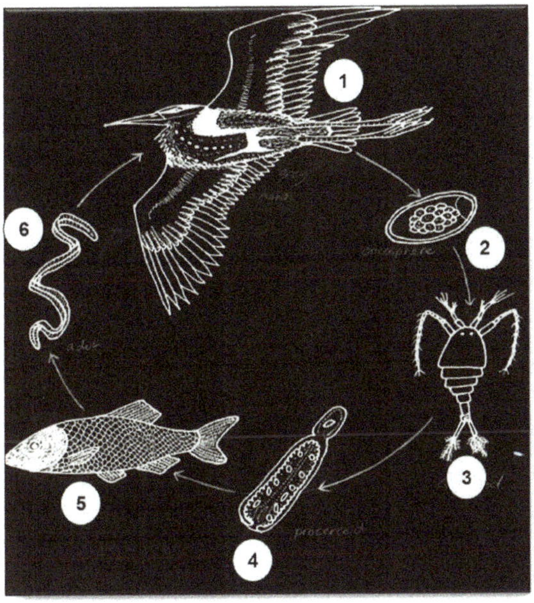

life history traits measurable physiological and phenotypic properties that affect the health of an animal and the population. Examples include brood size, developmental rate, and lifespan.

lifetime monogamy a pair is formed and individuals in the pair remain together throughout their life.

ligand a molecule recognized by a binding structure such as a receptor.

ligand valency the combined effects of the binding of multiple ligands.

ligase an enzyme that is used to join fragments of DNA together, for example in gene splicing.

ligate/ligation specific binding of receptors on the host cell with their appropriate antigen on the surface of pathogens.

ligation capture a method for capturing restriction enzyme-digested DNA molecules via the annealing of an oligonucleotide containing complementary sequences to adaptor oligonucleotides to the DNA molecules and to the adaptor oligonucleotides. The adaptors and DNA molecules are then ligated together, allowing for polymerase chain reaction amplification of only the ligated products.

light chain (L chain) the smaller of the two types of chains that comprise a normal immunoglobulin or antibody molecule. Light chains occur in two forms: kappa and lambda.

lignin a complex, insoluble polymer whose structure, while not well understood, gives strength and rigidity to cellulose fibres in the cell walls of woody plants. Lignin makes up a significant portion of the mass of dry wood and, after cellulose, is the second most abundant form of organic carbon in the biosphere.

Ligula intestinalis a helminth parasite, a cestode whose larval stages are found in the body cavity of a range of fish hosts. Fish infected with *Ligula* can demonstrate behavioural abnormalities.

likelihood ratio the likelihood ratio for a test result compares the likelihood of that result in patients with disease to the likelihood of that result in patients without disease.

likelihood ratio of a negative test the ratio of the probability of a false negative result if the disease is present to the probability of a true negative result if the disease is absent.

likelihood ratio of a positive test the ratio of the probability of a true positive result if the disease is present to a false positive result if the disease is absent.

limit of detection (alternatively, detection limit) the lowest quantity of an analyte that can be distinguished from the absence of that analyte (typically within a stated confidence limit, generally 1%).

limitation the classical negative density-dependent process thought to be important in regulating filarial larval infection in culicine mosquitoes, whereby L3 larvae output per microfilaria (Mf) ingested by mosquitos is initially positively linear but declines to approach a constant value with increasing Mf intake.

limited tissue response the timing and amplitude of the inflammatory response to injury. Limited indicates a delayed or slow response and a small or minimal area of occurrence.

lin-28 messenger RNA-binding protein that selectively binds to the primary and precursor forms of let-7, and inhibits the processing of pri-let-7 to form the hairpin precursor.

LINE (long interspersed nuclear element) an autonomous retrotransposon whose mobility is dependent on target-primed reverse transcription.

line/strain a distinct interbreeding population, usually maintained in the laboratory, and which is isolated from other such populations, often generated by inbreeding.

lineage two or more groups that share a common ancestor. *See* **monophyletic group**.

lineage paths sequences of parent–child steps linking individuals with length equal to the number of steps.

linear models for microarray data (LIMMA) a software package for the analysis of gene expression microarray data, especially the use of linear models for the assessment of differential expression.

linear predictor the linear combination of covariates and coefficients within a statistical model.

linin fibrils very delicate fibre like strands that may appear in the nuclei between the karyosome and the peripheral chromatin of some protozoa after staining, especially in some trophozoites of *Entamoeba histolytica*.

link function a function that relates the expected value of a probability distribution to the linear predictor within a statistical model. The natural logarithmic link function is typically used within statistical models for count data. For binomial models, where p is the probability of 'success', the logit link is often used, logit $(P) = \ln[P/(1-P)]$.

linkage the proximity of two or more markers (e.g. genes, RFLP markers) on a chromosome; the closer the markers, the lower the probability that they will be separated during DNA repair or replication processes (binary fission in prokaryotes, mitosis or meiosis in eukaryotes), and hence the greater the probability that they will be inherited together.

linkage analysis the identification of the correct structure of complex glycans from a sample (in our case *B. mandrillaris* cysts) using a range of known or predictable candidate structures and after performing different analytical methods such as chemical, enzymatic, and other analytical strategies, as well as MS and NMR spectroscopic methods.

linkage disequilibrium (LD) non-random association of allelic states between two or more genetic loci, not necessarily on the same chromosome, within a population. *See* **Mendelian inheritance**.

linkage map a map of the relative positions of genetic loci on a chromosome, determined on the basis of how often the loci are inherited together.

linkage phase (gametic phase) the information of combinations of DNA alleles in a diploid individual inherited from the mother or father.

Linnaean classification system the taxonomic nomenclature that was developed by the Swedish botanist Carl (Carolus) Linnaeus, in which distinct hierarchical groups, such as phylum, class, order, family, genus and species are defined. This system classifies and names organisms according to scientific principles. Each species has two names; the first indicates the organism's general type (genus), the second names the unique species. *See* **binomial classification**.

lipases hydrolytic enzymes that catalyse the degradation of fats by hydrolysing their ester bonds such as those between glycerol and fatty acid chains.

lipid molecules that prefer to be in chloroform rather than water (hydrophobic, or will not dissolve in water).

lipid droplets intracellular organelles that store neutral lipids, such as triacylglycerols and sterol esters (including cholesterol).

lipid extraction a procedure by which lipids are extracted from a mixture, based on their specific solubility in organic solvents. Usually after addition of solvents, a biphasic organic–water system is obtained and lipids are purified in the organic phase.

lipid profile proportions of each lipid class detected and quantified in a sample. A lipid profile is usually shown as a histogram.

lipid raft membrane microdomains characterized by a specific combination of glycosphingolipids, sterols and protein receptors, more ordered and tightly packed than the surrounding bilayer and involved in important biological functions. In the plasma membrane of parasites, lipid rafts are believed to be important for pathogenic processes.

lipophosphoglycan (LPG) eukaryotic surface molecule made up of lipid and polysaccharide (glycan) connected via a phosphodiester bond; LPG is commonly found in protozoan parasites such as *Leishmania* spp., in which it plays a role in modulating host immune response.

lipopolysaccharide (LPS) endotoxin derived from the cell wall of Gram-negative bacteria which presents inflammatory and mitogenic actions. LPS is composed of O antigen (or O polysaccharide), core oligosaccharide and lipid A.

liquid chromatography (LC) an analytical technique for separating ions or molecules dissolved in a solvent by interaction with a second solid or liquid phase as a result of differences in adsorption, ion-exchange, partitioning or size.

liquid chromatography–mass spectrometry (LC-MS) a method separating (LC) and identifying (MS) compounds in a liquid solution, based on their selective retention in the matrix of specific columns. HPLC allows the separation of lipids solubilized in organic solvents and can be coupled to a mass spectrometer (LC-MS) for chemical identification.

liquid hybridization a method for capturing fragmented DNA molecules via the annealing of biotinylated oligonucleotides to the DNA molecules. The binding of biotin to streptavidin beads allows for washing and removal of uncaptured DNA molecules, and subsequent elution of the captured DNA molecules.

lithotrophic a term used to describe microbial metabolism that uses inorganic substrates as electron donors.

live, attenuated vaccine a vaccine made from parasitic mutants that have reduced virulence and lacks the capacity to cause a disease.

liver cycle the cycle of the parasite that occurs in the liver where sporozoites will mature into merozoites and be released into the bloodstream for the blood cycle.

liver non-parenchymal cells cells of the liver, excluding hepatocytes. These include Kupffer cells, hepatic stellate cells, liver sinusoidal endothelial cells, and bile duct cells.

liver sinusoidal endothelial cells specialized cells that make up endothelial lining of the smallest blood vessels of the liver, the sinusoids.

liver sinusoids sinusoidal blood vessels in the liver.

liver-stage vaccines vaccines in this group aim to prevent liver infection by sporozoites and/or inhibit development of the parasite in the liver. These vaccines are 'preventative' in that they prevent the vaccinated individual from contracting disease.

lobular flogosis pulmonary inflammation involving the lobule (i.e. the morpho-functional lung unit).

local adaptation an adaptation in response to spatially heterogeneous selective pressures. Locally adapted organisms gain higher relative fitness in their native environment compared with foreign environments. Two frequently used approaches for studying local adaptation are common-garden experiments, where organisms are placed in a single environment, and one or a few aspects of the environment are manipulated and reciprocal transplant, where organisms are moved between habitats. The former approach is often used for studying host–parasite systems, but it fails to capture adaptation to the environment as a whole.

local contraction units multiprotein complexes that are similar to muscle sarcomeres and are used by cells to measure substrate rigidity.

localize determination of the original position (locus) of a gene or other marker on a chromosome.

loci (singular locus) chromosomal locations of genes or genetic markers.

locked nucleic acid (LNA) a synthetic RNA nucleotide in which the ribose group is modified with an extra bridge connecting the 2′ oxygen and the 4′ carbon. Incorporating LNAs into DNA or RNA oligonucleotides increases the specificity and sensitivity of detection in a number of assays.

locus expansion the duplication of genes in a given locus, usually in a tandem array.

locus-specific databases collections of DNA variants that have been reported in disease-associated genes.

LOD score or logarithm of odds (LOD) for linkage the common (decimal) logarithm of the likelihood ratio for the odds in favour of linkage over no linkage. It provides a statistical test of the null hypothesis of free recombination versus the alternative hypothesis of linkage. Allele-sharing LOD scores are derived from non-parametric linkage analysis.

LOD value (logarithmic (base 10) odd) an estimate of the strength of the association between a genotype (QTL) and a trait of interest compared with null hypothesis of no QTLs by phenotype association. A LOD score of 3 (1 in 1000) is often used as the significance level.

logic evaluators or logic gates devices that carry out logical operations using one or more signal inputs to produce a single logical output. For example, a four-input logic gate carries out logical computation using four inputs.

logistic regression a regression method for a binary outcome variable (disease status), which is transformed to an odds ratio, to test if the genotype can predict disease status in case–control data.

loiasis the disease caused by infection with the parasite *Loa loa.* The disease is also known as Calabar swelling or tropical eyeworm because adult worms migrate across the eye. Tabanid flies of the genus *Chrysops* transmit the infection.

London Declaration on Neglected Tropical diseases (LDNTDs) a high-level advocacy meeting that took place in London in January 2012 that brought together international donors, the pharmaceutical sector, and implementation agencies to pledge support for development, donation, and distribution of drugs to combat NTDs. A year later, the London Centre for Neglected Tropical Diseases was launched bringing together a new local partnership with Imperial College London (ICL), London School of Hygiene and Tropical Medicine (LSHTM), and the Natural History Museum (NHM) working within the UK Coalition against Neglected Tropical Diseases (http://ntd-coalition.blogspot.co.uk/).

long terminal repeats (LTRs) stretches of DNA sequences that are identical and repeat in hundreds to thousands of copies. LTRs were founds at the end of retrotransposons and act as insertion sites for viruses to insert their genetic material into the host genome.

longitudinal cords in nematodes, four cords (lines) that extend from the anterior to the posterior end (one dorsal, one ventral, and two lateral) which enclose the longitudinal nerves and, in the lateral cords, the longitudinal excretory tubules.

longitudinal data measurements or observations made repeatedly on the same unit (repeated measures) through time, e.g. multiple hookworm egg counts made from the same host at different times.

long-lasting insecticidal nets (LLINs) mosquito nets made from strong fibres impregnated with a quick-acting pyrethroid insecticide, which irritates or kills mosquitoes on contact, preventing them from finding openings for a period of up to 3 years.

long-lived plasma cells long-lived antibody-secreting B cells, also part of the memory B cell compartment.

long-range restriction mapping restriction enzymes are proteins that cut DNA at precise locations. Restriction maps depict the chromosomal positions of restriction-enzyme cutting sites. These are used as biochemical 'signposts', or markers of specific areas along the chromosomes. The map will detail the positions where the DNA molecule is cut by particular restriction enzymes.

long-term adaptive response the long-term survival of helminth parasites indicates their success in outmanoeuvring the immune system via achieving some form of balanced parasitism in which transmission is maintained and acute morbidity avoided. Helminths achieve this longevity despite the host immune response and their apparent inability to undergo antigenic variation. In some cases strong evidence exists for helminth-induced suppression of the host immune response. This probably increases worm survival but may also reduce the pathological affect of infection on the host tissues.

loop-mediated isothermal amplification (LAMP) a molecular method that amplifies DNA by using four specific primers and *Bst* DNA polymerase in less than 1 hour under isothermal conditions. The test has high specificity and is simple and easy to perform.

louping ill (LI, synonym **ovine encephalomyelitis, infectious encephalomyelitis of sheep, trembling-ill)** a tick-transmitted viral disease whose occurrence is closely related to the distribution of the primary vector, the sheep tick *Ixodes ricinus*, which requires an environment with a high relative humidity. LI is a disease of all ages of sheep and can infect a range of hosts (important disease of grouse). All stages of ticks, larva, nymph, and adult, acquire the virus by feeding on a viraemic host. LI can be controlled in endemic areas via the use of vaccination. Two doses of vaccine with an interval of 2 to 8 weeks between injections are recommended to achieve optimal protection to natural infection. Vaccination of pregnant ewes during the last trimester is advocated to ensure that lambs receive maximal levels of passively acquired antibodies and are protected during the initial critical months of life.

low-endemicity areas (LEAs) areas where the prevalence is $<10\%$ and the number of *Schistosoma* spp. eggs per gram of faeces (e.p.g.) is <100 (1–99) e.p.g. Carriers are asymptomatic/oligosymptomatic.

low-transmission areas (LTAs) areas that present a focal distribution of susceptible *Biomphalaria* spp. snails and schistosomiasis prevalence in the examined population of $\leq 25\%$, in addition to the predominance of a low intensity of infection ($\geq 75\%$) and low morbidity in that most individuals are asymptomatic.

LTR (long terminal repeat) a long sequence (usually 200 to 600 bp in length) repeated at each end of a retroviral DNA or retrotransposon and that is necessary for reverse transcription, integration and transcription. The LTRs include promoter and termination sequences for RNA polymerase II.

lumen the central cavity of an organ such as the lumen of the intestine. The space within a tubular organ such as the intestine.

lungworm a term generally used for a variety of different groups of nematodes, some of which use the lower air passages and the lung parenchyma as the final habitat, such as nematodes within the Metastrongyloidea superfamily. Others nematode species, such as Ascarid larvae just migrate through the animals' lungs or respiratory tracts, causing various degrees of pathological damage according to the nature and intensity of the host–parasite interactions.

Lyme borreliosis a tick-borne disease caused by bacteria of the genus *Borrelia*.

lymph the tissue fluid which drains into and through the lymphatic system.

lymph node small organs distributed throughout the body, which contain large numbers of naive T and B cells. Antigen presenting cells migrate from peripheral tissues to lymph nodes upon pathogen uptake in order to present the antigen to and activate T cells.

lymphadenopathy enlarged lymph nodes.

lymphatic filariasis a parasitic disease caused by microscopic, thread-like worms. The adult worms live in the human lymph system. Infection spreads from person to person by mosquito bites. When a mosquito bites an individual who has lymphatic filariasis, microscopic worms circulating in the person's blood enter and infect the mosquito. The microscopic worms pass from the mosquito through the skin, and travel to the lymph vessels. In the lymph vessels they grow into adults and lives for about 5–7 years. The adult worms mate and release millions of microscopic worms, called microfilariae, into the blood.

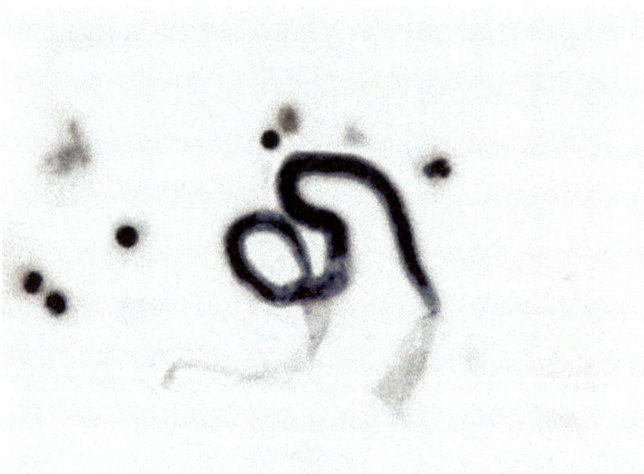

lymphocyte a type of white blood cell that is part of the immune system. There are two main types of lymphocytes: B cells and T cells. The B cells produce antibodies that are used to attack invading bacteria, viruses, and toxins. The T cells destroy the body's own cells that have themselves been taken over by viruses or become cancerous.

lymphocyte-activation gene 3 (LAG3) a member of the Ig superfamily, regulates expansion of activated primary T cells and development of the memory T cell pool. LAG3 is also an inhibitory molecule.

lymphokine-activated killer cells (LAK) soluble substances secreted by lymphocytes, which have a variety of effects on lymphocytes and other cell types.

lymphoproliferative responses cell-mediated proliferation in response to a specific antigen. It can be a measure of the cellular immunity of an individual to that antigen.

lymphotoxin (also called TNF-β) a T-cell derived cytokine which is cytotoxic for certain tumour cells and also has immunoregulatory functions.

lysosomes membrane-bound cellular organelles, containing several classes of hydrolytic enzymes involved in the digestion of phagocytosed material and the degradation of a large repertoire of cellular substrates.

lysozyme antibacterial enzymes present in phagocytic cell granules, tears and saliva, which digests peptidoglycans in bacterial cell walls.

M cells specialized epithelial cells of the follicle-associated epithelium of the gastrointestinal tract that are involved in the rapid uptake and presentation of particular antigens and microorganisms to immune cells of the lymphoid follicle, thereby inducing an effective immune response.

machine learning a field of computer science in which existing knowledge is used to train a program to recognize or to classify new instances within the same domain.

Macracanthorhynchus **(thorny-headed worm)** a worm belonging to the phylum Acanthocephala. The name thorny-headed worms comes from the fact that these worms have a spiny proboscis used as a holdfast in the gut. These vaguely resemble roundworms but in some ways are more similar to tapeworms as they lack a digestive tract. The body is flattened and has transverse grooves. The adult males are 10 cm and the females are up to 60 cm long. Hosts include pigs and wild boars (definitive) and dung beetles (intermediate). Life cycle starts by the beetle ingesting an egg. The egg hatches and the larva develops to a cystacanth stage, which encysts in the body cavity of the insect. When the pig eats the infected beetle, the larva is digested out and attaches to the wall of the small intestine and grows to the adult stage. Antemortem diagnosis is rare because eggs do not float in salt solutions. Faecal sedimentation test can be used. The egg has four shells, and is dark brown.

Macrocyclic lactones (MLs) antiparasitic (anthelmintic) drugs used as important broad spectrum anthelmintics to control parasitic nematodes. They are considered endectocides as their broad-spectrum of activity includes internal nematodes and external arthropod parasites, such as lice, certain Dipterans, mange mites, *Demodex*, and sometimes fleas. MLs comprise two subgroups, the avermectins including ivermectin (IVM), abamectin (ABA), doramectin, eprinomectin and selamectin and the milbemycins, including moxidectin (MOX) and milbemycin oxime. MLs have a dual mechanism of action. First, they potentiate the action of the inhibitory neurotransmitter GABA. This excess binding of GABA to its chloride-channel receptor results in a rapid influx of chloride ions, membrane hyperpolarization, and inhibition of neurotransmission in the worm. Second, they stimulate the permanent opening of ligand-gated chloride channels, especially glutamate-gated channels in the membranes of neurones in worms. This leads to the inhibition of nerves responsible for the control of muscles of nematodes. As a result, the worms show flaccid paralysis and cannot persist any longer in their environment within the host. Ivermectin (but not milbemycin) has an abnormal safety profile when administered to collie breed dogs. Signs of toxicosis in Collie breeds include mydriasis, depression, ataxia, and coma.

macrolides the macrolide antibiotic erythromycin and its analogue azithromycin have activity against *Toxoplasma gondii*. Also, spriromycin has activity against *T. gondii* and *Isospora* spp.

macronucleus in some ciliates (e.g. *Chilodenella*), the large kidney-shaped nucleus. Its function is not clearly understood but it is usually in association with a 'micronucleus'.

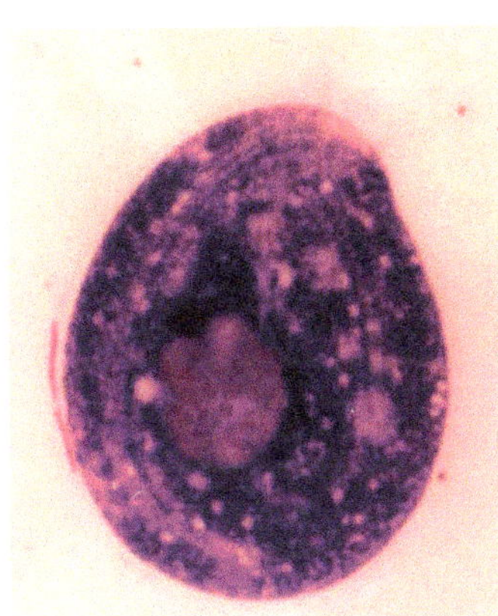

macroparasite macroscopically visible parasite, which as an adult produces eggs or other transmission stages that are released from the final host for further development and maturation. This includes helminths and arthropods.

macrophage large phagocytic innate immune cell of the mononuclear series found within tissues. Main functions are phagocytosis, and antigen presentation to T cells.

macropinocytic endocytosis a process by which cells internalize molecules using a non-selective uptake pattern.

macropinocytosis a process that involves the uptake of molecules into cells in a clathrin-independent manner.

macrorestriction map map depicting the order of and distance between sites at which restriction enzymes cleave chromosomes.

macroscopic pertaining to the state of being of large size; visible to the unaided eye.

maggot the soft-bodied, worm-like larva of a holometabolous insect, e.g. beetle grubs, fly larvae, especially ones living in decaying flesh or tissue debris.

magnetic activated cell sorting (MACS) selection of a population based on surface marker specific antibody beads retained by a magnet through a column and flushed to recuperate the population of interest.

magnetic resonance imaging (MRI) medical imaging using stimulation of water molecules by a powerful magnetic field to obtain a map of these molecules in the body.

MAINE-seq (MNase-assisted isolation of nucleosomes followed by next-generation sequencing) a procedure that relies on the enzymatic activity of micrococcal nuclease (MNase) to digest the linker DNA between nucleosomes in native chromatin, resulting in mono-nucleosomal DNA fragments (because the DNA protected by the nucleosome remains intact).

major allergen denotes an antigen against which more than 50% of a selected group of patients (allergic to the allergen source in question) react.

major histocompatibility complex (MHC) a group of genes that control several aspects of the immune response. They code for markers located on the surface of all body cells and are recognized by the body as 'self' (belonging to the body). These genes define a person's tissue type and are used to determine whether a transplant would be compatible.

major histocompatibility complex class II (MHC II) genes MHC II genes encode peptide transporters expressed by antigen-presenting cells (macrophages, dendritic cells, B cells) that facilitate recognition of antigenic peptides by T cells; in humans, the MHC II genes are *HLA-DR*, *HLA-DP*, and *HLA-DQ*. Expression of MHC II genes is up-regulated upon exposure to γ interferon (IFN-γ) and inhibited by exposure to interleukin 10 (IL-10).

major surface proteases (MSP) MSP is a zinc metalloprotease of kinetoplastids that was first described in *Leishmania* and called GP63. It is the most abundant surface glycoprotein of *Leishmania* with a molecular weight of 63 kDa. The three subspecies of *T. brucei* contain genes for three different MSPs (MSP-A, MSP-B, and MSP-C), which are very similar to each other and all have approximately 30% sequence homology to GP63.

malaria one of the world's most common infectious diseases, caused by a parasite that is transmitted to humans by a female mosquito's bite. The discovery of this parasite in mosquitoes earned the British scientist Ronald Ross the Nobel Prize in Physiology or Medicine in 1902. In 1907, Alphonse Laveran received the prize for his findings that the parasite was present in human blood.

malaria control the reduction of the malaria disease burden to a level at which it is no longer a public health problem.

malaria elimination defined by the WHO as the reduction to zero of the incidence of infection caused by human malaria parasites in a defined geographical area as a result of deliberate efforts. Continued measures to prevent re-establishment of transmission are required.

malaria parasite metabolic pathways (MPMP) a manually constructed set of pathways for *Plasmodium falciparum* based on the presence of at least three to four enzymes that act consecutively in the pathway and have annotations in the genome.

malaria pre-elimination defined as reduction of malaria burden at an annual malaria incidence of less than 1 case per 1000 people at risk. This requires setting up high quality-controlled systems for reaching the elimination phase.

malaria receptivity a term used to describe the degree to which infection occurs. In areas of high receptivity, the presence of abundant anopheline vectors and the existence of other ecological and climatic factors favour parasite transmission.

malaria-specific mortality rate the mortality rate attributed to malaria.

male-killing the selective killing of male embryos.

MALT mucosa-associated lymphoreticular tissue.

mammalian the group of vertebrates that have internal development of the embryo, mammary glands that can produce milk, live-born young with a body covering of hair or fur, a four-chambered heart, a well-developed cerebral cortex and the ability to maintain a constant body temperature a permanent set of teeth.

mammalian haploid cells mammalian cells that harbour only one copy of the genome.

mammillated having many small nipple-like protrusions.

management of anthelmintic resistance proper understanding of the modes of action of the antiparasitic drugs is fundamental to prevent development of parasite resistance in the target pest(s). Using antiparasitic drugs with the same modes of action contributes to this problem by killing the susceptible parasites and leaving only those with resistance to the entire class of the drugs that work through similar mechanisms. Development of resistant parasite can be avoided or delayed by rotating pest control chemicals that work through different modes of action. Alternatively, better results are obtained if two drugs with different mechanisms are combined. Because resistance to insecticides is often due to metabolic inactivation, the situation for ectoparasites is more complex.

mange contagious skin disease caused by mites (e.g. *Sarcoptes* spp., *Chorioptes* spp., *Psoroptes* spp.). clinical signs and location of skin lesions vary depending on the aetiological agent and include pruritus, papules, crusts, excoriation, and lichenification, which may complicate with secondary bacterial infections.

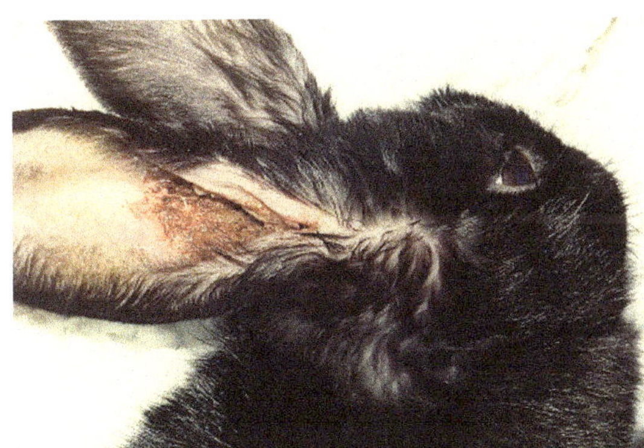

mange mites the parasitic mites of the families sarcoptidae and psoroptidae, known as 'mange mites', generally give rise to well-defined dermatoses. Mange mites live on the surface of the skin (*Chorioptes* and *Psoroptes*) or may be superficial burrowers (*Sarcoptes*). the entire life cycle takes place on the host and comprises the egg, the larval stage, two nymphal stages (protonymph and tritonymph) and the adult mite. under optimal conditions the whole life cycle may be as short as 11 days (*Psoroptes ovis*); however, on average it may take about 2–3 weeks.

Manhattan plot a scatterplot used to present GWAS results. In a Manhattan plot, DNA co-ordinates of SNPs are displayed on the x axis, whereas the negative logarithm of the *P*-value of an association with a specific trait or disease is displayed on the y-axis. The name of the plot is derived from its similarity with the Manhattan skyline, a profile of buildings with a range of heights.

manipulation hypothesis hypothesis that claims that many species of parasites evolved an ability to increase the probability of transmission from their intermediate to definitive hosts by modifying the phenotype, typically the behaviour, of the infected host.

manipulative pathogens parasites that induce phenotypic changes (morphology, behaviour, physiology, and so on) in their hosts that increase the probability of their transmission from one host to another and/or ensure that propagules will be released in an appropriate location.

mannose binding protein (mannose binding lectin) member of the collectin family of calcium-dependent lectins, and an acute phase protein. it functions as a stimulator of the classical pathway of complement activation, and as an opsonin for phagocytosis by binding to mannose, a sugar residue usually found in an exposed form only on the surface of microorganisms.

mannose receptor (MR) a C-type lectin carbohydrate binding protein that serves as a broad-range pattern recognition receptor via detection of mannose on the surface of pathogens. MR is predominantly expressed in macrophages but is also found on neutrophils and dendritic cells.

mannose-binding lectin 2 (MBL-2) belongs to the collectin protein family and is an important element in the innate immune system. MBL-2 recognizes mannose and N-acetylglucosamine on many microorganisms and is capable of activating the classical complement pathway.

MAP kinases protein kinases, including c-Jun N-terminal kinase (JNK), p38 and extracellular signal-regulated kinase (ERK), functioning in post TLR signalling by activating the transcription factor AP-1 (activating protein 1).

MAPK (mitogen-activated protein kinases) serine-threonine kinases that respond to extracellular stimuli and regulate various cellular activities, such as gene expression, mitosis, differentiation, proliferation, cell survival and apoptosis.

mapping is the process of making a representative diagram cataloguing the genes and other features of a chromosome and showing their relative locations. Cytogenetic maps are made using photomicrographs of chromosomes stained to reveal structural variations. Genetic maps use the idea of linkage to estimate the relative locations of genes. Physical maps, made using recombinant DNA technology, show the actual physical locations of landmarks along a chromosome.

mapping population the group of related organisms used in constructing a genetic map.

mapping quality $\log_{10}$ transformed measure of the probability that a read is incorrectly mapped multiplied by 10.

marginal effects the average effect of a locus across all other loci and environmental effects.

marginal zone region at the interface between the red pulp and the white pulp of the spleen.

margination leucocyte adhesion to the endothelium of blood vessels in the early phase of an acute inflammatory reaction.

marker is a DNA sequence with a known physical location on a chromosome. Markers can help link an inherited disease with the responsible genes. DNA segments close to each other on a chromosome tend to be inherited together. Markers are used to track the inheritance of a nearby gene that has not yet been identified but whose approximate location is known. The marker itself may be a part of a gene or may have no known function.

Markov chain Monte Carlo (MCMC) a stochastic algorithm central to Bayesian statistical inference which samples parameter values from the posterior probability distribution by combining information from the likelihood of the observed data and the prior probability distribution of the parameters.

marsupial a mammal whose distinguishing features include the birth of young at an early fetal stage of development, and generally, a pouch (marsupium) in which further development of the fetus occurs.

MASCOT a search engine that uses mass spectrometry and tandem mass spectrometry data to identify proteins using DNA and protein sequence databases.

mass cytometry a technology that combines flow cytometry with mass spectrometry, which enables simultaneous detection of dozens of antigens in thousands of single cells using antibodies tagged with isotopically pure metal reporters. Based on current isotope availability, up to ≈100 proteins or protein modifications can be detected simultaneously in thousands of single cells using mass cytometry.

mass drug administration refers to the administration of an antimalarial drug or drug combination to the whole of a population at risk whether or not they are infected, usually with the aim of interrupting transmission. When mass drug administration is given on more than one occasion, it overlaps with intermittent preventative treatment.

mass selection trait selection based on phenotype of a group of parents. Parents of next generation are simply mixed to form the next generation so no measure of specific correlation of parent–offspring is possible. This is in contrast to family selection.

mass spectrometry (MS) analytical technique that measures the mass-to-charge ratio of charged (ionized) molecules. A mass spectrometer is composed of an ion source, a mass analyser and an ion detector. The mass analyser can be a quadrupole (Q), a time-of-flight analyser (TOF) or an ion trap. TOF has a very high mass accuracy. MS data are eventually represented by a spectrum composed of peaks corresponding to mass-to-charge ratios (*m*/*z*). MS peaks are specific for fragments of ionized lipid molecules. MS sensitivity allows the detection of picograms to femtograms of lipids and other classes of molecules.

mass testing and treatment (MTAT) mass testing for infection followed by treating all infected persons in a targeted contiguous area or population, irrespective of whether they are symptomatic. MTAT aims to reduce the size of the infectious reservoir in the targeted area.

mass-to-charge ratio (*m*/*z*) charged particles move in electric and magnetic fields following a path which is strictly dependent on their mass-to-charge ratio. Mass spectroscopy peaks therefore correspond to the *m*/*z* quantities of the charged fragments striking the detector. Neutral fragments cannot be detected. When the fragment is positively charged, this quantity is positive and vice versa. When containing only one charge, the m/z quantity corresponds to the mass of the fragment.

mast cell a tissue cell with abundant granules, which resembles the blood basophil. Both these cell types bear high affinity Fc receptors for immunoglobulin E (IgE), which when cross-linked by IgE and antigen cause degranulation and the release of a number of mediators including histamine and leukotrienes.

mast cell and/or degranulation cross-linking of two FcεRI via the antigen–IgE complexes on the surface of mast cells or basophils elicits a cascade of events, by releasing inflammatory mediators. These can elicit symptoms such as urticaria, angioedema or anaphylaxis, but also symptoms such as in rhinitis, asthma or gastrointestinal reactions.

matching-allele model genetic model of infection assuming that, for each gene conferring resistance to the host, there is a corresponding gene in the parasite with an equal number of alleles. For each allele of the host gene, only one 'matching' allele of the corresponding parasite gene enables infection to occur.

mate-guarding a strategy designed to both preserve access to a mate and to prevent the encroachment of intrasexual rivals. Mate-guarding can vary from simple vigilance to violence and sexual coercion. It is commonly expressed by male coercion of females via mate grasping, forced copulation, sexual harassment or punishment, and typically occurs when the sex ratio is male biased.

maternal immunity the transfer of protective antibodies from a mother to its offspring; in chickens this occurs through the egg yolk that is absorbed by the newborn chick.

mate-switching a change of mating partner. Although mate-switching diverges from social monogamy it does not necessarily diverge from genetic monogamy. Mate-switching can take the form of 'unfaithfulness' or 'divorce': unfaithfulness is when an individual leaves its initial partner occasionally to pair with another individual but then returns to the initial partner, whereas divorce is a definitive change of partner.

mating system the ways in which animal societies are structured in relation to sexual behaviour. The mating systems generally recognized in animals are monogamy, polygamy and promiscuity (or polygynandry).

mating types many protists and fungi exist in two or more mating types, such that a cell can only mate with members of a different mating type. Unlike male and female sexes, different mating types are usually not distinguishable by morphology.

matrix-assisted laser desorption/ionization time-of-flight mass spectrometry (MALDI-TOF ms) a method in which gas-phase ions are formed by the interaction of a pulsed laser with a solid matrix, accelerated to a constant kinetic energy and their m/z determined by measuring the time they take to reach a detector.

mature oocysts containing sporozoites, or rarely sporocysts freed from oocysts, become the infective stage for an intermediate host.

Maurer's clefts one of two main features of the exomembrane system. A single membrane flattened cisternae found adjacent to the membrane of infected RBCs. Maurer's clefts are characterized by a translucent lumen and an electron-dense coat of variable thickness. Maurer's clefts are an intermediate compartment for protein export from the parasite to the host cell membrane.

maxicircle DNA the approximately 22 kb component of kinetoplast DNA that encodes mRNA and ribosomal RNA genes; there are tens of copies of these DNAs, and the gene-encoding region is homogeneous in sequence.

maximum likelihood (ML) estimation a framework for estimating parameters of a statistical model by conditioning the probability of the observed data (the likelihood) on unknown parameter values using a probability distribution.

maximum likelihood estimators estimates of unknown parameters obtained by maximizing the likelihood for the observed data given a statistical model.

MBL (mannose-binding lectin) it binds to mannan-derived molecules present in various microbial aggressors. It participates in the lectin pathway of complement activation.

MBOAT (membrane bound O-acyl transferase) multipass transmembrane enzymes that have two common active site domains – a hydrophobic region containing a conserved histidine, and a hydrophilic region containing a conserved asparagine; MBOATs acylate cholesterol/diacylglycerol, amino acids, and lysophopholipids.

McDonald–Kreitman (MK) test a test of the neutral theory of molecular evolution for protein coding regions. This test compares the ratio of non-synonymous to synonymous changes within species with the ratio of non-synonymous to synonymous changes between species.

mechanical vector an object, either animate or inanimate, which transports a parasite to its host. this differs from a biological vector in that the biological vector contributes actively to the growth and/or development of the parasite.

mechanosensation neuronal activity in response to touch or mechanical pressure.

mechanotransduction the process by which physical properties of the cellular environment are converted into changes in cell signalling and cell state.

meconium ileus obstruction of the gut that usually develops *in utero* in the ileum of the small intestine and that is highly suggestive of cystic fibrosis.

mediator a multiprotein complex that functions as a key component of the RNA polymerase II transcriptional machinery. It functions as a bridge between the enzyme and regulatory proteins and is involved in both activation and repression of messenger RNA production.

Medicines for Malaria Venture (MMV) product development partnership organization created in 1999 with the mission to discover, develop, and deliver safe, efficient, and inexpensive new antimalarials in endemic countries such as those in Africa, Asia, and South America. Currently, 70 staff members work in tight collaboration with more than 300 academic and industrial partners around the world to achieve such a mission. For several years now, and in conjunction with many members of the antimalarial community, the objective is to fulfil the eradication agenda.

medulla inner (central) region of an organ.

megabase (Mb) unit of length for DNA fragments equal to 1 million nucleotides and roughly equal to 1 cM. *See* **centimorgan**.

megacolon enlargement of the colon.

mega-oesophagus enlargement of the oesophagus.

megakaryocyte a bone marrow precursor of platelets.

Meglumine antimoniate this compound selectively inhibits the leishmanial enzymes required for glycolytic and fatty acid oxidation. It is used alone or in combination with allopurinol to treat leishamaniosis. On their own, antimony compounds are rarely successful in completely eradicating the parasite in infected dogs. In contrast, combinational therapy seems to be the most recommended option for treating dogs with confirmed clinical illness because it can yield a longer period of clinical remission than treatment with either drug alone. Extreme caution should be exercised in patients with cardiac, hepatic or renal insufficiency.

***Megninia ginglymura* (feather mite)** spends its entire life cycle on the host's plumage, staying mainly on feather surfaces, but sometimes invading the quills. Affected birds often show no clinical signs, but infestations have frequently been associated with significant damage in commercial poultry farms.

Mehliś gland belongs to the reproductive system of flatworms and represents a large unicellular gland surrounding the ootype. In schistosomes it can be part of the ootype itself. It has been suggested that Mehliś gland secretions are likely to contribute to egg formation and/or facilitate egg passage to the uterus.

meiosis the process of division of the nuclear material of a cell that, by the final division, yields a gamete with a haploid number of chromosomes. The central event in heredity. Type of cell division in which the number of chromosomes in the daughter cells is halved. Thus, they are in the haploid state. Two successive divisions occur in the process, giving four daughter cells. The first division takes place in four stages: prophase, metaphase, anaphase and telophase. The second division has three stages: metaphase, anaphase and telophase. In animals, meiosis occurs in the formation of gametes, e.g. eggs and sperm. *See* **mitosis**.

meiotic drive a distortion in the transmission ratio of a particular gene or chromosomal rearrangement at meiosis.

meiotic recombination a process that takes place during meiosis and generates a haploid product with a genotype that differs from both haploid genotypes that constituted the meiotic diploid cell.

meiotic sex chromosome inactivation (MSCI) a process leading to the transcriptional silencing of the X and Y chromosomes during the meiotic phase of spermatogenesis.

melanization part of the innate immune response of insects and other arthropods in the defence against pathogens. Pathogens are coated by a layer of the black melanin pigment and killed. Melanization is also a mechanism of wound healing, covering injuries with melanin.

melanization reaction a branch of insect immunity that involves the rapid synthesis and deposition of a brown–black pigment, melanin, at the site of infection and injury. Melanin possesses cytotoxic activity towards microorganisms, restrains the invasion of microbial pathogens into the body cavity of the host and assists in wound healing. Melanization is triggered after the recognition of microbial compounds, such as β-1,3-glucans, lipopolysaccharides and peptidoglycans.

melarsomine an organic arsenical compound used to treat adult heart-worms and is used solely in dogs. Melarsomine denatures proteins/enzymes by binding to the sulfhydryl groups of cysteine residues. Mild localized oedema may occur following injection, liver toxicity, nephrotoxicity, thromboembolic pneumonia. Overdose may result in distress, restlessness, pawing, salivation, vomiting, tachycardia, tachypnea, dyspnoea, abdominal pain, hind limb weakness, and recumbency. Severe overdose cases terminate in circulatory collapse, coma, and death. Compared to thiacetarsamide, it is more efficacious and less likely to be associated with hepatotoxicity. It is administered by deep IM injection in the lumbar epaxial musculature in dogs. It should not be given intravenously or subcutaneously.

membrane semipermeable biological barrier consisting of lipids, proteins, and small amounts of carbohydrate. Membranes control the flow of chemical substances (e.g. nutrients, protons, ions, and wastes) in and out of cells or cellular compartments. They also serve as structural supports for systems of membrane-embedded proteins that mediate important biological processes such as photosynthesis and cellular respiration.

membrane attack complex (MAC) complex of complement components C5b–C9 which inserts as a pore into the membrane of a target cell leading to its lysis or apoptosis.

membrane blebs spherical membrane bulges formed by the combination of internal hydrostatic pressure and weak points of either the actin cortex or its linkage to the plasma membrane. Blebs rapidly fill with cytoplasm but initially lack F-actin. Bleb growth is slowed by actin polymerization under the membrane.

membrane-feeding assay an experimental method for testing the efficacy of antibodies generated against both parasite and mosquito antigens in blocking transmission of the pathogen through the insect vector. To conduct the assay, infected blood is mixed with antibodies and placed in a special water-jacketed apparatus from which mosquitoes feed on the blood–antibody mixture through a thin membrane.

memory (in immunology) a characteristic of the acquired immune response of lymphocytes whereby a second encounter with a given antigen produces a secondary immune response; faster, greater and longer lasting than the primary immune response.

memory cells T and B lymphocytes formed as a result of exposure to antigen (primary immune response). They have the ability to mount a secondary immune response which is enhanced to antigen as compared to lymphocytes that had not previously encountered antigen.

memory T cells are long-lived, capable of antigen-independent maintenance and can mediate a robust recall response upon antigen re-encounter.

Mendelian and complex diseases Mendelian disorders are the result of alterations in one or several genes that can be traced in family pedigrees, complex diseases arise as a consequence of the combined effect of multiple genetic determinants, which may vary between individuals, and environmental factors. The term ‘complex disease’ is also used to refer to those diseases where the phenotype cannot be easily predicted from the genotype. This is proposed to be due to the interaction between genes (epistasis), modulation by environmental factors or stochastic processes, or epigenetic changes. It is important to note that some of these factors, such as incomplete penetrance and variable expressivity, can also make it difficult to predict the phenotype of Mendelian disorders based solely on genotype.

Mendelian inheritance a hereditary process where genetic traits are passed from parents to offspring and are explained in terms of chromosomes separating, independent assortment of genes and the homologous exchange of segments of DNA. there are three modes of Mendelian inheritance: autosomal dominant, autosomal recessive and x-linked inheritance. named after Gregor Mendel, who first studied and recognized the existence of genes and this method of inheritance by breeding experiments with different varieties of peas. *See* **recessive gene**.

Mendelian loci genetic loci that have alleles with discrete effects on the phenotype and that follow Mendel's laws of segregation and independent assortment.

meningitis infection or inflammation of the meninges, the membranes that enclose the brain and the spinal cord.

meningoencephalitic stage second stage of African trypanosomiasis in which parasites cross the blood–brain barrier and are found in the central nervous system. Symptoms include sleep disturbances and neurological manifestations.

merogony asexual reproduction, commonly by multiple fission of the parasite nucleus followed by simultaneous cellular division producing daughter cells (merozoites); asexual schizogony. *See* **asexual reproduction**.

meront asexual multinucleate stage that forms during merogony (an asexual schizont). *See* **schizont**.

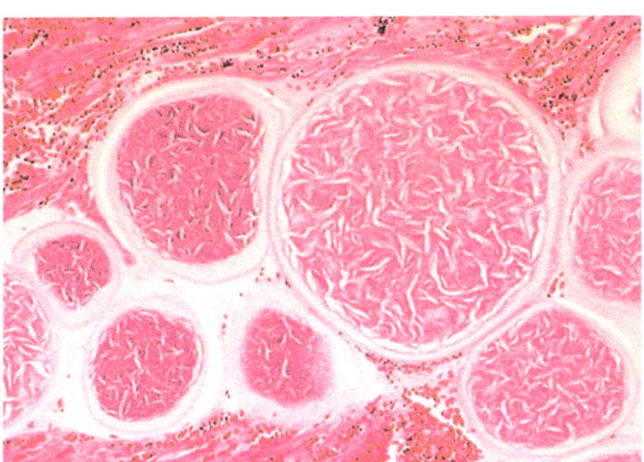

MEROPS a curated database that classifies all known peptidases (http://merops.sanger.ac.uk). MEROPS also includes information on peptidase inhibitors, substrates, specificity and extensive summaries and links to the literature.

merosomes merozoite-filled vesicles that bud-off infected hepatic cells into the bloodstream.

merozoite surface protein 1, 19-kDa fragment (MSP1-19) 19-kDa fragment of the 200-kDa *Plasmodium falciparum* merozoite protein that coats the surface of the parasite and is processed and cleaved into several fragments before erythrocyte invasion.

merozoites merozoites are formed by asexual reproduction in the life cycle of *Plasmodium*. They use specialized apical organelles to invade host erythrocytes and initiate intraerythrocytic development.

Meselson effect a process by which asexual diploid and tetraploid organisms accumulate mutations in the alleles of the different sets of chromosomes in the cell in the absence of sexual or ameiotic recombination.

mesenteric veins are those found in the mesentery, which drain blood away from the intestine.

mesenteric venules a small blood vessel that receives blood from the capillaries and transports it to the mesenteric veins.

mesentery a sheet of tissue or membrane that enfolds and supports an internal organ by attaching it to the body wall or another organ. adults of schistosomes live and female worms pass their eggs in the mesenteric venules associated with the intestine and urinary bladder.

Mesocestoides cestodes, which belong to tapeworms. The life cycle of *Mesocestoides* sp. has yet to be fully clarified, but it is known that two intermediate hosts are required for its completion, and carnivores and birds are considered to be definitive hosts. Coprophagous arthropods such as Oribatid mites may serve as the first intermediate host. In the first intermediate host, oncospheres (first larval stage) develop into cysticercoids (second larval stage). A tetrathyridium (third larval stage) is formed when an infected mite is eaten by a second intermediate host, such as rodents, amphibians, reptiles, and birds. The final adult form develops when the second intermediate host is ingested by the definitive host. The adult form of *Mesocestoides* sp. develops within the intestines of the definitive host, such as dogs, cats, and wild carnivores, approximately 2–3 weeks after ingestion of the second intermediate host.

messenger RNA (mRNA) RNA molecule that specifies the amino acid sequence of a protein. It is the intermediary molecule between DNA and ribosomes. mRNA takes encoded specifications from the cell's DNA, processes the message and takes it to the ribosomes, where the amino acids are assembled into polypeptides, which are then folded into proteins. *See* **genetic code**, **transcription**.

meta-analysis statistical synthesis of the results from several studies that address the same question.

metabiosis the dependence of one organism upon another for its existence;

metabolic control analysis a method for calculating the degree of sensitivity of the flux through a pathway to variations in the concentrations of each of its enzymes. Metabolic control analysis requires detailed knowledge of the kinetic parameters for each reaction.

metabolic engineering optimizing genetic and regulatory processes within cells to increase the cells' production of a certain substance.

metabolic flux the rate of turnover of metabolite molecules through a metabolic pathway.

metabolic flux analysis (MFA) a method for measuring all the metabolic fluxes of an organism's central metabolism; ^{13}C-labelled substrate is taken up by an organism, and the distribution of ^{13}C throughout the metabolic network enables the quantification of labelled metabolite pools.

metabolic reconstruction a method for predicting an organism's complement of metabolic enzymes and the network of reactions that it supports; the prediction is derived from a combination of genome sequence analysis, experimental data and reasoning based on network structure.

metabolic resistance partial or complete resistance to the harmful effects of insecticides or other alien compounds provided by detoxifying enzymes, of which cytochrome P450 monooxygenases are a well-known superfamily that is important in both insects and humans.

metabolism a collection of all biochemical reactions (anabolic and catabolic) that an organism uses to obtain the energy and materials it needs to sustain life. An organism uses energy and common biochemical intermediates released from the breakdown of nutrients to drive the synthesis of biological molecules.

metabolites small molecules (<500 Da) that are the substrates, intermediates, and products of enzyme-catalysed metabolic reactions.

metabolome the complete profile of small-molecule metabolites, such as amino acids, nucleotides, antioxidants, organic acids, vitamins, hormones, drugs and food components that are found within a cell, tissue, organ or entire organism.

metabolomics type of global molecular analysis that involves identifying and quantifying the metabolome – all metabolites present in a cell at a given time.

metacentric chromosome a chromosome with the centromere in the middle.

metacercaria (ae) the infective stage of a fluke trematodes enclosed in a protective cyst that resists adverse environmental conditions. This stage develops from the cercaria and is infective for the definitive host.

metacestode immature tapeworm which develops from the hexacanth embryo (oncosphere) and grows in the intermediate host (mammal), but one not yet sexually mature. Cattle may harbour *cysticercus bovis*, the metacestode of *Taenia saginata* (the beef tapeworm of humans) in their muscles. likewise, *Taenia solium* (the pork tapeworm of humans) occurs in pigs as *C. cellulosae*.

metacyc database of metabolic pathways developed by the biocyc consortium.

metacyclic form the *T. brucei* developmental stage that travels from the salivary gland of the tsetse fly into the mammalian host; cellular division is arrested in preparation for transmission to the mammalian host.

metacyclic promastigotes the virulent form of *Leishmania* promastigotes that arise following metacyclogenesis, a developmental process that occurs as the parasites move progressively from the midgut to the foregut and proboscis of the sand fly insect vector.

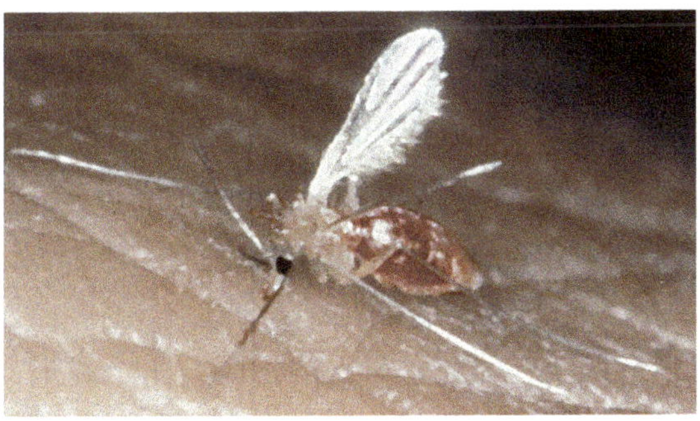

metacyclogenesis a process by which trypanosomes transform into mammalian infective, metacyclic trypomastigotes.

metadata data that describe specific characteristics and usage aspects (e.g. what data are about, when and how data were created, who can access the data, and available formats) of raw data generated from different analyses.

metagenome genetic material recovered directly from environmental samples.

metagenome/metatranscriptome the total genomic DNA or RNA transcripts of all organisms within a community.

metagenomics the study of a collection of genetic material (genomes) from a mixed community of organisms. Metagenomics usually refers to the study of microbial communities.

metalloprotease a proteolytic enzyme that is defined by the presence of an essential active-site metal ion, which is most often zinc.

metallothioneins a group of low-molecular weight proteins with high-affinity binding of metals thereby reducing their toxicity; occur in many animal and plant taxa; might be induced by factors other than metals as well.

metamorphosis of arthropods most of arthropods hatch from eggs in a form differs markedly from those of adult and must undergo morphological and developmental changes (i.e. metamorphosis) to reach adult stage. Such arthropods are known as *metabola*, e.g. *Cyclops*. Others emerge from eggs resembling adults except for minor differences are known as *ametabola*. *Metabola* are further divided into two groups: *holometabola* (complete metamorphosis), where each stage differs markedly from other in morphology, feeding habits and types of food, e.g. fly and fleas; *hemimetabola* (incomplete metamorphosis), in which no clear changes in feeding neither habits nor morphology between different stages can be detected, e.g. ticks, lice.

metaomics high-throughput, global analysis of DNA, RNA, proteins, or metabolites isolated directly from a community of organisms living in a particular environment.

metaphase a stage during the process of cell division (mitosis or meiosis). Usually, individual chromosomes cannot be observed in the cell nucleus. However, during metaphase of mitosis or meiosis the chromosomes condense and become distinguishable as they align in the centre of the dividing cell. Metaphase chromosomes are used during the karyotyping procedure that is used to look for chromosomal abnormalities.

metapopulation a group of spatially separated populations of the same species that interact at some level. The concept has been mostly applied to species in fragmented habitats.

metapopulation dynamics and relevance for control if populations are genetically diverse and stable for neutral genetic markers, this implies relatively large population sizes. If populations are not genetically stable, this might imply local extinction followed by recolonization from neighbouring locations, or periodic size contractions followed by re-emergence from residual pockets. The term 'sink-source' is used to reflect these dynamics. In other words, the dynamics of any single population is dependent upon migration from neighbouring populations as well as temporal changes within a single population, which might include local extinction and recolonization.

metaproteomics high-throughput, global analysis of proteins isolated directly from a community of organisms living in a particular environment. Metaproteomics can reveal which genes are actively translated into functional proteins by a community.

metastrongyloids the parasitic nematodes belonging to the Metastrongyloidea superfamily. These parasites are characterized by life cycles typically requiring an intermediate mollusc host for developing to the infective stage. The vast majority of metastrongyloids affect the lungs and associated sites of animals, with few exceptions.

metatranscriptome transcriptome of a group of interacting organisms or species.

metatranscriptomics high-throughput, global analysis of RNA isolated directly from a community of organisms living in a particular environment. Metatranscriptomics can reveal which genes are actively expressed by a community.

metazoa the subkingdom Metazoa includes all organisms that are not protozoa. Metazoa encompasses two groups of major importance parasitologically: the helminths ('worms') and the arthropods (insects, ticks, and the like).

metazoonosis an infection caused by a pathogen (e.g. some *Leishmania* spp.) that can be transmitted from dogs to humans via an arthropod vector.

methyl-β-cyclodextrin (MβCD) chemical commonly used to chelate cholesterol from membranes, thereby disrupting lipid rafts.

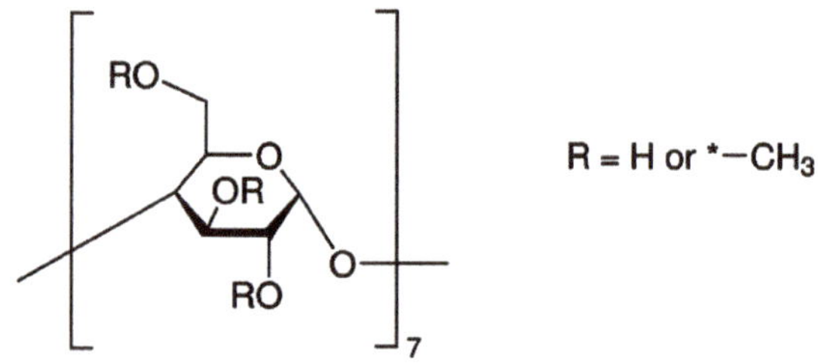

metronidazole 5-nitroimidazole is one of the most widely used antiprotozoal drugs (trade name Flagyl®). It is probably unique in having activity not only against parasitic protozoa but also against anaerobic bacteria. Metronidazole is effective against *Giardia* sp., *Trichomonas vaginalis* and *Entamoeba histolytica* among others. Metronidazole is a prodrug that is well absorbed from the gastrointestinal tract and reaches high tissue concentrations, making it effective against tissue and gastrointestinal protozoa. Once metronidazole enters the protozoan, its nitro group is reduced probably by the low redox enzyme pyruvate:ferredoxin oxidoreductase (PFOR), which results in the generation of toxic intermediates that disrupt the parasite's DNA synthesis (among many other actions). Metronidazole is generally well tolerated, but may cause gastrointestinal upset, has an antibuse effect and at high doses, neurotoxicity manifested by ataxia, tremors, and weakness may become apparent.

MHC (major histocompatibility complex) a genetic region encoding molecules involved in antigen presentation to T-cells. Class I MHC molecules are present on virtually all nucleated cells and are encoded mainly by the H-2K, D and L loci in mice and by HLA-A, B and C in man, whilst class II MHC molecules are expressed on antigen-presenting cells (primarily dendritic cells, macrophages and B-cells) and are encoded by H-2A and E in mice and HLA-DR, DQ and DP in man. Allelic differences are associated with the most intense graft rejection within a species.

MHC (major histocompatibility complex) class I heavy chain MHC class I molecules are trimeric structures composed of a transmembrane heavy chain non-covalently associated with a light chain (β_2-microglobulin) and an 8–11 amino acid peptide. The heavy chain consists of an extracellular part divided into three domains (α1, α2 and α3) followed by a transmembrane domain and a cytoplasmic domain containing conserved serine and tyrosine residues. The distal α1 and α2 domains are highly polymorphic and form a cavity where the 8–11 amino acid peptides binds.

MHC class I molecule a molecule encoded to genes of the MHC which participates in antigen presentation to cytotoxic T ($CD8^+$) cells.

MHC restriction the necessity that T-cells recognize processed antigen only when presented by MHC molecules of the original haplotype associated with T-cell priming.

microaerophilic a term used to describe microbial metabolism that requires oxygen (O_2) concentrations to be very low; for example, microaerophilic ferrous iron Fe^{2+}-oxidizing bacteria function when the O_2 concentration is <50 µm.

microaggregates small (50–250 µm in size) mineral–organic matter complexes in soils that physically protect organic matter from degradation.

microarray analytical technique used to measure the mRNA abundance (gene expression) of thousands of genes in one experiment. the most common type of microarray is a glass slide onto which DNA fragments are chemically attached in an ordered pattern. as fluorescently labelled nucleic acids from a sample are applied to the microarray, they bind the immobilized DNA fragments and generate a fluorescent signal indicating the relative abundance of each nucleic acid in the sample.

microarray feature an element of an array, a micrometre-sized spot where interaction between the printed antigen and serum proteins takes place in the case of antibody profiling microarrays.

microarray probe a single-stranded DNA complementary to a specific sequence. The probes are either synthesized directly on the chip or attached to it. Probes might be either short (e.g. ≈25-mers on Affymetrix oligomer microarrays) or long (e.g. a few hundred nucleotides on cDNA microarrays).

microarray technology a developing technology used to study the expression of many genes at once. it involves placing thousands of gene sequences in known locations on a glass slide called a gene chip. a sample containing DNA or RNA is placed in contact with the gene chip. complementary base pairing between the sample and the gene sequences on the chip produces light that is measured. areas on the chip producing light identify genes that are expressed in the sample.

microbial genetics the study of genes and gene function in bacteria, archaea, and other microorganisms. often used in research in the fields of bioremediation, alternative energy, and disease prevention. *See* **bioremediation**, **biotechnology**.

microbiome the collection of the genomes of the microorganisms in a particular habitat.

microbiota the total collection of microbial organisms within a community, here generally referring to the resident microbes of the human body.

microcolonies groups of progeny cells that are generated from the growth and division of single cells.

microcontact printing also known as micropatterning. A form of surface patterning, usually with fluorescent-labelled extracellular matrix proteins.

microcytosis erythrocyte volume (and size) is smaller than normal, reflected by a reduced mean corpuscular volume (mcv).

microdroplet PCR massively parallel PCR amplification of target sequences in microdroplets. The process involves the preparation of template and PCR mix in picolitre volume and primer droplets, combination of individual template and primer pair droplet, pooling the fused droplets for thermal cycling, and releasing of PCR amplicons for purification and sequencing.

microenvironment the local environment of a single cell within a population and their relative positioning to each other, such as the local crowding of cells, the amount of neighbours, whether cells face empty space on one site and cells on another site, and whether cells are solitary.

microfilariae the larval filarial nematodes are known as microfilariae and are found in the blood or in the tissue beneath the skin. This facilitates transmission by the insect when it comes for its next feed.

microfluidics technology platforms that deal with the behaviour, precise control, and manipulation of fluids that are geometrically constrained to a small, typically submillimetre, scale. A technology in which small fabricated chambers are used to isolate and image single cells.

microgametocyte in relation to malaria, the male gametocyte which produces a number of microgametes. In relation to *Toxoplasma* and *Sarcocystis*, the male stage of the parasite found in the epithelial cells of the intestine. The nucleus undergoes many divisions leading to the formation of microgametes.

microglia specialized macrophages found within the CNS that play a role in innate immune defences, development, and maintenance of the surrounding environment.

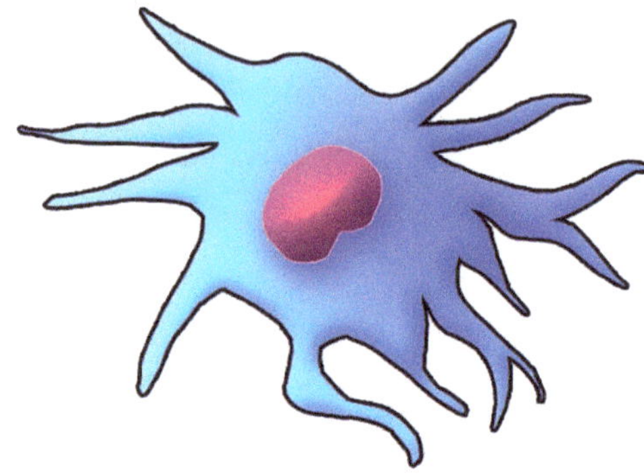

microglial cells resident macrophages of the CNS.

microinjection a technique for introducing a solution of DNA into a cell using a fine microcapillary pipet.

micrometre an instrument for measuring units equal to a micron (0.001 mm).

micromorphology body structures which can only be seen by means of magnification.

micronemes a class of secretory organelles in the apical complex, which release proteins onto the parasite surface during motility or invasion. Secreted proteins include surface-destined proteases in addition to adhesins containing domains that bind to the host-cell surface.

micronuclei chromosome fragments that are not incorporated into the nucleus at cell division.

micronucleus in ciliates, a small nucleus closely associated with the macronucleus. The macronucleus and micronucleus perform differently than do the nuclei in most other protozoa during division.

microorganism any unicellular prokaryotic or eukaryotic organism, sometimes called a microbe.

microparasite a small-sized parasite that replicates and grows into an adult within the same host. Includes viruses, bacteria and protozoans.

microparasite model an epidemiological model for host–parasite dynamics for which the proportion of infected hosts, rather than the number of parasites per host, is the unit to be tracked. Typically, hosts are classed as susceptible, infected, or recovered; an example is *Hydramoeba hydroxena* on hydras.

microparticles (MPs) small (<0.1 µm) vesicles that are released from cells and circulate in the extracellular milieu. MPs consist of a plasma membrane surrounding a small amount of cytosol. The membrane of endothelial MPs contains receptors and other cell-surface molecules. Under normal physiological conditions, low levels of MPs are constantly shed into the blood from ECs that line the blood vessels. The levels of endothelium-derived MPs circulating in the blood may increase in some vascular and infectious diseases.

micropyle in coccidia, a pore in the cyst wall closed by a plug of material which is more easily dissolved than the cyst wall and through which the structures formed in the cyst emerge.

microRNA (miRNA) miRNAs are genome-encoded 20–24 nucleotides (nt) duplex-structured small RNAs with 5′-phosphate and 3′-hydroxyl groups with 2 nt overhangs. These are excised by the cropping activity of Dicer on the primary miRNA (pri-miRNA) transcript that can adopt an imperfect hairpin-like structure. One of the duplexes serves as a guide strand, which is referred to as miRNA (and the antisense strand to miRNA is known as miRNA*), and is selectively loaded onto RISC and serves as a sequence-specificity determinant in recognizing the target transcript. Based on complete or partial complementarity between miRNA and its target transcript, RISC can cause either target mRNA degradation or inhibit protein synthesis.

microRNA (miRNA) sponge RNA molecules with tandemly arrayed miRNA antisense sequences that can sequester miRNAs from their endogenous targets. Stably expressed miRNA sponges are valuable to achieve spatiotemporal miRNA loss-of-function.

microRNAs (miRNA) endogenous non-coding RNAs (≈22 nt) that regulate gene expression at transcriptional and post-transcriptional levels; produced from genes or introns, processed from long primary transcripts. miRNAs target and bind complementary sequences in the 3′-untranslated region of mRNAs, leading to gene silencing via translation repression and/or mRNA degradation. Mature miRNAs associate with Ago clade of Argonautes.

microRNome (miRNome) defined as the full complement of miRNAs in the genome.

microsatellite a polymorphism characterized by a variable number of tandem repeats often defined by the numbers of repeats in a row of at least two or more nucleotides. Useful markers for family linkage studies and determining the unique DNA 'fingerprint' of an individual for forensic medicine. Also known by other names, including simple sequence repeat (SSR) or simple tandem repeat (STR).

microsatellite instability (MSI) somatic variation of STRs (microsatellites) associated with phenotypic changes such as cancer, often due to mutations in DNA repair genes.

microsatellite locus a 2–6 bp repeat motif in a DNA stretch that varies in length between alleles. Their high variability, genome ubiquity, codominance, and biparental inheritance make them useful for population and individual level studies of both sexes.

microscopy terms limitations imposed by sample tilt geometry because of the limited tilt range possible in electron tomography, information from high tilt angles, normally above 70°, is not possible. This causes an artefact in the tomographic reconstruction, known as the 'missing wedge.' thus, the 3D volume obtained will contain some elongation and smearing in the axial plane.

microspora microspora are small intracellular parasites that differ significantly in structure from the Apicomplexa. These parasites are characterized by the structure of their spores, which have a complex tubular extrusion mechanism (polar tubule) used to inject the infective material (sporoplasm) into host cells. Microspora is a green alga.

microsporidia unicellular eukaryotes fungus classified as an ascomycetes; intracellular parasite; reproduction through spores. Ubiquitous, spore-forming, intracellular parasites infecting invertebrates and vertebrates. Some of them are important opportunistic pathogens in humans, including three species of genus *Encephalitozoon*. Intraspecies genetic variation with a different range of hosts is known in *Encephalitozoon cuniculi* distinguishing four genotypes. Recently, *E. cuniculi* is observed in pet animals, mainly *E. cuniculi* genotype in pet rabbits. Microsporidial vacuoles with typical spores were found in brain and kidney tissues and *E. cuniculi* DNA in all tested organs. The ITS of ribosomal RNA gene showed 100% homology with *E. cuniculi* genotype III previously identified in dogs, tamarin colonies from zoos, swine, birds and humans. Pet lemmings could represent a new potential source of the infection for their breeders. Microsporidia are ubiquitous, spore-forming, intracellular parasites infecting a wide range of invertebrates and vertebrates. Nearly 100 genera and more than 1200 species were described to date. These organisms can cause economically important diseases in insect, fish and mammals. Some microsporidia have a zoonotic potential, 14 species were described in humans to date with *Enterocytozoon bieneusi*, *Encephalitozoon intestinalis*, *Encephalitozoon hellem* and *Encephalitozoon cuniculi* as the most important opportunistic pathogens.

microstratification overlap mapping (MOM) the concept of defining the overlapping distribution of infections or co-endemicity in a particular geographic area at a finer level of spatial resolution than mapping of infections has previously achieved or been considered necessary.

microtubule quartet four specialized subpellicular microtubules that extend from the basal body to the anterior of the cell and subtend the region of plasma membrane where the flagellum attaches to the cell body. These four microtubules constitute part of the flagellum attachment zone, are associated with a subdomain of the smooth endoplasmic reticulum and are antiparallel to the other subpellicular microtubules.

microvilli microscopic cellular membrane extensions that increase the surface area of absorptive cells, for example, enterocytes in the small intestine.

microvillous membrane the apical membrane (maternal facing) of the polarized syncytiotrophoblast that contains many nutrient transporters.

midges biting flies belonging to the order of Nematocera, which include three families of relevance for cattle: Culicidae (gnats), Simulidae (blackflies) and Ceratopogonidae (biting midges). All of them are haematophagus and can act as transmitter of diseases. Midges like all flies are holometabolous insects. The requirements on the environmental conditions are strongly dependent on the genus. Midge females require blood for the egg production. While Culicidae and *Culicoides* are mainly active in the dusk and during night, *Simulium* has its blood meals during the day. Culicidae (represented by the three genera *Aedes*, *Culex* and *Anopheles*), need still water for their life cycle. Females place their eggs on the water, while larvae and pupae are located under the water surface. Several generations of Culicidae may develop between spring and autumn. *Simulidae* attach their eggs to stones or plants in rapidly flowing oxygen-rich water. The eggs and life stages of *Culicoides* are adapted to water with rich plant and dung environment. Of major importance are mass attacks of Simulidae in spring on pastures near to streams or rivers, when the warm temperatures allow a fast development of the hibernating eggs or larvae. Mosquitoes are pests. The stab is painful and causes itching. The animals become anxious and may show local skin swelling, especially when exposed to mass attacks. Special attention has to be paid to large clusters of blackflies. Allergic reactions to toxins of the salivary gland may cause generalized petechial bleeding and oedema formation. In some cases death may occur due to a cardio-circular collapse. Diseases which might occur in Europe and which are transmitted by mosquitoes: Bluetongue (Orbivirus, transmitted by the biting midges *Culicoides* sp.) and

Onchocerciasis (-osis) (*Onchocerca gutturosa*, transmitted by *Simulium* sp.). Many other viruses can be transmitted, especially arboviruses, of which the importance in many cases is not known, yet). Cattle on pasture are mainly affected. Animals may show pruritus, petechiae and/or local inflammatory skin reactions.

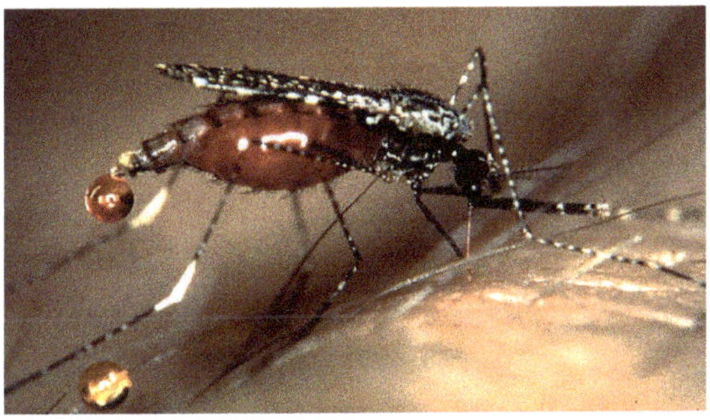

midgut epithelium (MGT-E) single layer of cells lining the mosquito gut. The cells are lying on a basal lamina that separates them from the haemolymph and contain microvilli on the anterior side facing the midgut lumen. The main function of the midgut epithelium is secretion of digestive enzymes and absorption of nutrients.

midgut lumen interior cavity of the mosquito midgut where the blood meal is contained and digested.

migration in population genetics, migration means the (permanent) movement of genes into or out of a population. Thus, a 'migrating' warbler does not cause any migration (in the genetic sense) by moving from breeding grounds in Wyoming to wintering grounds in Mexico and then returning to breed in the same Wyoming locale. We refer to the process of genes moving among populations as gene flow.

migration inhibition factor (MIF) a lymphokine that inhibits the motility of macrophages in culture.

migratory endoparasite commonly, nematode parasites that migrate through host plant root tissue to access multiple distinct feeding sites.

milk spot in the liver migrating *Ascaris suum* larvae in the pig's liver causes an inflammatory reaction, tissue destruction and haemorrhage. This is followed by an intense infiltration of eosiniphils and collagen production. These lesions are visible at necropsy on liver surfaces as whitish areas and are commonly called 'milk spots' since they resemble splashes of milk.

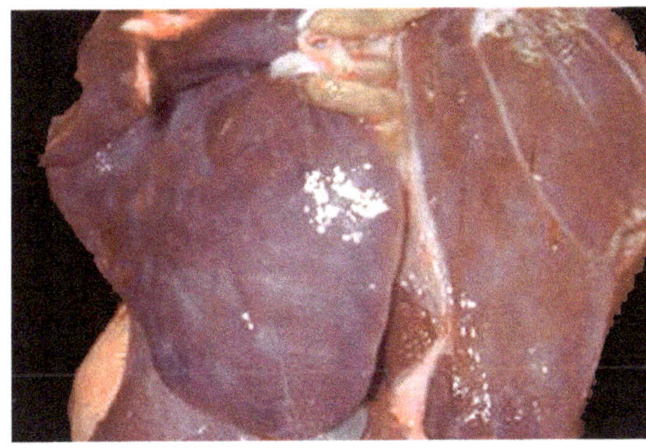

Miltefosine (hexadecylphosphocholine) the first oral drug used for treatment of leishmaniasis. Miltefosine belongs to the class of alkylphosphocholine drugs, which are phosphocholine esters of aliphatic long-chain alcohols. The dosage for treatment of visceral leishmaniasis is based on the patient's body weight (b.w.): 100 mg/day (28 days) for b.w. >25 kg and 50 mg/day (28 days) for b.w. <25 kg. For the treatment of PKDL, the same dose has been used but the duration increased to 12 weeks. miltefosine has a dual effect on leishmaniasis because it causes mitochondrial dysfunction, inhibition of cytochrome c oxidase, and increased generation of reactive oxygen species resulting in an apoptotic-like death in *Leishmania* parasites; in parallel, it raises the proinflammatory cytokines and molecules in *Leishmania*-infected macrophages and in patients with human leishmaniasis.

mimicking molecule compounds produced by the parasite and exposed on its body surface. These compounds enable the parasite to become antigenically closely related to the host and thus avoid inducing host immune responses.

mimotopes these are peptides that resemble structural features of epitope sequences and elicit immune responses.

miniexon or spliced leader short sequences that are joined to the 5′ ends of pre-mRNAs by trans-splicing. They are found primarily in primitive eukaryotes (i.e. protozoan and nematodes).

minimal inhibitory concentration (MIC) the lowest concentration of an antibiotic that, under a defined set of experimental conditions, inhibits visible growth of a bacterial culture following an overnight incubation.

minimal inoculum for resistance (MIR) in the laboratory, incubation of antimalarials or drug candidates with various amounts of parasites (inocula) can lead to recrudescence after several days. This adaptation to growth in the presence of the drug depends on the capacity of the parasite to accumulate mutations leading to a modification of its drug targets or drug efflux pumps. The occurrence of mutated parasites correlates with the number of genomes (parasites) in the inoculum. The lower the inoculum allowing the appearance of a resistant mutant, the more the drug is assessed as inducing resistance. The MIR is the minimum necessary to produce resistance and is used to attribute a risk level to the molecule that is tested (a MIR of 10^5 parasites is a high risk, 10^7 a medium risk, and 10^9 a low risk).

minimal selective concentration the lowest concentration of an antibiotic that results in the selection of a resistant mutant in a population over an isogenic susceptible strain.

minimum efficacy threshold the minimum impact of a molecule, prototype, or strategy measured by entomological endpoints that is required for decision making whether further investment in development and testing of that molecule, prototype, or strategy is warranted.

minor histocompatibility antigens these antigens, encoded outside the Major Histocompatibility Complex, are numerous, but do not generate rapid graft rejection or primary responses of T cells *in vitro*. They do not serve as restricting elements in cell interactions.

miracidia (singular **miracidium**) the first larval stage of a trematode, hatched from an egg, ciliated and often free-swimming.

miracidia hatching a parasitological technique based on the strong positive phototropic behaviour of miracidia. A schistosome-infected faecal suspension (or homogenized infected-tissue, for example liver or intestine) is placed in a flask, which is further placed in the dark; the miracidia hatch and swim to an illuminated collecting container, and are then trapped in this compartment. The miracidia are easily recovered with a pipette.

mismatch (MM) probe an Affymetrix probe that is designed to be complementary to a target sequence except for a one-base mismatch at the central position.

mismatch-repair system a strand-specific DNA-repair system that is present in most organisms; it recognizes and repairs erroneous DNA replication, recombination and damage.

missense mutation is when the change of a single bp causes the substitution of a different amino acid in the resulting protein. This amino acid substitution may have no effect, or it may render the protein non-functional.

missing enzyme or pathway hole an enzyme that is presumed to be present in a genome on the basis of the results of metabolic reconstruction, but for which a gene has yet to be identified.

mitochondrial control region a segment of the mitochondrial genome that is thought to contain the origin of replication of one strand, and the origin of transcription of both strands. The control region is normally located in the major non-coding portion of the genome. The control region frequently contains both hypervariable domains and conserved sequence blocks. Some parts of the control region may fold into hairpin loops that likely act as signals for polymerases.

mitochondrial DNA (mtDNA) the genetic material of the mitochondria – the organelle that generates energy for the cell. The DNA in mitochondria is different from that in the nucleus. Many scientists believe that this DNA is the remnant of a bacterium that invaded the cell in very early evolution. mtDNA is typically passed on only from the mother during sexual reproduction, as only the nucleus of the sperm enters the egg upon fertilization. Therefore, little change occurs in the mtDNA from generation to generation.

mitochondrial DNA (mtDNA) haplogroups groups of related mtDNA sequences. The faster coalescence times of mtDNA than nuclear markers make them useful for shallow divergence studies. The maternal inheritance allows studying the evolutionary processes of female lineages. The comparison of biparental nuclear and uniparental mtDNA regions allows the study of sex-biased phenomena.

mitochondrial haplotype the genetic constitution of a genetic locus, a chromosome or a genome is referred to as a haplotype. In the absence of recombination, a whole mitochondrial genome represents a single haplotype. This makes mitochondrial sequences useful tools for studying the evolutionary history of populations.

mitochondrion (plural **mitochondria**) endosymbiotic organelle with its own small mitochondrial chromosome. Generally plays a major role in energy conversion electron transport and proton motive force. The cell's power plant.

mitogen a substance that stimulates nonspecifically the proliferation of clones of lymphocytes.

mitogen-activated protein kinase (MAPK) pathway a signal transduction pathway involving several kinases that is responsive to numerous external stimuli (growth factors, differentiation factors and other cellular conditions).

mitonuclear incompatibilities genetic incompatibilities between nuclear and mitochondrial genomes. The opportunity for mitonuclear incompatibilities arises because mitochondrial gene products must interact with nuclear gene products for respiration to occur (the core genes for oxidative phosphorylation (OXPHOS) are encoded by mitochondria while the remaining ones are encoded by the nucleus). Hence, both groups of genes must co-evolve. Other incompatibilities are conceivable because mitochondrial genes may code for functions other than respiration and ATP production.

mitophagy the process by which mitochondria are targeted to and degraded by the autophagy pathway. In many settings, the critical targeting step is mediated by PINK1 and Parkin proteins, which couple recognition of damaged mitochondria, particularly those with loss of membrane potential, to the autophagy machinery.

mitosis the usual type of cell division in which the parent nucleus splits into two identical daughter nuclei, which contain the same number of chromosomes and identical genes to that of the parent nucleus. Also termed karyokinesis. *See* **meiosis**.

mitotic division nuclear division in which definite chromosomes and a spindle, associated with the presence of centrosomes. in some instances the nuclear membrane persists while in others the nuclear membrane is not evident as seen in higher species.

mitotic recombination recombination that takes place in cells that are dividing only by mitosis. A mitotic crossover causes all loci distal to the crossover to become homozygous. Mitotic gene conversion makes individual genes or short segments of chromosomes homozygous.

mixed lymphocyte culture an *in vitro* test for cellular immunity in which lymphocytes from two individuals are cultured together and when these cells are expressing allogeneic MHC a proliferative response is generally observed.

mixed lymphocyte reaction (MLR) a T-cell proliferative response induced by cells expressing allogeneic major histocompatibility complex.

mixing the coming together of individuals originating from different habitats or patches after a period of within-patch selection. Mixing may result in competition between individuals for the colonization of new patches. In the context of infection, mixing may occur when transmission-stage pathogens disperse from one host to infect another.

mixing models the isotopic composition of a mixed sample is determined by the isotopic signature and the contribution of the original sources by applying mass balance equations. Using the simplest form, it is usually only possible to determine $n+1$ sources for n isotopes; thus, if data are available for two isotopes, for example carbon and nitrogen, it is possible to determine the contribution of three food sources if the isotopic value of all three food sources and the isotopic value of the target organism are known and distinct.

mixotrophic a term used to describe microbial metabolism that uses an organic substrate as a carbon source and an inorganic compound as electron donor (i.e. having both autotrophic and heterotrophic capabilities).

model mathematical or other (e.g. engineering) representation used in computer simulations to calculate the evolving state of dynamic systems.

model ecosystem a specific type of ecosystem that is widely studied in great detail by a community of researchers to provide insights into the processes controlling the behaviour of other ecosystems.

model organism is one studied widely by a community of researchers. Biological understanding obtained from model-organism research is used to provide insights into the biological mechanisms of other organisms. Model organisms include the bacteria *Escherichia coli*, the yeast *Saccharomyces cerevisiae*, and the mustard weed *Arabidopsis thaliana*.

model uncertainty the range of simulation outputs as a result of using different models or changing the assumptions of the model.

model variant one of the 14 versions of the open malaria transmission models that make different assumptions about immune decay, transmission heterogeneity, and treatment access.

model-based clustering analyses analyses that group individuals into genetic clusters from their multi-locus genotypes and allele frequencies, and assign each individual to the detected clusters. Genetic clusters are defined based on assumptions on the definition of a population (e.g. a panmictic group).

modelling use of statistical and computational techniques to create working computer-based models of biological phenomena that can help to formulate hypotheses for experimentation and predict outcomes of research.

molecular biology the molecular basis of biological activity. The interaction of biomolecules in various cell systems, including the interaction between nucleic acids, proteins and their biosynthesis and regulation.

molecular clock the assumption of an approximately constant rate of nucleotide substitutions for a particular gene. With a constant molecular clock the genetic difference between any two species is proportional to the time since these species last shared a common ancestor.

molecular clock hypothesis proposes that molecular change is linear with time, and constant over different taxa and in different places. If that is so, then the sequence difference between homologues in different taxa can be used to estimate time since divergence.

molecular genetics the study of macromolecules important in biological inheritance.

molecular machine a highly organized assembly of proteins and other molecules that work together as a functional unit to carry out operational, structural, and regulatory activities in the cells.

molecular mimicry the use of proteins by a pathogen that are similar to host proteins to mimic those proteins in order to interfere with biological signalling in the host and to benefit the pathogen.

molecular population genetics a branch of evolutionary biology focused on understanding the evolutionary and demographic processes that shape genetic variation at the nucleotide level. Since its inception, there has been considerable focus on identifying genes that have been the targets of natural selection and the form of selection shaping nucleotide variation.

mollusca a member of the phylum Mollusca that acts as an intermediate host for the class Trematode.

monandrous mating system a female mates with only one male during the breeding season.

monecious both male and female reproductive organs are present in the same individual. This term is synonymous with hermaphroditic.

monepantel a new molecule that kills sheep nematodes resistant to all other anthelmintic groups. Monepantel paralyses worms by attacking a unique kind of acetylcholine (Ach) receptor – Hco-MPTL-1 – present only in nematodes. This drug is approved for use in NZ, Australia and some European countries, but not in the US.

monoclonal antibody an homogeneous antibody derived from a single B-cell clone and therefore all bearing identical antigen-binding sites and isotype.

monocytes circulating leucocytes with a single kidney-shaped nucleus, specialized in ingesting large foreign particles and cell debris. Monocytes migrate to tissues and organs where they can differentiate onto macrophages.

monogamy a mating system where one male has a mating relationship with only one female. Monogamy can be considered at different temporal levels (i.e. lifetime monogamy versus serial monogamy) and, independently, can include or exclude mate change. *See* **genetic monogamy**, **social monogamy**.

monogenean members of the class Monogenoidea (phylum Plathelminthes); hermaphroditic flatworms; oviparous or viviparous; usually with single-host life-cycle; mainly external parasites of fish and other aquatic vertebrates.

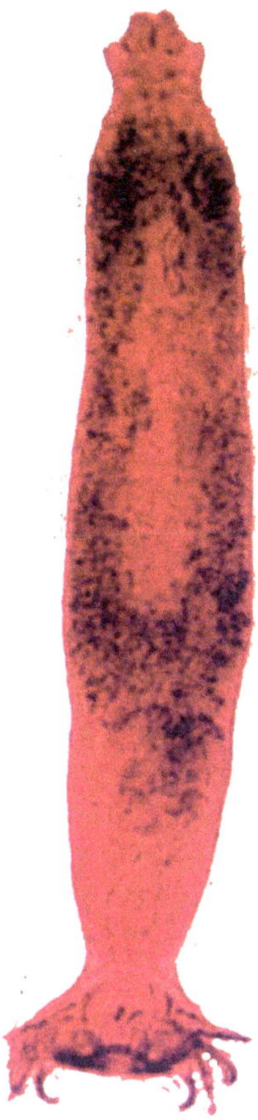

monogenetic a single generation constituting a complete life cycle.

monokaryotic having a single nucleus.

monokines cytokines secreted by macrophages and monocytes.

monomorphic having a single form. The term is used to refer to those isolates of *Trypanosoma brucei* that produce only a single morphotype during the mammalian bloodstream stage of the life cycle; that is, they do not exhibit the long slender-to-short stumpy form transition. Monomorphic forms generally occur as a result of prolonged laboratory cultivation and tend to produce an acute, highly virulent infection in mice, which is marked by the absence of the multiple waves of parasitaemia that are typically seen in infections with field isolates.

mononuclear phagocyte system a system that comprises blood monocytes and tissue macrophages.

monophyletic group (clade) evolutionary assemblage/lineage of taxa that originates from a common ancestor and all of its descendants.

monostable system a system for which only one stable state exists.

mono-unsaturated molecules, such as fats, with only one double bond in their chemical structure. Some plant oils and margarines, avocados, olives, nuts and seeds contain mostly monounsaturated fats.

monoxenus parasite one host parasite – requires only one host for the completion of its life cycle and development, e.g. *Eimeria*.

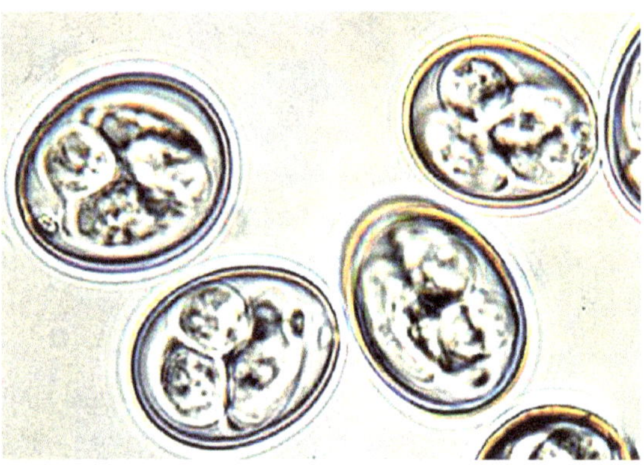

monozoic a condition in members of the tapeworm subclass Cestodaria in which there is only a single set of reproductive organs.

Montpellier 1 (MON 1) the multi-locus enzyme electrophoresis (MLEE) typing system standardized by a reference research group of Montpellier (France), the name of a *Leishmania infantum* MLEE genotype that is frequently recorded in the Old and New Worlds.

moral standing to say that a group of organisms has moral standing is to say that their well being must be given some consideration. It does not decide the question of whether they have the same moral standing as people (and thus have 'human' rights).

Moran's *I* coefficient of correlation that measures the correlation of the values of a variable, defined for a set of locations, between pairs of locations located at a given geographical distance. Moran's *I* ranges from –1 (strong negative spatial autocorrelation, i.e. nearby populations will be dissimilar) to 1 (strong positive autocorrelation, i.e. nearby populations will share more genetic similarity than expected by chance); a Moran's *I* of 0 is expected in the case of a random pattern.

morbidity associated with schistosomiasis is typically due to direct lesions by schistosome eggs and concurrent immunopathology to tissue granulomata. Manifestations for urogenital schistosomiasis include (macro)haematuria and (micro)haematuria where there is visual blood in the urine or cryptic amounts as detected by reagent strips, respectively. Urine reagent strips also detect proteinuria, the main biological component being albumin, which can be graded into various levels of albuminuria.

morphological species concept a species is defined by being sufficiently morphologically distinct from all others. The definition can include multivariate tests of the statistical distance between species centroids in relation to intraspecific variation about the centroids.

morphological voucher a preserved specimen archived in a collection facility (e.g. a museum). In DNA barcoding, vouchering (the preservation of morphological vouchers) is standard practice for specimens from which DNA barcode sequences were obtained.

morphometry measurement of form, size, and shape of organisms.

mortality events multiple host deaths occurring at the same time and as a result of parasite infection.

morula stage the cleaving stage of an egg in which it forms a mulberry-like, solid mass of cells.

mosaic aneuploidy the number of copies of a given chromosome varies between chromosomes, or between different species, or between different strains of a given species.

mosaic hybrid zone a region with a patchy distribution of parental and hybrid genotypes, as opposed to a smooth and continuous geographical transition from one parental species to the other through an area of hybridization.

mosquito cycle sexual stage in the mosquito gut where two gametocytes (male and female) will fuse to give an oocyte.

mosquito larviciding the treatment of breeding sites with insecticides that specifically target the larval life stage of mosquitoes.

most recent common ancestor (MRCA) although the ancestries of two alleles may both pass through the same individual, they pass through different alleles with probability 0.5, in which case that individual is not the MRCA of the alleles.

motif a particular sequence of DNA bps. The number of possible motif sequences increases with motif length.

motif activity response analysis (MARA) a method for inferring DNA-binding-motif activity and for linking motifs to promoters. MARA models promoter expression as a linear function of motif activity and the number of functional binding sites.

mo(u)lting (=ecdysis) shedding of an external covering such as integument or exoskeleton; in arthropods and nematodes, shedding the external covering allows the parasite to expand in its new skin.

mo(u)lt the act of shedding the outer body covering, e.g. skin, cuticle, feathers, exoskeleton, etc.

mouse model is a laboratory mouse used to study some aspect of human physiology or disease. A variety of different model organisms are used in this regard, but mice are especially useful because they share mammalian features with humans and suffer from many of the same diseases. A large number of mouse models have been created to target specific human diseases using selective breeding and genetic engineering.

MPC the minimum parasiticidal concentration is the lowest concentration of antimalarial drug in the blood that provides maximal inhibition of parasite multiplication.

MRE11 meiotic recombination 11, a conserved member of the RAD52 epistasis group that forms the MRX/N complex (MRX in yeast, comprising MRE11, RAD50 and XRS2, or MRN in mammals, comprising MRE11, RAD50 and NBS1). This protein is involved in general homologous recombination, in DSB repair and in telomere stability.

mRNA isoforms different mRNAs produced from the same precursor mRNA.

mRNAs these protein-coding RNA species are transcribed from the coding DNA strand and contain an ORF, a 5′ UTR and a 3′ UTR. *3′-Untranslated region (UTR*) is a section of mRNA following the coding region that contains several types of regulatory sequences that affect mRNA stability and translation, including miRNA-target sites.

MS2 tagging a technique that uses the MS2 coat protein of the MS2 bacteriophage to detect RNA molecules owing to the high binding affinity of MS2 for a specific RNA stem–loop structure. Upon insertion of the MS2 recognition sequence in the RNA of interest, the intracellular localization of the RNA can be monitored in live cells using GFP-tagged MS2.

mtDNA (mitochondrial DNA) sequencing of mtDNA is a widely used technique in systematics. The mostly maternal, clonal transmission of mtDNA provides both opportunities and problems for phylogenetic analysis.

mucocutaneous leishmaniasis a disease caused by metastasis of some *Leishmania* spp. to facial mucosal tissues, leading to ulceration and destruction of mucosal tissues in the nose, mouth, and throat.

mucosa-associated lymphoid tissue (MALT) lymphoid tissue present in the surface mucosa of the respiratory, gastrointestinal and genitourinary tracts.

mucosal immunity immune responses expressed at mucosal surfaces. These include BALT (bronchial associated lymphoid tissues, e.g. retropharyngeal lymph nodes, Harderian gland) and GALT (gut-associated lymphoid tissues, e.g. lymph nodes draining the gut, Peyer's patches, lymphoglandular complexes).

mucosal vaccine immunization through oral, nasal or vaginal routes.

mucous membrane any of several moist surfaces in the body of vertebrates in which there are mucus-secreting or goblet cells; examples are the orbit of the eye, nasal passages, inside the mouth.

Muellerius capillaris adult worms are found in the parenchyma of the lungs. It has an indirect life cycle, requires a snail or slug intermediate host. Goats are most susceptible and can die from pneumonia. Worms are found in granulomatous foci in the parenchyma.

Muller's ratchet the process by which the number of deleterious mutations increases irreversibly in non-recombinant asexual populations due to the impossibility to recover the less mutated class.

multidrug resistance (MDR) resistance to multiple, structurally divergent drugs.

multifactor dimensionality reduction a data-mining algorithm that can reduce a high-dimensional multi-locus model of multifactorial classes (that is, single-nucleotide polymorphism genotype combinations) into a one-dimensional model of one variable of either high-risk (potential interacting) or low-risk classes based on the ratio of cases and controls in each class. The algorithm uses cross-validation iteratively to define the best classification.

multigenic involving the action, and potentially interaction, of multiple genes.

multi-locus enzyme electrophoresis (MLEE) standard MON classification is based on 15 enzymatic systems analysed by starch gel electrophoresis. The same enzyme from different samples might show different mobility in the gel (isoenzyme or electromorphs; strictly allozymes). Zymodemes are strains sharing the same isoenzyme profiles for all 15 enzymes. MLEE markers are co-dominant. *See* **allozyme**, **isozymes**.

multi-locus genotype (MLG) the composite genotype made up of all individual genotypes recorded at each locus under survey in a given individual.

multi-locus microsatellite typing (MLMT) microsatellites are stretches of short nucleotide motives (1–6 bp) that are repeated tandemly, which present allelic length variation. They mutate fast, however, 10–20 independent markers have to be analysed for each strain owing to homoplasy. Microsatellite-containing sequences are amplified and length polymorphisms are detected using PAGE, MetaPhor agarose gel electrophoresis or, preferably, capillary sequencers. A multi-locus microsatellite profile is compiled for each sample from the fragment length measured for the microsatellite markers used and analysed directly or after the calculation of genetic distances using different types of population genetic analyses. Microsatellite markers are co-dominant.

multi-locus sequence typing (MLST) a highly standardized method that relies on the comparison of partial sequences (usually 700 bp) of a defined set of housekeeping genes (7–10). Similarly to MLEE, alleles are scored as identical or not, regardless of how many different polymorphic loci they have. Strains sharing the same allele combinations for the set of genes tested are referred to as sequence types. MLST markers are co-dominant and are amenable for population and phylogenetic analyses. Alternatively, the data can be analysed as sequences.

multi-marker test a statistical method that measures the strength of association between a trait and multiple SNP markers.

multiple infection an infection in a single host that involves two or more parasite genotypes or species.

multiple myeloma plasma cell malignancy resulting in high levels of monoclonal immunoglobulin in serum and of free light chains (Bence-Jones protein) in urine.

multiplex polymerase chain reaction a variant of the polymerase chain reaction test in which more than one target sequence is amplified using more than one pair of primers.

multiplexing a methodology that allows the sequencing of several individuals in a single run by using barcodes in order to assign each read obtained from the sequencing run to a specific individual. Every individual from a single sequencing run will be given a unique barcode (or a unique combination of barcodes if barcodes are fixed to both forward and reverse primers; this allows to multiplex a larger number of samples); each read from the same individual will therefore share the same barcode(s) which will be different from the barcodes of the other individuals, allowing to track the individual of origin of each read.

multipotent a chromosome form during the first meiotic division where more than two homologous chromosomes synapse.

multivariate readouts phenotypic readouts consisting of multiple features of the cellular activity, state and microenvironment.

murine pertaining to mice.

museum vouchers a specimen and all its associated data that are preserved in a museum. The voucher is proof of the existence of a species or object at a particular time and space and ensures a defined reference point for future research.

mutagen an agent that causes a permanent genetic change in a cell. Does not include changes occurring during normal genetic recombination.

mutagenicity the capacity of a chemical or physical agent to cause permanent genetic alterations.

mutation any heritable change in DNA sequence. The process by which a gene undergoes a change in the base sequence. Some mutations result in the gene no longer coding for the correct protein, or producing a reduced amount of the protein. *See* **polymorphism**.

mutational load a reduction of the average fitness of the members of a population because of the accumulation of deleterious mutations in the population at any given time.

mutational space all possible mutations that can confer a specific phenotype. This can vary from one to several mutations, depending on the system that is studied.

mutational target size the fraction of the genome in which new mutations can potentially cause variation for a trait. For most complex traits this is large, thus suggesting that many loci can influence trait variation.

mutators individuals that have a higher mutation rate than the species-average mutation rate.

mutualism a facultative relationship occurs between two organisms when each member of the symbiotic relationship benefits the other. Relationship between ungulates and their ciliates; lichens.

mutualistic an interaction that benefits both parties. This term is typically used for interactions between multiple species.

myasthenia gravis an autoimmune disease in which autoantibodies react with nicotinic acetylcholine receptors leading to severe muscle weakness.

mycelia sterilia fungal mycelia producing neither asexual nor sexual spores.

myeloid cells leucocytes from the mononucleated phagocyte system (granulocytes and various monocyte, DC, and macrophage subsets) characterized by the common expression of the αM integrin CD11b and differing ontogeny. M1-type/M2-type myeloid cells refer to a pathogenic insult-dependent continuum of activation states ranging from inflammation induction to tissue-healing function.

myeloid differentiation primary response gene (88) (MyD88) an adapter protein involved in the signalling cascade of almost all Toll-like receptors.

myeloma a tumour of plasma cells, generally secreting a single species of immunoglobulin.

myeloma protein monoclonal antibody secreted by myeloma cells.

myiasis invasion of healthy body tissues by the larvae of dipteran (two winged) true flies, and is characterized as cutaneous (subdermal tissue), gastrointestinal, nasopharyngeal, ocular or urinary, depending on the region invaded.

myositis inflammation of the muscle.

myristoylation the addition of the saturated 14 carbon fatty acid, myristic acid, to a glycine residue via amide linkages.

myzocytosis a predatory mode whereby a cell is able to penetrate the cell surface and draw in the cytoplasmic contents of a prey or host cell.

myzozoa a clade of avleolates comprising the most recent common ancestor of apicomplexans and dinoflagellates and all of its descendants.

A B C D E F G H I J K L

M **N** O P Q R S T U V W

X Y Z

NAbs natural antibodies (constitutive antibodies involved with innate immunity.

N-acetylglucosamine (GlcNAc) and glycosamine (GlcN) amino sugars differing by the presence of an acetyl group on carbon two and the nitrogen-containing side chain.

NADPH-oxidase (Nox) and the related dual oxidase (Duox) play various biological and pathogenic roles via regulated generation of ROS.

Naegleria fowleri commonly referred to as the 'brain-eating amoeba' or 'brain-eating amoeba', a free-living amoeba, (single-celled living organism). It can cause a rare and devastating infection of the brain called primary amebic meningoencephalitis (PAM). The amoeba is commonly found in warm freshwater lakes, rivers, hot springs and soil. *Naegleria fowleri* usually infects people when contaminated water enters the body through the nose. Once the amoeba enters the nose, it travels via olfactory nerves to the brain where it causes primary amebic meningoencephalitis (PAM), which is usually fatal. Infection typically occurs when people go swimming or diving in warm freshwater places, like lakes and rivers. In very rare instances, *Naegleria* infections may also occur when contaminated water from other sources (such as inadequately chlorinated swimming pool water or heated and contaminated tap water) enters the nose. You cannot get infected from drinking water contaminated with *Naegleria*.

naive lymphocyte a mature T- or B-cell which has not yet been activated by encounter with antigen.

NANOGA a transcription factor that is crucial for the self-renewal of undifferentiated embryonic stem cells.

nanoRNAs (nRNAs) small (2–6 nucleotides) oligoribonucleotides of unknown origin that can prime the transcription of some mRNAs.

Nano-SIMS an imaging technique that uses a nanoscale secondary ion mass spectrometer (nano-SIMS) and cells labelled with stable isotopes of carbon and/or nitrogen to identify areas of active growth and follow nutrient fluxes between cells.

nanotechnology is the science of manipulating matter on the atomic and molecular scales to solve problems. Nanotechnology is a developing applied science that has the potential to make significant contributions to many fields, including engineering, computer science, and medicine.

narrow-sense heritability (h2) the proportion of variation due to the additive effects of genes.

national control programmes (NCPs) disease intervention programmes operating at a national level where a formal commitment in terms of policy, finances, staffing, and timetabling of a country to adopt and scale-up preventive chemotherapy (PC) to a country-wide application. Measurement and assessment of progress towards reductions in selected disease targets are an obligatory requirement of the monitoring and evaluation component.

native organisms that have not been recently introduced into an ecosystem.

native elongating transcript sequencing (NET-seq) a deep sequencing technology used to identify nascent transcripts that are associated with an elongating RNA polymerase.

natural alloantibodies antibodies against foreign antigens found in the serum of normal, unimmunized individuals. Most probably arise as a result of exposure to cross-reacting bacterial antigens (e.g. anti-A and B red blood cell antigens)

natural defences set of traits primarily involved in the prevention or control of infections (behavioural defences, structures, chemical defences, immunity). These traits can be constitutive or inducible, and are plastic.

natural host this term implies that the host is the usual one and is synonymous with typical host.

natural killer (NK) cell non-T, non-B lymphocytes found in normal individuals and capable of killing some tumour cells and some virus-infected cells independently of bound antibody to the target cell.

natural population a population not in artificial breeding or captivity. One might question the use of this term for invasive species closely associated with human habitats, not their ancestral habitats. However, it has become rather 'natural' for such species to share human habitats. Here it is used synonymously with field population.

natural selection one of the conclusions drawn by the British naturalist Charles Darwin (1809–82) from the theory of evolution: certain organisms with particular characteristics are more likely to survive and hence pass on their characteristics to their offspring, i.e. survival of the fittest. Thus the characteristics of a population are controlled by this process.

nauplius (plural **nauplii**) two life-stages of free-living non-feeding planktonic larvae that hatch from the egg and moult to become copepodids.

NCBO (National Center for Biomedical Ontology) a consortium of leading biologists, clinicians, informaticians and ontologists who develop innovative technology and methods that enable scientists to create, disseminate and manage biomedical information and knowledge in computer-processable form.

near-clade occasional bouts of genetic exchange explain that the genetic clusters observed in a predominantly clonal population cannot be equated to real clades (between which genetic exchange is completely absent). We have coined the term 'near-clades' to designate such genetic clusters, which are observed in many pathogen species.

necrophagy the consumption of dead and decaying organic matter.

necroptosis a proinflammatory form of cell death, which is dependent on receptor-interacting protein kinases 1 and 3 (RIPK1 and RIPK3), and is inhibited by caspase-8.

necrosis a passive, uncontrolled form of cell death characterized by cellular swelling, membrane lysis, and release of intracellular contents.

necrotic enteritis an intestinal disease that in chickens is caused by the bacterium *Clostridium perfringens* that causes severe inflammation and lesions in the infected portion of the intestine. The disease is exacerbated by concurrent coccidial infection.

necrotizing haemorrhagic encephalitis a degenerative brain disease, characterized by marked oedema, numerous haemorrhages, necrosis of blood vessel walls, demyelination of nerve fibres, and infiltration of the meninges with neutrophils, lymphocytes, and histiocytes. Typical signs are severe headache, fever, and vomiting; seizures may occur, and the patient may rapidly lose consciousness.

necrotizing orchitis inflammation of the testes, initially characterized by swelling and pain, that results in tissue necrosis and leads to infertility.

negative frequency-dependent selection selection operating in which fitness of an allele is inversely proportional to its frequency. This is an important form of balancing selection that tends to maintain allelic polymorphism within a population.

negative predictive value (NPV) the probability that a negative result accurately indicates the recycling endosomes.

negative regulatory factor (Nef) expressed by HIV after infection and is also considered a virulence factor.

negative selection selection against an allele that has a deleterious effect on fitness (also called purifying selection as it will tend to remove the allele from the population).

negative-feedback loop a circuit in which a molecule represses its own production and/or induces its own consumption.

neglected disease a disease for which there is a disproportionately low level of investment and research into the development of treatments relative to the size of the affected population. Neglected diseases typically affect developing countries, and the at-risk populations have insufficient purchasing power to allow the profitable development of therapeutics.

neglected parasitic infections (NPIs) a group of five parasitic diseases that have been targeted by the US CDC as priorities for public health action based on the (i) number of people infected, (ii) severity of the illnesses and (iii) ability to prevent and treat them. These infections are considered neglected because relatively little attention has been devoted to their surveillance, prevention, and/or treatment. They include Chagas disease, cysticercosis, toxocariasis, toxoplasmosis, and trichomoniasis. Anyone, regardless of race or economic status, can become infected although minorities, immigrants, and people living in poor or disadvantaged communities appear to be most at risk.

neglected tropical diseases (NTDs) a group of several tropical diseases that collectively cause significant mortality and morbidity, often in impoverished rural communities.

neighbour-joining method phylogenetic reconstruction method based on an algorithm that minimizes the total length of the tree. This method does not explore or compare possible topologies.

nemathelminthes nematodes, also called roundworms, are generally relatively small. They have cylindrical bodies covered by a 'tough skin' called a cuticle. The sexes of roundworm are separate, and these organisms have a complete digestive system. Nematodes may be intestinal parasites or may infect the blood and tissue. The nematode life cycle has three stages: egg, four-larval, and adult. The parasitic animal nematodes are endoparasites. However, the first three larval stages can develop either in the external environment or within the intermediate host. The larval stage can be subdivided into: first stage larva (ready to hatch). This larva sheds it external cuticle to grow to the next stage. The shedding of this outer layer is called moulting. Second stage larva eventually moults into the next stage. The third stage larva is also known as the infective stage because it is infective for the definitive host. Fourth stage larva occurs when the third stage moults within the definitive host. It eventually develops into the sexually mature adult.

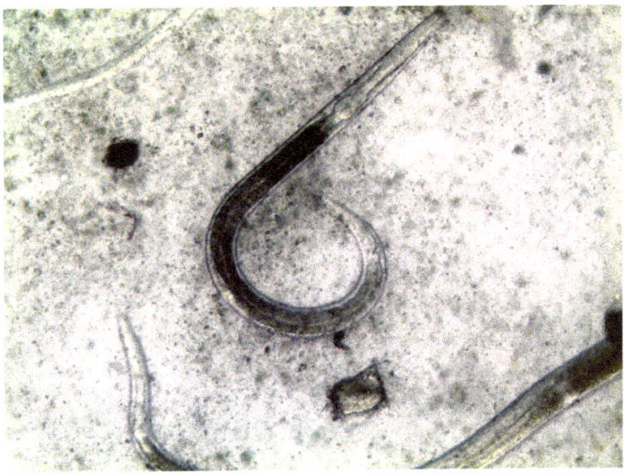

nematode cuticle it is the nematode exoskeleton that provides structural integrity, environmental protection and plays a role in hydrostatic skeleton function. It mainly consists of collagens, lipids, carbohydrates and glycoproteins. it is produced by the hypodermis, an ectodermal cell layer that surrounds the body of the animal. The nematode cuticle is permeable and can grow between moults (unlike arthropod exoskeleton).

nematodirosis is common in temperate zones. It is of special significance to lambs in UK especially at the Scottish border and northern England. *Nematodirus* spp. inhabit small intestine of ruminants. Common species include *N. battus* (sheep and calves occasionally), *N. fillicolis* and *N. spathiger* (sheep), and *N. helvetianus* (cattle). *N. battus* life cycle is a little different to common Trichostrongyle life cycle, e.g. (1) preparasitc phase (L1–L3) occurs within egg shells (2) larvae take months to develop. (3) eggs are extremely tough and resistant to freezing and drought, viable up to two years on a pasture. (4) hatch is stimulated by cold period followed by a mean day/night temperature of ≈10°C, but larva three (L3) are susceptible to climate and need to be ingested quickly. Therefore, this is usually the first nematode species to peak on pasture in spring. Nematodirosis is characterized by strong age-related immunity (immunity occurs after about 3 months of age), increased eosinophils and mast cells at site of infection, and villus shedding (important rejection mechanism when infected with species which are closely associated with villi).

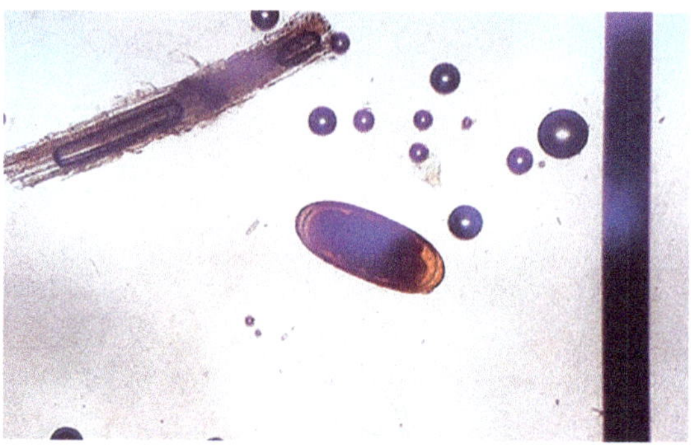

neoblasts pluripotent adult stem cells with capacity for indefinite self-renewal, totipotency; drive long-term tissue homeostasis; endow planarians with ability to regenerate. Neoblast-like stem cells occur in tapeworms and schistosomes. Characteristics include roundish-to-ovoid cells with a high nuclear-to-cytoplasmic ratio, large nucleolus, exhibit chromatoid bodies (RNA-protein bodies near nuclei), and numerous free ribosomes.

Neodermata the subphylum of the platyhelminthes that includes the classes monogenea, trematoda and cestoda. All members of the Neodermata are parasites.

neofunctionalization the process by which a duplicated gene gains a different but often related function from its parent gene. It was hypothesized that the gene duplication event creates redundancy so that the new gene copy can accumulate mutations that then allow it to gain a new molecular function.

neonatal FcR (FcRn) FcRn is unrelated to classical FcRs and binds to a different region in the antibody Fc fragment. Structurally, it is related to the family of MHC class I molecules and is responsible for regulating IgG half-life. Recent research has also shown FcRn to be involved in regulating antigen presentation.

neonicotinoids (or chloronicotinyl guanidines): imidacloprid, nitenpyram, and dinotefuran neonicotinoids act like agonists on postsynaptic nicotinic acetylcholine receptors, mainly in motoneurons, inducing nerve membrane depolarization, which causes spastic paralysis in insects. they are derived from the observation of the insecticidal activity of nicotine. They target insect-specific receptors, which explain their wide safety margin in mammals.

neoplastic disease disease associated with a tumour.

neospecies a newly derived species.

Neospora caninum an intracellular protozoan that infects many domestic and wild animals. Dogs are known as a definitive host of *N. caninum* and involved in transmitting infections to intermediate hosts by shedding oocysts.

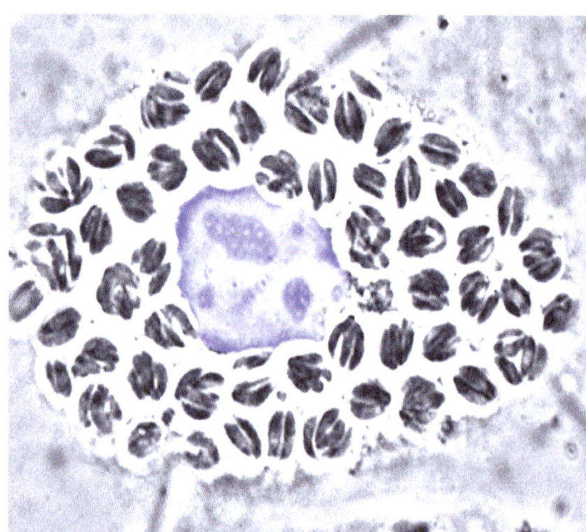

nested PCR modification of PCR intended to reduce non-specific binding in products due to the amplification of unexpected primer binding sites. A nested PCR method for detecting *B. mandrillaris* from the environment (water and soil samples) was developed in 2011 and targets the 18S rDNA gene of this amoebic pathogen.

net-transporting process that describes only the net movement (i.e. not the exchange) of a substance across a membrane.

network diagram of relationships allowing lateral connections among groups.

network architecture a pattern of links connecting the nodes of a network.

network motif a recurrent pattern of interconnections in a network that appear more frequently than expected by chance in random networks of similar size.

neuroborreliosis a manifestation of Lyme borreliosis with involvement of the central nervous system, typically presenting in humans as a polyradiculitis (inflammation of several nerve roots) weeks to months after the tick bite.

neurocysticercosis (NCC) infection of the brain by a cysticercus; it can be detected by CT scan. The presence of the larvae can induce seizure or mass effect, perturbing the normal function of the brain according to their position in the brain.

neuroendocrine crypt cells are cells that are distributed throughout the intestinal epithelium and that secrete peptide hormones in an endocrine or paracrine manner from dense core or neurosecretory granules.

neuroexocytosis nanomachine a molecular machine of nanometre dimensions that is used for the release of neurotransmitters.

neurogenesis the formation of new neurons.

neuronal ablation experiment in which subsets neurons are killed using either a high-powered laser or cell-specific expression of pro-apoptotic genes. Behavioural phenotypes of ablated animals can be used to assign functions to specific neurons.

neuropeptides polypeptides that act to modulate neuronal activity. In *Caenorhabditis elegans* large families of neuropeptides include insulin-like peptides, FLPs (FMRFamide-like peptides), and NLPs (neuropeptide-like proteins).

neuroschistosomiasis *Schistosoma* spp. involvement of the central nervous system affecting the spinal cord (*S. mansoni*, *S. haematobium*) or encephalus (*S. japonicum*). Nerve damage is induced by egg deposition and an inflammatory immune response. Clinical manifestations can develop in both acute and chronic phases of infection.

neurotransmitter an endogenous chemical that transmits signals across a synapse from a neuron to a postsynaptic cell.

neurovascular unit the other name for cells and structures forming the blood–brain barrier.

neutral alleles allelic variants at a locus that have no effect on the fitness of the bearer.

neutral evolution a neutral mutation is one that has no adaptive significance to the organism or is without a phenotypic effect. Such changes are therefore not subjected to selective forces. The theory of neutral evolution states that, at the molecular level, the vast majority of changes are neutral or nearly so and that evolution is primarily determined by random genetic drift, rather than by natural selection.

neutral evolutionary pressures non-selective forces, such as demography (e.g. population expansions and reductions) and stochastic events (e.g. genetic drift). Most of the variation in the genome is shaped by neutral evolutionary pressures.

neutral loss scan a tandem mass spectroscopy mode used when a molecular ion, produced after fragmentation of a lipid class, loses a specific uncharged fragment. Because uncharged fragments cannot be detected, the specific loss of the corresponding neutral fragment is recorded. The lipid class can thus be characterized in a complex mixture by selecting only the initial molecular ions that lose this specific neutral fragment. Phosphatidylserine species can, for instance, be detected by constant neutral loss scans of m/z 87.

neutral process the process of accumulation of mutations (and therefore new alleles in a given population) that do not affect the fitness of individuals. Because neutral alleles are not subject to natural selection, their frequency in a population is governed by the processes of mutation and random genetic drift.

neutralization an immunological test to detect antibodies that bind to a virus and reduce its titre in an infectivity assay. For orthobunyaviruses, neutralizing antibodies are elicited by the viral glycoproteins and are used for the identification of specific viruses.

neutralizing antibodies are antibodies that block infectivity (e.g. of a virus), usually by binding to the foreign particle (the virion) and incapacitating it in some way.

neutrophil the major circulating phagocytic polymorphonuclear granulocyte. Enters tissues early in an inflammatory response and is also able to mediate ADCC.

neutrophil extracellular trap (NET) a network of extracellular fibres composed of DNA and histones generated by activated neutrophils that bind pathogen.

never-edited mRNA transcripts that forgo kinetoplastRNA editing.

New World the western hemisphere.

next generation' sequencing (also known as massively parallel sequencing) a term used to describe a class of DNA-sequencing platforms that employ highly parallel approaches to sequence millions or billions of bps in a single run. Relative to traditional capillary sequencing, next-generation sequencers generate data dramatically more quickly and inexpensively, at the cost of shorter read lengths.

NF-κB nuclear factor kappa B, a family of transcription factors that includes the proteins RelA (p65), RelB, c-Rel, p50 and p52. These factors play a key role in regulating immune responses. It is one of the transcription factors involved in inflammatory signalling pathways utilized by TLRs.

niche multidimensional space within which a species makes a living.

nitenpyram even though nitenpyram has the same mode of action, it is a systemic and not a contact insecticide. It is administered orally to carnivores, as a tablet. It is rapidly absorbed, with a peak plasma level occurring within 30 minutes to 2 hours after administration. It is excreted in the urine (half-life of 4 hours in dogs and 8 hours in cats) and disappears within 2 days. While feeding on blood, fleas ingest nitenpyram and die within the next 15–30 minutes. It is therefore a short-acting, systemic insecticide with rapid action.

nitric oxide (NO) nitric oxide produced by nitric oxide synthetases can function as a cellular signalling molecule and at higher concentrations as an antimicrobial agent. NO is able to activate guanylate cyclases that typically act downstream of G-protein-coupled receptors in signalling cascades.

nitrogen fixation process carried out by certain species of bacteria and archaea in which atmospheric nitrogen (N2) is converted to organic nitrogen-containing compounds that can be used by other organisms.

nitrogenase enzyme that catalyses the conversion of atmospheric nitrogen (N_2) to nitrate in nitrogen-fixing bacteria and archaea.

nitrogenous base a nitrogen-containing molecule having the chemical properties of a base. DNA contains the nitrogenous bases adenine (A), guanine (G), cytosine (C), and thymine (T).

nitroreductase (NTR) enzymes a family of FMN- or FAD-dependent enzymes capable of metabolizing nitroaromatic compounds. NTRs utilize NADH or NADPH as a reductive cofactor. The NTRs are subdivided into two classes depending upon their reaction mechanism.

N-methyl-cyclooctadepsipeptide synthetase (PFSYN) enzyme catalysing synthesis of PF1022 type cyclooctadepsipeptides from the precursors leucine, (R)-lactic acid, (R)-phenyllactic acid, and S-adenosyl-(R)-methionine in the presence of ATP and MgCl2.

N-myristoyltransferase (NMT) enzymes that catalyse the transfer of myristate from myristoyl-CoA to an N-terminal glycine residue.

N-nucleotides non-templated nucleotides added to the junctions between antibody (and T-cell receptor) variable (V), diversity

(D) and joining (J) gene segments during gene rearrangement.

nod-like receptor a family of cytoplasmic pattern recognition receptors involved in sensing the presence of pathogens.

nodulation the binding of multiple haemocytes to aggregations of bacteria.

noise propagation stochastic expression of a target gene resulting from gene expression noise in upstream regulatory genes, within a given circuit.

non-allelic homologous recombination (NAHR) the genetic exchange between similar sequences, such as segmental duplications, as opposed to truly homologous sequences on paired chromosomes. Also known as 'ectopic' recombination.

non-artemisinin therapy an antimalarial medicine that does not contain artemisinin or any of its derivatives.

non-clinical animals (or clinically healthy animals or healthy infected dogs) animals infected by one or more canine vector borne disease-causing pathogens that, in contrast to clinical animals, do not exhibit apparent clinical signs and haematological or biochemical abnormalities. The classification of these animals is based upon both their history and physical examination.

non-coding DNA sequences do not code for amino acids. Most non-coding DNA lies between genes on the chromosome and has no known function. Other non-coding DNA, called introns, is found within genes. Some non-coding DNA plays a role in the regulation of gene expression.

non-coding RNAs (ncRNAs) functional RNA molecules that do not encode proteins (unlike messenger RNAs). These often associate with chromatin, where they scaffold the chromatin and recruit repressive histone modifier complexes.

non-competitive cross species co-hybridization (CSH) cross-species co-hybridization of one or more RNA samples derived from a single target species to a microarray that represents a different species.

non-disjunction the failure of homologous chromosomes to segregate properly during the first meiotic division, or the failure of sister chromatids to segregate properly during the second meiotic division or during mitosis. Non-disjunction produces aneuploid cells.

non-equilibrium population a population in which the equilibrium assumptions of panmixia and/or constant size are violated.

non-malarial febrile illness (NMFI) infectious diseases in patients who present with undifferentiated fever and require malaria RDTs/microscopy – but in whom these tests are negative (i.e. other causes of malaria-like symptoms).

non-recommended therapy a treatment that is not endorsed by government guidelines for treatment of uncomplicated malaria.

non-selective and selective autophagy there are three primary types of autophagy: microautophagy, macroautophagy and a mechanistically unrelated process, chaperone-mediated autophagy that only occurs in mammalian cells. Both micro and macroautophagy can be selective or non-selective and these processes have been best characterized in yeast. The most distinguishing feature of macroautophagy is the formation of the double-membrane bound phagophore and autophagosome. In contrast, during microautophagy the cargoes are sequestered by direct invagination or protusion/septation of the yeast vacuole membrane. Non-selective autophagy is used for the turnover of bulk cytoplasm under starvation conditions, whereas selective autophagy specifically targets damaged or superfluous organelles, including mitochondria and peroxisomes, as well as invasive microbes; each process involves a core set of machinery, as well as specific components, and accordingly is identified with a unique name – mitophagy for selective mitochondria degradation by autophagy, pexophagy for peroxisomes, xenophagy for microbes, etc.

nonsense mutation is the substitution of a single bp that leads to the appearance of a stop codon where previously there was a codon specifying an amino acid. The presence of this premature stop codon results in the production of a shortened and likely non-functional, protein.

nonsense-mediated decay a cellular mRNA quality control mechanism which facilitates the detection and degradation of premature termination (nonsense) mutations thus preventing the expression of erroneous proteins.

non-synonymous a nucleotide difference in a gene that alters the encoded protein sequence.

non-synonymous and synonymous genetic variants non-synonymous genetic variants are variants that can alter the amino acid sequence of a protein. By contrast, a synonymous genetic variant also resides in a coding exon but has no effect on the amino acid sequence.

non-synonymous mutation a mutation in an exon of a protein-coding gene coding that results in a change in the amino acid. Synonymous mutations do not result in modification to the sequence of amino acids.

non-synonymous nucleotide substitution a change in nucleotide sequence that alters the encoded amino acid.

non-synonymous substitutions (missense) a substitution that alters a codon and causes a change in amino acid residue.

normal host this ambiguous term is often used as a synonym for typical host.

northern blot a laboratory technique used to detect a specific RNA sequence within a mixture after separation by gel electrophoresis. The RNA fragments are transferred out of the gel to the surface of a membrane. The membrane is exposed to a DNA probe labelled with a radioactive or chemical tag. If the probe binds to the membrane, then the complementary RNA sequence is present in the sample.

nosocomial infection an infectious disease contracted while under medical care, usually acquired in a hospital.

notifiable disease disease or condition that is required by law to be notified to the State health department.

notification the process of reporting a notifiable infectious disease.

novel transcripts previously unidentified RNA transcripts containing a spliced leader and a poly(A) tail. They comprise protein-coding and non-protein-coding RNAs.

nuclear factor of activated T cells (NFAT) is a family of transcription activators that play a significant role in T cell activation and differentiation. NFAT was first identified as an inducible DNA-binding protein able to bind the IL-2 promoter. NFAT is also involved in other immune cell functions (dendritic cells, B cells).

nuclear fusion fusion of two nuclei. This results in the combination of genetic material from two nuclei into a single nucleus.

nuclear magnetic resonance (NMR) a technique used to study molecular structure by analysing the absorption of electromagnetic resonance at a specific frequency in atoms subjected to strong magnetic field.

nuclear membrane a double membrane that encloses the cell nucleus. It serves to separate the chromosomes from the rest of the cell. The nuclear membrane includes an array of small holes or pores that permit the passage of certain materials, such as nucleic acids and proteins, between the nucleus and cytoplasm.

nuclear pore complex (NPC) a complex of ≈30 proteins that form a pore at the nuclear envelope and mediate transport of molecules between the nucleoplasm and the cytoplasm of eukaryotic cells.

nuclear transfer technology a method of cloning a living organism. The process involves removing the nucleus of an egg cell and replacing it with a nucleus from any cell of the organism being cloned.

nuclear-encoded polymerase (NEP) a viral-like, nuclear-encoded RNA polymerase commonly found in plant chloroplasts and/or mitochondria that is encoded on the nuclear genome.

nucleic acid an important class of macromolecules found in all cells and viruses. The functions of nucleic acids have to do with the storage and expression of genetic information. DNA encodes the information the cell needs to make proteins. A related type of nucleic acid, called RNA, comes in different molecular forms that participate in protein synthesis. *See* **deoxyribonucleic acid**.

nucleolar organizing region a part of the chromosome containing rRNA genes.

nucleolus is a region found within the cell nucleus that is concerned with producing and assembling the cell's ribosomes. Following assembly, ribosomes are transported to the cell cytoplasm where they serve as the sites for protein synthesis.

nucleophile the term is used as shorthand for the amino acid providing the functional group that makes an initial attack on the carbonyl carbon of a peptide bond. In canonical cysteine peptidases, the nucleophile is the sulfur of the -SH functional group of the active site cystei.

nucleoplasm the protoplasm of the nucleus not including the karyosome, plastin and chromatin.

nucleopore one of a series of small holes found in the nuclear membrane. The nucleopore serves as a channel used for transporting nucleic acids and proteins into and out of the cell nucleus.

nucleosome a basic unit of DNA packaging in eukaryotes that consists of section of DNA (≈166 bp) wrapping around a histone core. Nucleosome structure helps to condense DNA into smaller volume. Nucleosomes are subunits of chromatin.

nucleosome occupancy the amount of nucleosomes present at a particular genomic region at a specific time. Nucleosome occupancy is influenced both by (i) the density of nucleosomes (e.g. the length of the linker between adjacent nucleosomes), and (ii) the stability of these nucleosomes (e.g. the frequency/probability with which the nucleosome is displaced from the DNA).

nucleotide the subunit of nucleic acids, DNA and RNA, that consists of a five-carbon sugar, a phosphate group and a nitrogenous base. The bases used in DNA are adenine (A), cytosine (C), guanine (G), and thymine (T). In RNA, the base uracil (U) takes the place of thymine. *See* **bp**, **deoxyribonucleic acid**, **ribonucleic acid**.

nucleotide oligomerization domain (nod)-like receptors a family of multidomain cytosolic pattern recognition receptors (PRRs) molecules detecting bacterial molecular patterns. C-terminal leucine rich repeats regions are involved in pattern-associated molecular patterns (PAMP) recognition.

nucleotide-binding domain and LLR-containing protein (NLR) resistance proteins' that monitor or guard host proteins that are targeted by pathogen effector proteins. NLRs induce an effector-triggered immunity defence response upon perceiving an effector-mediated modification to the host target protein.

nucleotide-binding oligomerization domain receptors (NLRs) also called Nod-like receptors, are intracellular sensors of PAMPS and DAMPs able to co-operate with toll-like receptors (TLRs) to regulate or potentiate the inflammatory and apoptotic response.

nucleus a spheroid body within the protoplasm of a cell, distinguished from the rest of the cell by its dense structure and presence of chromatin. The nucleus controls growth, cell division, and other activities of the cell; and contains DNA, the basic substance controlling the genetic characteristics of the cell or organism.

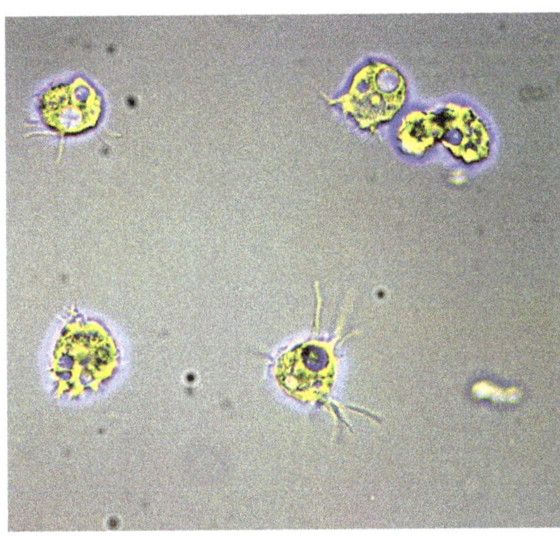

nude mice mouse strain, which is T-cell deficient due to a homozygous gene defect (nu/nu) resulting in the absence of a thymus (and also lack of body hair).

nuisance biting arthropods are arthropods that do not transmit pathogens but affect quality of life owing to nuisance biting.

null completely missing.

null allele an allele that is not amplified although present. The non-amplification of the allele can occur because of the presence of mutations in the flanking regions preventing the binding of the primers.

null cells an early population of lymphocytes that lacks T-and B-cell differentiation antigens, so they cannot be assigned to the B nor to the T lineage.

null mutants defective gene mutants where the genes are non-functional.

number needed to harm (NNH) the number of patients for whom there is one additional patient who experiences a harmful outcome.

number needed to treat (NNT) the number of patients with a particular condition who must receive an intervention to prevent the occurrence of one adverse outcome.

nurse cell is formed from mammalian striated muscle cells upon infection with larvae of the parasite *Trichinella*. It constitutes an intracellular niche fulfilling larval metabolic requirements and protecting it against host defence mechanisms.

nymph the pre-adult stage of an insect, which has hemimetabolic development in a terrestrial environment.

A B C D E F G H I J K L

M N **O** P Q R S T U V W

X Y Z

obligate endosymbionts beneficial symbionts that have intracellular lifestyle in specific host cells and depend on the host for survival.

obligate parasite a parasite that requires a specific host in order to complete its life cycle.

observational study (non-experimental study) changes or differences in one characteristic (e.g. whether or not people received a specific treatment or intervention) are studied in relation to changes or differences in other(s) (e.g. whether or not they died), without the intervention of the investigator examples include case control and cohort studies.

observed scale measurement of a binary phenotype in terms of whether the participant exhibits the phenotype or not.

occasional transmission the direct or indirect transmission of an otherwise vector-dependent parasite or pathogen arising under natural but rare circumstances, e.g. as the result of a scavenging or predation event or through aggressive interactions during the breeding season.

occult blood blood in such small quantities that it is not readily detectable except by chemical means.

occult infection an infection that does not display detectable abnormalities at the physical examination but might present haematological alterations consistent with a chronic, 'hidden' infection. This infection is not detectable by usual diagnostic methods and might be recognized by secondary clinical manifestations at first.

octopamine a biogenic amine which acts as a neurotransmitter in some invertebrates; in plant parasitic nematodes it stimulates pumping of the feeding tube/pharynx.

octoploid having eight complete sets of homologous chromosomes in a cell.

odds ratio the ratio of the odds that an outcome occurs given a set of covariates compared to the odds that the outcome occurs in their absence. for example, the odds of observing (by Kato–Katz) hookworm eggs one week after treatment with albendazole divided by the odds of observing hookworm eggs immediately before treatment.

oedema abnormal accumulation of fluid in or around cells, tissues, or serous cavities of the body due to unbalance of fluid homeostasis. They may be observed in limbs and ventral parts of the body. Alveolar and interstitial oedemas in lungs are characterized by fluid spilling into the air spaces. *See* **bottle jaw**.

oesophagostomosis infection associated with *oesophogostomum* species. They are called 'nodular worm' because larvae developing in the colonic mucosa cause an inflammatory response leading to nodules in the wall of the large bowl which interfere with digestion and absorption. life cycle include the following sequential events; adult males and females are present in colon lumen, eggs are passed in the faeces; larvae develop on the ground through L1 to L2 to L3; infective stage, L3 is ingested by the host; larval development to the L4 stage occurs in the colonic submucosa; larvae may arrest here for up to 2 months; larvae emerge, moult again in the colonic lumen and develop to adults. prepatent period is 30 day. Infection of sows is most important and heavy infections can lead to weight loss, reduced litter sizes and reduces litter weaning weights. Diagnosis can be accomplished by faecal flotation for detection of strongyle-type egg.

***Oestrus ovis* infestation** is common in sheep, and occurs when the adult *O. ovis* fly deposits larvae around the nostrils of the sheep. the larvae migrate up the nasal passages into the sinuses, where they develop for weeks to months before coming out to pasture to pupate on the ground. Infestation of the nasal cavity with the larvae (bots) irritates the tissues, causing rhinitis, sneezing, mucopurulent nasal discharges, and decreased airflow.

Secondary bacterial infection and pneumonia may develop. Endoscopy is helpful in making a diagnosis. Ivermectin is used in the late summer to prevent the build up of heavy infestations and again in the winter to kill overwintering larvae.

off-target reads the sequencing reads that do not align to the target region.

Old World the eastern hemisphere, comprising that part of the world that was known before the discovery of the Americas. It includes Europe, Asia, and Africa.

olfactory aversive learning a change in olfactory preference following conditioning with an aversive stimulus. in the case of pathogen avoidance, infection and cellular damage act as the unconditioned stimulus, and bacterial odours act as the conditioned stimuli that become aversive following training.

oligoclonal a few different clones, or the product of a few different clones.

oligogenic a phenotypic trait produced by two or more genes working together.

oligomycin inhibitor of the mitochondrial ATP synthase.

oligonucleotide short nucleic acid polymer, typically with 50 or fewer bases. *See* **nucleotide**.

oligotrophic environments that are characterized by low concentrations of growth-requiring nutrients and, consequently, low microbial biomass. Such habitats dominate the upper ocean of the large subtropical gyres. An oligotroph is the term used to describe an organism that is adapted to growing in habitats with low-nutrient conditions.

omega (ω) the ratio between the rate of non-synonymous and synonymous substitution; ω >1 is evidence for positive selection assuming that synonymous mutations do not impact fitness.

omics term for a range of new high-throughput biological research methods (e.g. genomics, transcriptomics, proteomics, and metabolomics) that systematically investigate entire networks of genes, proteins, and metabolites within cells.

Onchocerca cervicalis a widespread filarial nematode of horses. Adult worms live in the ligamentum nuchae, producing microfilariae which migrate in connective tissue to the upper dermis, especially the skin of the ventral midline. *Culicoides* spp. act as a vector. This parasite causes fibrotic, caseous and calcified lesions in the ligamentum nuchae. The conditions known as 'poll evil' or 'fistulous withers' are no longer thought to be associated with this parasite.

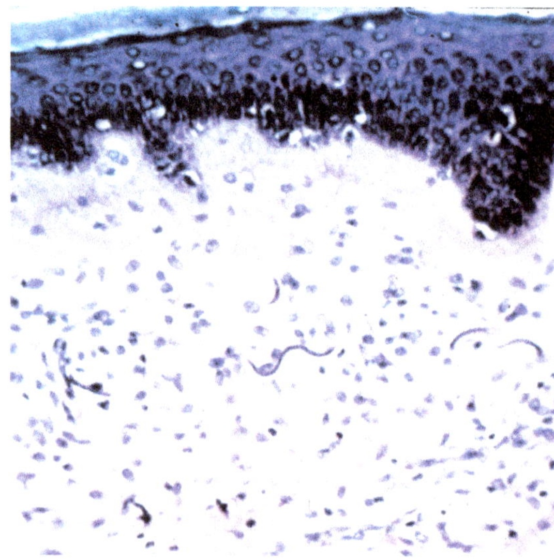

Onchocerca ochengi a nodule-dwelling filarial nematode parasite of cattle in Africa. Adult worms reside in intradermal nodules. Females are sedentary; males may move between nodules. The first stage larvae (microfilariae) disperse in the skin of the host waiting to be taken up by the intermediate host.

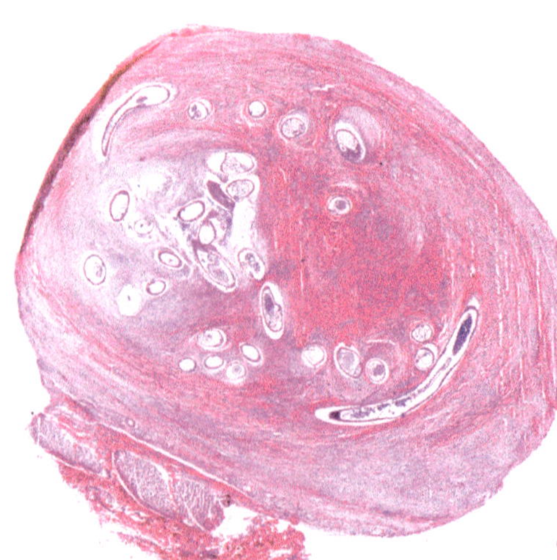

Onchocerca volvulus is a filarial nematode that is transmitted to humans by blackflies of the genus *Simulium*, cause a disease known as river blindness.

onchosphere the six-hooked (hexacanth) embryo that is contained in the egg membranes of members of the subclass cestoda.

oncofetal antigen antigen whose expression is normally restricted to the fetus, but which may be expressed during malignancy in adults.

oncogene is a mutated gene that contributes to the development of a cancer. In their normal, unmutated state, onocgenes are called proto-oncogenes, and they play roles in the regulation of cell division. Some oncogenes work like putting your foot down on the accelerator of a car, pushing a cell to divide. Other oncogenes work like removing your foot from the brake while parked on a hill, also causing the cell to divide.

oncomiracidium a free-swimming, ciliated larva of members of the class Monogenea.

oncotic pressure the force exerted by proteins to draw tissue fluids into the blood.

oncovirus a virus associated with cancer.

one health approach an interdisciplinary approach for combating threats to the health of animals, humans, and the environment they share on earth.

one-allele mate choice substitution of a single 'assortative mating' allele that causes all individuals, in any environment, to mate assortatively. A biologically plausible example would be if all individuals mate where they were raised. This causes assortative mating because individuals subjected to the same forces of selection become mates.

one-carbon mitochondrial metabolism exchange of one carbon molecules at different levels of oxidation between folate intermediates catalysed by enzyme complexes loosely attached to the inner mitochondrial membrane. The glycine cleavage system (GCV), serine hydroxymethyltransferase (SHMT), and 5,10-methenyltetrahydrofolate dehydrogenase multienzyme complex (MTHFD) are their main components.

oocyst first structure formed immediately following syngamy (fusion of gametes) in apicomplexan protists; oocyst matures, sporozoites develop within sporocysts (except cryptosporidia where sporozoites develop free in the oocyst). This is the resistant spore phase of apicomplexans that can remain viable outside a host for long periods of time.

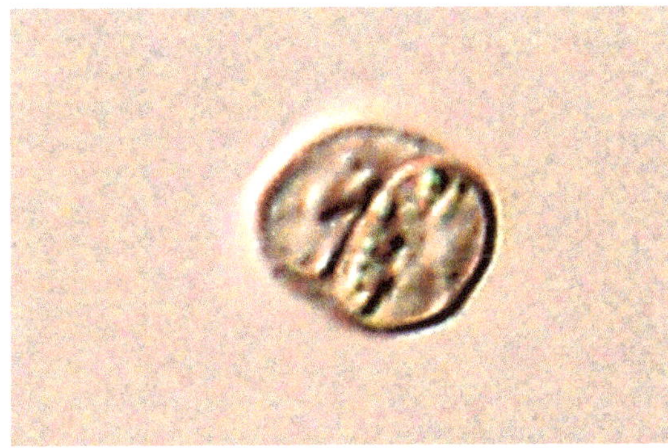

oogonial cells small, immature cells in the anterior small part of the bulb-shaped ovary of mature *Schistosoma mansoni* females. Bigger primary oocytes occur in the expanded posterior part of the ovary.

ookinete the motile zygote stage of *Plasmodium* which is seen in the midgut of the mosquito shortly after syngamy.

oomycetes a distinct phylogenetic lineage of fungus-like (filamentous) eukaryotic microorganisms that can reproduce both sexually and asexually. They are closely related to photosynthetic organisms such as brown algae and diatoms.

open chromatin regions correspond to nucleosome-depleted regions of the chromatin, allowing the binding of proteins to the chromatin. these regions can be identified by assays such as DNase-seq or faire (formaldehyde assisted isolation of regulatory elements)-seq.

open malaria modelling platform the name of the project in C++ programming language located at https://github.com/SwissTPH/openmalaria/wiki including implementation of the open malaria simulator that supports several models of malaria, providing a general way for comparing, fitting, and evaluating different model structures, and for quantitative prediction of effects of different interventions and integrated malaria control programs.

open reading frame expressed sequence tags (ORESTES) a portion of a DNA molecule that, when translated into amino acids, contains no stop codons. The genetic code reads DNA sequences in groups of three bps, which means that a double-stranded DNA molecule can read in any of six possible reading frames – three in the forward direction and three in the reverse. A long open reading frame is likely part of a gene.

operational taxonomic unit (OTU) a cluster of organisms similar at a defined sequence identity threshold (e.g. 95%) within some region (e.g. the whole 16s gene), used in the absence of a named species, genus, etc.

operculated eggs of some trematode and cestode helminths have a cap-like structure at one end through which the embryo (larva) emerges. Protrusions surrounding the base of the operculum, called shoulders, are present in the eggs of some worm species.

operculum or cap-like structure at one or both ends of certain worm eggs, i.e. *Trichuris*.

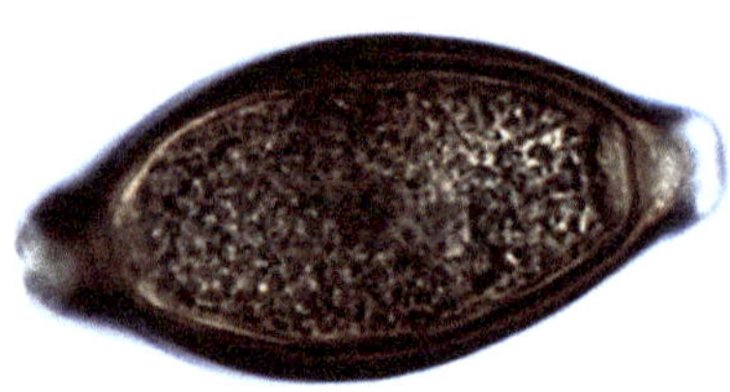

operon functional units of genomic DNA that contain a cluster of genes under the control of a single promoter.

opisthaptor the highly specialized, posterior hold-fast of members of the class Monogenea.

opisthomastigote a morphotype characterized by cells with long and narrow flagellar pocket extending from a post-nuclear (posterior) kinetoplast to the anterior end of the cell.

opisthomorph a morphotype similar to choanomatigotes but with a posterior kinetoplast.

opportunistic infectious agents infectious agents, usually pathogens, that are able to establish infections in particularly susceptible (immunocompromised) hosts, or through unusual routes, such as eyes, which are relatively poorly immunologically defended. These can include organisms that are obligate parasites or those that are sapronotic, for example *Toxoplasma gondii* (obligate) and *Histoplasma capsulatum* (sapronotic).

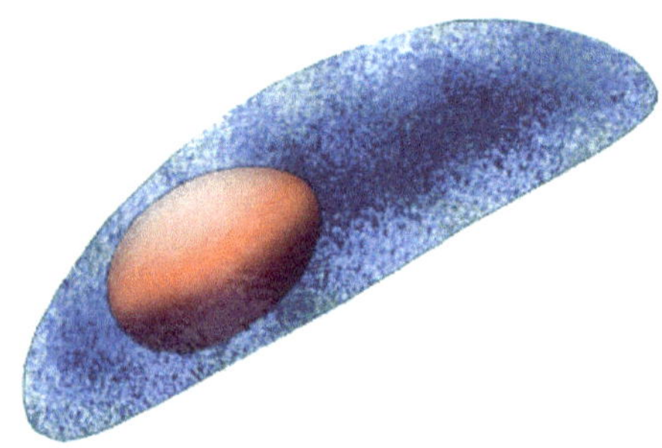

opportunistic parasite an organism that is not typically a parasite but may become parasitic under specific conditions. Normally free living amebae of the genus Naegleria, *Acanthamoeba*, and *Balamuthia* are usually free-living but may become opportunistic parasites.

opportunistic pathogen organisms of low virulence that are unable to cause disease in a healthy individual, but may invade and cause disease in an individual whose immunological defences are impaired.

opsonin proteins or protein complexes that bind to antigens on the surface of foreign particles enhancing phagocytosis. Opsonins form part of the immune response of the host. Receptors on immune cells then bind to opsonins, resulting in indirect detection of pathogens.

opsonization the coating of a particle with antibody and/or complement (opsonins) that leads to enhanced phagocytosis of the bacterium by phagocytic cells.

opsonizing antibodies promote uptake of parasites by phagocytic cells (e.g. monocytes, macrophages, neutrophils) expressing the appropriate Fc receptors on their surface; are predominantly IgG1 and IgG3.

optical trap also known as laser tweezers. An appliance that provides force from a highly focused laser beam to hold or move objects such as microspheres.

OPU (operational phylogenetic unit) a group of sequences that appear as a monophyletic clade that is meaningfully separated from the remaining sequences in a genealogical reconstruction.

organ a collection of tissues that structurally form a functional unit specialized to perform a particular function. One's heart, kidneys, and lungs are examples of organs.

organelle a subcellular, membrane-bounded structure that has one or more specific jobs to perform in the cell, much like an organ does in the body. Among the more important cell organelles are the nuclei, which store genetic information; mitochondria, which produce chemical energy; and ribosomes, which assemble proteins.

organism a living thing that contains DNA and is capable of metabolism and replication by itself; for example, bacteria, plants and animals.

organophosphates (OPs) a large group of drugs with insecticidal, acaricidal, and helminthicidal properties. Commonly used organophosphates include chlorpyrifos, coumafos, cythioate, fentrothion, dimpylate (diazinon), dichlorvos, fenthion, malathion, phosmet, tetrachlorvinphos, and trichlorfon. These products are effective on flies, fleas, ticks, mites, and lice. These drugs work by binding to and inhibiting the enzyme acetylcholinesterase (AChE), which degrades acetylcholine (ACh). ACh thus accumulates at the neural synapse, resulting in continuous neuronal stimulation and paralysis of the parasite. The binding of AChE by an OP results in phosphorylation and inhibition of the enzyme. The phosphorylated enzyme is capable of reactivation, but this reaction proceeds so slowly that the OPs are considered irreversible AChE inhibitors. The OPs have a low margin of safety. They should not be used with other agents that inhibit AChE or block neuromuscular transmission, and care should be taken when used in combination with CNS depressants. OPs should not be used in young, sick, or pregnant animals or in greyhounds, whippets, Persian cats, and certain breeds of cattle (Chianina, Charolais, Simmental, Brahman) due to high sensitivity.

organotypic cultures *ex vivo* cultures of tissue or tissue sections that have been obtained from HIV-negative donors in which physiological infection by HIV-1 can be simulated. When the tissue architecture of these organs is preserved, such cultures are referred to as organotypic.

origin recognition complex (ORC) the multi-subunit complex found in eukaryotes that binds to origins. The ORC is normally considered to be composed of six subunits, Orc1–6. Interacts with Cdc6, which is evolutionarily related to the Orc1 subunit.

ornate coloured; patterned. Some ticks are referred to as ornate ticks based on the presence of colourful markings distinct to certain species, such as *Dermacentor variabilis*, also known as the American dog tick or Wood tick.

Ornstein–Uhlenbeck processes mathematical models that describe random variations of a quantitative variable towards an optimum value. This model is widely used to describe evolutionary processes, as it can incorporate both genetic drift and natural selection.

orphan (hypothetical) gene putative ORFs, in ESTs or genomic sequences, without any resemblance to previously determined protein coding sequences.

orphan genes genes without detectable sequence similarity in the genomes of other organisms. According to the strictest definition these are genes which do not encode any previously identified protein domains. There is no general agreement or rule, but usually proteins which do not show any sequence similarity in BLASTP searches with cut-off values E <10–5 or E <10–10 are referred to as 'orphan' or 'novel'. There is increasing evidence from genomic and transcriptomic sequencing that 'orphans' represent genes which have a narrow phylogenetic distribution (homologous 'orphans' can be present in closely related species but are not present in more distantly related species or in other genera).

orthochromatic erythroblast erythrocyte progenitor that is the last stage before enucleation.

orthogonally acting RNA elements RNA elements that have distinct target-binding specificities and do not exhibit crosstalk.

orthologous term used to describe the relationship between genes or features present in different species but sharing a common evolutionary descent, e.g. the genes encoding α-haemoglobin in humans and α-haemoglobin in chimpanzees are orthologous.

orthologous genes homologous genes related through a speciation event. They are commonly assumed to share a similar function.

orthologues homologous genes in different organisms that are believed to have evolved from a common ancestor; they usually fulfil similar functions in each organism. This is in contrast to paralogues, which are genes in the same organism that have arisen by gene duplication and have typically diverged in function as well as sequence.

orthology describes genes in different species that derive from a common ancestor. Orthologous genes may or may not have the same function.

OrthoMCL a method for the identification of orthologous groups across multiple eukaryotic taxa using a Markov Cluster algorithm to group (putative) orthologues and paralogues.

oslerosis the disease caused by nematodes belonging to the genus *Oslerus*. Felid oslerosis is caused by *Oslerus rostratus*, a rare metastrongyloid infecting the domestic cat and wild felids. Little information is available on clinical pictures in cats, although a mild bronchitis has been described in experimental infections. No information is available on the treatment of the cat infection.

osmiophilic body a membrane-bound vesicle that is predominantly found in female gametocytes.

osmotic fragility cell lysis in a solution of a permeant solute. The kinetics of cell lysis have been used to quantify the permeability of the solute.

outbreak *See* **epidemic**.

outgroup taxon phylogenetically outside the clade of interest (the ingroup). When one uses an outgroup in phylogenetic inference, the ingroup is implicitly assumed to be monophyletic.

overall disease risk the lifetime probability that an individual will be affected by a disease.

overdispersion the occurrence of variance that is greater than expected based on a simple probability distribution.

overdominant selection a balancing selection that favours the heterozygotes.

ovigerous female a female louse carrying two conspicuous uniseriate egg strings containing a total of 100–1000 eggs.

oviparous egg-laying organisms where the embryos have developed little or not at all. Seen, for example, in the nematodes *Toxocara canis* and *Ancylostoma caninum*.

ovoviviparous a species in which eggs hatch in utero within the adult female and the freed young motile forms are released from the female worm, e.g. *Strongyloides stercoralis* and *Dirofilaria immitis*.

ovum (-a) a female reproductive/germ cell with a haploid number of chromosomes preceding combination with the male gamete and the addition of an embryonic membrane and other shell layers.

oxadiazines (indoxacarb) an insecticide molecule that binds to a specific receptor, leading to a voltage-dependent blockade of Na^+ channels, causing inhibition of nerve activity and lethal paralysis. Indoxacarb is used in the form of spot-ons for dogs and cats to control fleas. A combination with permethrin in dogs has been developed to add tick elimination.

oxidation loss of one or more electrons from a chemical substance.

oxidative phosphorylation (OXPHOS) the metabolic process that mitochondria perform to produce ATP by oxidation of organic molecules (mostly glucose) to H_2O and CO_2, when O_2 is available.

***Oxyuris equi* (pinworm)** a nematode that occurs in the caecum, colon, and rectum of horses. The fertilized female migrates down the gut and crawls onto the peri-anal area, where she lays her eggs onto the skin in yellow clusters. *Oxyuris equi* provokes irritation of the perianal region of horses during egg laying, causing them to rub and bite their tails. This can result in hair loss and sometimes physical damage to the tissues of the area. *Oxyuris equi* has no zoonotic potential and does not infect other domestic animals. Humans serve as definitive hosts of *Enterobius vermicularis*, a pinworm that is unique to primates.

padlock probes oligonucleotide probes consisting of target-complementary homology arms connected by linker sequences. Upon recognition of the target, the extremities of the probe are ligated, and the circle created is amplified through rolling circle amplification (RCA). The resulting product is detected using fluorescence *in situ* hybridization (FISH) probes. Padlock probes and RCA combine high specificity of detection with high signal amplification, and they have been used in numerous applications *in vitro* and *in situ*.

paediatric formulation antimalarial drug packaged specifically for children.

PAF (platelet activating factor) an alkyl phospholipid released by a variety of cell types including mast cells and basophils, which has immunoregulatory effects on lymphocytes and monocytes/macrophages as well as causing platelet aggregation and degranulation.

paired-end reads DNA fragments that were sequenced from both ends, yielding pairs of reads that are separated by a defined distance that is dependent on the library preparation protocol.

PAMP/DAMP pathogen or danger associated molecular patterns, molecular motifs associated with groups of pathogens or non-infectious stimuli, e.g. cellular debris from dying cells, that are recognized by cells of the innate immune system.

pan genome the entire set of genes carried by a group of strains being examined (that is, essentially the total gene pool). Mathematically, the pan genome is the union set of the genomes from the strains under study.

panmictic a population where mating is random with all individuals can mate freely without genetic restriction.

panmixia absence of any differentiation among subpopulations (because of high levels of gene flow, creating effectively one single large population with no internal structure). The adjective is panmictic.

PaO_2/F_iO_2 ratio of partial pressure of arterial oxygen/inspired fraction of oxygen. This ratio reflects the efficiency of oxygen transfer from the inspired air to the arterial blood and is used to define ALI (PaO_2/F_iO_2 <300 mmHg) and ARDS (P_aO_2/F_iO_2 <200 mmHg).

papilla (-ae) a nipple-like structure.

papule/popular small, solid nodule in the skin, which may be intensely pruritic.

papulonodules a papule is a circumscribed, solid elevation of skin with no visible fluid and indurations, whereas nodules are larger solid areas of skin, with more than 0.5 cm area and indurations. Patients who have both papules and nodules are considered as papulonodules.

parabasal (parabasal body) in flagellates, a heavy fibre (rhizoplast) present in some flagellates. It supports the cytostome in *Chilomastix* and is the basal fibril of the undulating membrane in *Trichomonas*. The function of parabasals in *Giardia* is unknown.

parabasalid a member of a group of primarily flagellated protists, most of which form commensal or parasitic relationships with animals; includes the trichomonads.

paracellular the opposeite of transcellular, this term describes the transport through the intercellular space between the columnar intestinal epithelial cells. Such transport is relevant mainly for small molecules, but some mononuclear phagocytes establish tight junctions to extend cellular processes in a paracellular manner.

paracellular diapedesis migration of cells between endothelial or epithelial cells including transient openings of their cellular junctions.

paracortex the region located between the cortex and medulla of lymph nodes in which T cells predominate.

paracrine a term used to describe the activation of cellular receptors on cells adjacent to the cell that produces the ligand. *See* **autocrine**.

paracrine signalling a form of cell signalling in which cells secrete molecules to signal to nearby cells in order to bring about changes in differentiation or behaviour.

paraflagellar rod the extra-axonemal structure of kinetoplastids. the paraflagellar rod lies alongside the axoneme and functions in flagellar motility.

parallel analysis of RNA structure (PARS) a genome-wide measurement of RNA secondary structure based on deep sequencing of RNA fragments generated with structure specific enzymes.

parallel speciation parallel evolution of traits underlying reproductive isolation in independent populations evolving along similar environmental gradients.

paralogues homologous genes related through a gene duplication event.

parameter a quantity whose value varies with the circumstances in which it is applied.

parametric bootstrap a parametric model is fit to the data, and then samples are drawn from this model in order to quantify confidence in a particular estimated parameter.

parametric versus non-parametric statistical tests parameters are quantities that are constant for particular distributions but that take on different values for different members of families of the same kind of distribution. For example, the population mean and variance are parameters of the normal distribution. One or more population parameter is estimated when using a parametric test. Non-parametric tests (sometimes called 'distribution-free' tests) have weaker distributional assumptions and do not involve parameter estimation. When parametric assumptions are reasonable (or when the parametric statistic is robust with respect to their violation), parametric tests are generally statistically more powerful than their non-parametric counterparts.

paramphistome infections have been associated with significant morbidity, caused chiefly by the activity of juvenile flukes in the intestine of the ruminant final host. Most cases have been reported in tropical and subtropical areas. However, recent reports of an apparent increase in the incidence of rumen fluke and its geographical range in Europe have renewed interest in a parasites previously considered of little significance in temperate regions. Moreover, the identity of rumen flukes present in the British is currently being revised.

paramutation the heritable change of function of an allele that is induced by another allele.

parapatric abutting of geographical ranges of populations, such that only the members of populations at the contact edges of geographical ranges encounter one another.

paraphyletic containing the descendants of a common ancestor that retain shared primitive characters, but omitting descendants that have lost those characters.

paraphyly description of a group of organisms that does not include all the descendants of a common ancestor, as opposed to monophyletic groups, which include a common ancestor and all its descendants. Organisms of monophyletic and paraphyletic groups share characteristics that are identical-by-descent (i.e. that have been inherited from a recent common ancestor).

parasexual recombination of DNA that does not involve meiosis. Examples include the fusion of nuclei within a fungal heterokaryon or bacterial conjugation.

parasitaemia the presence of parasites in the bloodstream.

parasite control reduction of infection incidence, prevalence, or morbidity to a locally acceptable level. Requires continued interventions and might constitute the first phase of a national intervention program.

parasite density number of red blood cells (RBCs) infected by a given parasites divided by the total number of RBCs, expressed in percentage. Pertains only to those parasites that infect RBCs.

parasite elimination reduction of the incidence of infection to zero in a defined geographic area. Requires continued measures to prevent re-establishment of transmission.

parasite eradication permanent reduction to zero of the worldwide incidence of infection. Once accomplished, intervention measures will no longer be required. Ultimate goal of the global programme.

parasite evolution the constant change of a parasite population in the face of selection pressure.

parasite fitness basic reproductive ratio: the expected number of secondary infections arising from a single individual during the entire infectious period, in a population of susceptible individuals.

parasite nomenclature the system of naming organisms including parasites was established in 1735 by Carolus Linnaeus. Scientific nomenclature assigns each organism two names; the genus name is the first name and is always capitalized; the specific (species name) follows and is not capitalized. The parasite is referred to by both the genus and specific species, and both names are underlined or italicized. By custom, after a scientific name has been mentioned once, it can be abbreviated with the initial of the genus followed by the species name.

parasite pathogenesis is the mechanism by which a parasite replicates in the host, and in doing so, injuries the host and produces disease.

parasite persistence the presence of parasites even after clinical cure. It is generally seen in tryponosomasis and leishmaniasis.

parasite tolerance ability of an organism to reduce the fitness costs of a particular parasite burden.

parasite-derived factors factors that can contribute to parasite establishment and pathogenesis. For example, *Acanthamoeba* encephalitis adhesins such as mannose-binding protein, laminin-binding protein, ecto-ATPases, neuraminidases, proteases, elastases, phospholipases and glycosidases.

parasites are organisms that are metabolically and physiologically dependent on other organisms, their hosts, for survival and development. although all parasites are considered as eukaryotic, some are unicellular and others are multicellular. they range in size from tiny protozoa as small as 1 to 2 μm in diameter (= the size of many bacteria) to arthropods and tapeworms that can measure several metres in length. the life cycles of parasites are complex; with some parasites establish a permanent relationship with hosts and others going through a series of developmental stages in a succession of hosts.

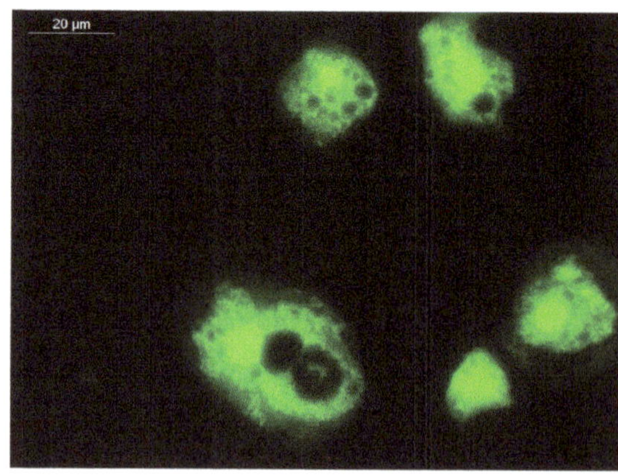

parasitic castrator a parasite that blocks reproduction by its host, either completely or in part, by specifically targeting reproductive organs in a manner that is independent of the intensity of infection.

parasitic gastritis in pigs the nematodes *Hyostrongylus rubidus* – a trichostrongyle parasite of swine stomachs. Three stomach worms occur in pigs: *H. rubidus*, *Ascarops strongylina*, and *Physocephalus sexalatus*. *H. rubidus* is the most common and the most pathogenic of the three. Infections occur in indoor husbandry systems. *H. rubidus* – has a direct life cycle. Parasitic development is similar to that of *Ostertagia* spp. in ruminants. L3 larvae burrow into the gastric glands and moult to L4 and to adults in the gastric glands. Larvae can arrest in the gastric glands as a L4. Adult males and females reside in the stomach. Fertilized female produce strongyle-type egg in faeces.

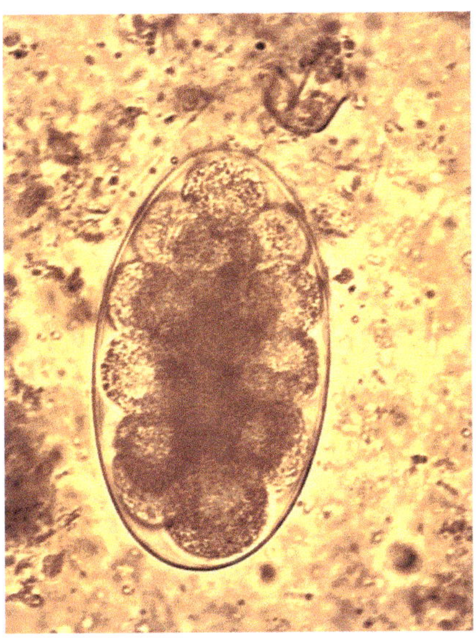

parasitic gastroenteritis (PGE) gastrointestinal nematodes infect small ruminants causing a condition called PGE. The GI nematodes of greatest importance in small ruminants are members of the nematode of order strongylida and most belong to the superfamily Trichostrongyloidea. *Haemonchus contortus* is the most important strongylid nematode (abomasums) worldwide. A second important trichostrongylid abomasal nematode that contributes to PGE is *Teladrosagia circumcincta* (formerly *Ostertagia circumcincta*). *Nematodirus* is a third important genus that affect small intestine of sheep seriously especially in northern Europe. *Trichostrongylus* is a fourth important genus of trichostronglid nematode (small intestine) that contributes to PGE.

parasitic phase of life cycle the development and maturation to adult males and females occurring in the definitive host of a parasite life cycle.

parasiticide a chemical or chemotherapeutants used to kill parasites. Can be called a medicine, pesticide, or drug in different countries.

parasitism obligatory association between two distinct organisms (the parasite and the host), where the parasite lives at the expense of the host and at the same time may cause damage (pathological lesions or disease condition) to the host.

parasitogenic due to parasites.

parasitoidan animal which is parasitic in one stage of the life cycle and subsequently free-living in the another stage, as the parasitic Hymenoptera or Trombiculid mites.

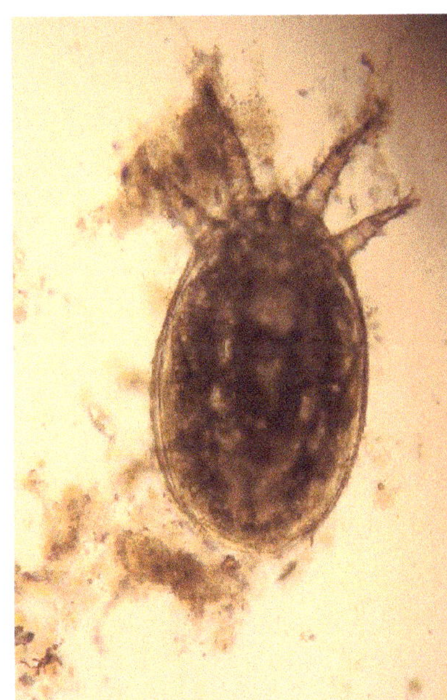

parasitological pertaining to or emanating from parasitology.

parasitological detection (pd) direct methods used for detection of parasite evaluative forms in biological samples.

parasitological examination inspection of and test for the presence of parasites. Procedures include examination of faeces for protozoa, worm eggs or larvae and for tapeworm segments, skin scrapings for arthropod parasites, blood samples for haemoprotozoa, microfilariae, for plasma pepsinogen levels, and examination of gross specimens.

parasitologist a person skilled in parasitology.

parasitology the scientific study of parasites and parasitism, and the host–parasite relationships. Traditionally this area of study has focused on parasites belonging to the protozoa, helminths and arthropods.

parasitophorous vacuolar membrane (PVM) a membrane surrounding the intracellular parasite, derived originally from the host cell membrane at the time of invasion. With parasite development, the lipids and proteins that make up this membrane change dramatically.

parasitophorous vacuole (PV) a vacuole formed by invagination of the plasma membrane of the host cell during invasion. the newly formed PV membrane surrounding the parasite can be transient (e.g. *Babesia*) or persist for the duration of the life of parasites within the host cell (e.g. *Toxoplasma* and *Plasmodium*).

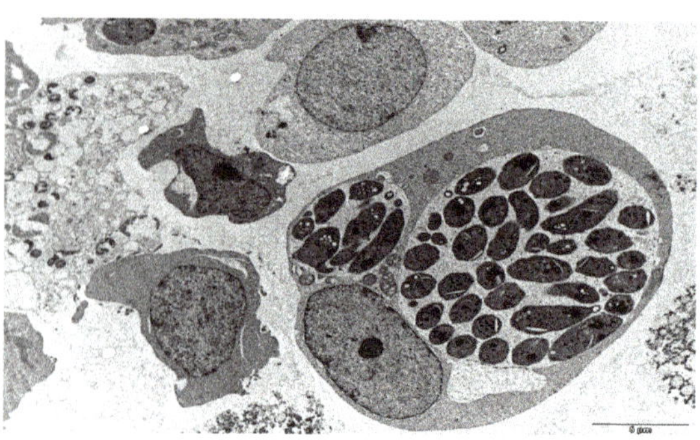

parasitosis a disease caused by a parasitic infestation.

parasitotropic having affinity for parasites.

paratenesis the passage of an infective stage larva by a transport (paratenic) host to the definitive host. The larva does not undergo essential development within the transport host, but is maintained in its infective stage from one season of transmission to another.

paratenic host an atypical (substitute) intermediate host that ingested and harbours the stage infective for the definitive host, which remains active and unchanged (a collector or storage host). If a suitable definitive host ingests the paratenic host or a part of it containing the infective stage, the parasite can grow to maturity.

paratope the antigen combining site on an immunoglobulin that is complementary to an epitope (part of the antigenic determinant).

paratransgenesis genetic modification of a symbiont to affect the vector competence of its host. Symbionts need to be transformed easily without affecting their fitness, they need to spread across host populations, and they need to reside in host organs affected by pathogens.

paratransgenic fly a fly that carries a genetically modified symbiotic organism. The expression of products in symbiotic bacteria can confer beneficial traits upon the paratransgenic fly, such as antiparasite properties. This mode of gene expression is an alternative approach to transgenic insects, where foreign genes are expressed in insect cells.

parenchyma the soft, undifferentiated tissue composing the general substance of the body of some invertebrates, e.g. members of the phylum Platyhelminthes.

parenteral vaccine vaccines administered through a non-mucosal surface (intravenous, intramuscular, etc.).

Paromomycin an aminoglycoside antibiotic used in the treatment of amoebiasis and cryptosporidiosis. It is poorly absorbed orally and has little activity against intestinal bacteria. Due to the potential to cause nephrotoxicity, it should not be used in patients with renal disease.

parous rate the proportion of the vector species that has already laid a batch of eggs.

parsimonious interpretation preference for the least complex explanation for an observation (i.e. that requiring the fewest unproven caveats or assumptions).

parthenogenesis asexual reproductive mode that does not involve fertilization of the female gamete. The development of eggs of certain parasite species without the fertilization by the union of male and female gametes, e.g. production of viable eggs by parasitic adult females in intestinal infections of *Strongyloides stercoralis*.

partial hepatectomy surgical removal of part of the liver.

partially methylated domains (PMDs) large contiguous regions of the genome (mean length ≈153 kb) that display an intermediate level of DNA methylation (average <70%).

partially sequenced regions a common approach in empirical population genetics, in which polymorphism data (plural; datum is singular) are generated for a series of small fragments spanning a large genomic region.

partner choice the preferential interaction with individuals who are more co-operative of the same or a different species.

PAS (phagophore assembly site) a perivacuolar location or compartment where the nucleation of the phagophore initiates. Most components of the macroautophagy core machinery locate at least transiently at the PAS in yeast; however, a mammalian equivalent of the PAS has not been identified.

passeriform birds the largest order of birds that contains more than half of all bird species. This order contains, e.g. finches, sparrows, wrens, robins, warblers, and blackbirds.

passive agglutination the agglutination of inert particles by antibodies directed against antigens bound to their surface.

passive case detection (PCD) the detection of malaria cases based on the spontaneous presentation of ill patients to fixed health services.

passive immunity the transfer of immunity against an infectious agent to a naive animal by providing antibodies, that were produced in another animal, either orally or via injection.

passive immunization direct administration of immune response products (e.g. antibodies or stimulated immune cells) obtained from an appropriate donor(s) to a recipient patient.

passive transmission a form of transmission in which the host/parasite contact is more or less accidental. Ordinarily, the host comes to the parasite, and the intimate contact necessary for infection is provided by the host, e.g. amoebiasis.

patch clamp technique an electrophysiological technique for recording the very small *currents* that flow through single ion channels. It is used to study single or multiple ion channels in cell membranes.

patent period the interval during which eggs are passed by the definitive host.

pathobiont symbionts that are able to cause disease when the environmental conditions are altered.

pathogen an organism or agent that causes disease (e.g. bacteria, viruses, parasites and fungi).

pathogen effector any microbial secreted molecule providing a fitness benefit to the pathogen during host colonization, such as by suppressing the host immune system, by extracting nutrients from the host or even by suppressing the growth of microbial competitors in the host environment. Effectors are proteins or secondary metabolites and act in the plant apoplast or inside host cells.

pathogen-associated molecular pattern (PAMPs) molecules characteristic of pathogens, such as the cell-wall component of Gram-negative bacteria, lipopolysaccharide (LPS).

pathogenesis the causing of disease or toxic response in another organism.

pathogenesis-related (PR) proteins a heterogeneous group of proteins induced in plants by pathogen infection and exogenous chemicals.

pathogenic giving origin to disease or symptoms of disease.

pathogenicity the ability of a pathogen or parasite to cause disease in a host.

pathogenicity gene a gene whose direct or indirect product has a role in pathogenic interaction with the host.

pathognomonic characteristic clinical sign for a particular disease.

pathology the study of a disease, especially structural and functional changes caused by disease.

pathway series of molecular interactions that occur in a specific sequence to carry out a particular cellular process (e.g. sense a signal from the environment, convert sunlight to chemical energy, break down or harvest energy from a carbohydrate, synthesize ATP, or construct a molecular machine).

pathway hole a reaction known to occur in a pathway, but for which no gene has been assigned.

PCR correction the use of PCR to compare parasite genotypes and adjust drug efficacy results when the post-treatment parasite represents a new infection.

PCR fingerprinting based on the amplification of genomic DNA with single non-specific primers that anneal to mini- and micro-satellite DNA sequences. The polymorphic DNA fragments are separated on agarose gels and the resulting banding patterns are analysed as described for DNA fingerprinting.

PCR-restriction fragment length polymorphism (RFLP) following PCR amplification of a locus, the amplicon is treated with a restriction endonuclease. If the recognition site for this enzyme is present in the amplicon, two or more restriction fragments are generated. Thus, sequence variation between individuals at the recognition site(s) can be detected by electrophoresis.

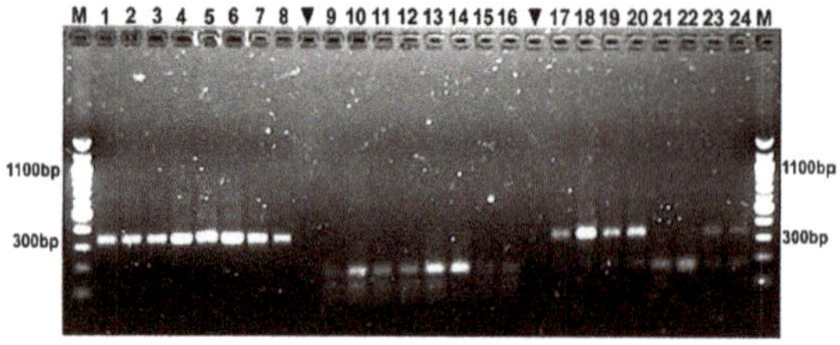

PCR-RFLP different amplified targets are digested with a defined set of restriction enzymes. The restriction fragments are separated by PAGE or on agarose gels. The evaluation of the resulting banding patterns is as for DNA fingerprinting.

pedigree a set of individuals connected by parent–child relationships.

pedigree clustering clustering of inherited allelic variants at a locus in families. Robust tests are used to control for this clustering when using multicase families for allelic association testing by transmission disequilibrium test (TDT) or case/pseudo-control logistic regression analysis (i.e. to be sure that association between an allele and disease is real and not just because of the clustering of alleles in families).

pedogenesis a process in the molluscan intermediate host which involves the production of new individuals by a single larval form in which one trematode egg may eventually develop into hundreds of adults.

pelagic organisms that live in the water column of marine systems.

pellicle a double external membrane.

pellicular membrane surface membrane of the cell body (excludes flagellum and flagellar pocket).

penetrance the proportion of individuals carrying a particular variation of a gene that also express an associated phenotype trait of interest; thus, genetic variants carrying a high risk of susceptibility to a particular disease or resistance to a particular insecticide have high penetrance for that trait.

penetrative transmission transmission by penetration through the surface tissues of the host by the helminth larva.

penitrem a fungal neurotoxin that specifically inhibits BK channels including SLO-1.

pentastomida the pentastomids (synonym tongue worms) are respiratory endoparasites of reptiles, birds, and mammals. Pentastomida shares features with both Arthropoda and Annelida, but is usually regarded as its own phylum. Adult pentastomids are white and cylindrical or flattened parasites that possess two distinct body regions: an anterior cephalothorax and an abdomen. Diagnosis may be made by finding the eggs in the faeces or respiratory secretions. There is currently no effective treatment for pentastome infestations. Infestations can be controlled through good hygiene and feeding parasite-free prey. Humans can act as intermediate hosts, becoming infected by having their hands contaminated from the faeces or saliva of the reptile, and accidentally ingesting the eggs. In some cases infected patients develop localized inflammation.

pepsinogen inactive precursor of the enzyme pepsin that contributes to peptic digestion in the abomasum. Infection with *Ostertagia ostertagi* in cattle results in elevation of abomasal pH, inhibition of the normal conversion of pepsinogen to pepsin, and leakage of pepsinogen into the bloodstream. Serum pepsinogen concentration in calves is used as an indicator of abomasal damage induced by *O. ostertagi* infection.

peptide an organic compound composed of two or more amino acids linked together chemically by peptide bonds. A component of a polypeptide. *See* **polypeptide**.

peptide exit tunnel the site, formed by proteins and ribosomalRNA, where the nascent polypeptide chain leaves the ribosome.

peptidoglycan recognition protein (PGRP) protein used in the innate immune response of insects to recognize surface peptidoglycans of pathogens and parasites.

percutaneous infection occurs when the free-living waterborne stage of the schistosome life cycle, known as cercariae, penetrate human skin and transform into schistosomulae.

perfect gold standard a test which has both a specificity and sensitivity of 1 or 100%. A perfect gold standard accurately predicts disease or infection status in every individual and in every stage of infection. In practice, a perfect gold standard is rarely available.

perfect match (PM) probe an Affymetrix probe that is designed to be complementary to a target sequence.

perforin a cytolytic protein found in the granules of cytotoxic T lymphocytes and NK cells. The component C9 of the complement system is also a perforin.

performance (provision, utilization and coverage) measures of program activities among intended recipients.

periarteriolar lymphoid sheath (PALS) the lymphoid tissue which forms the white pulp of the spleen.

pericentric inversion the inversion of chromosomal material that includes the centromere.

pericyte vascular mural cell embedded within the endothelial basement membrane of blood microvessels.

peridomestic mammal reservoir mammals (e.g. dogs) that live in and around human habitations, acting as a reservoir of infection.

period of communicability the time during which an infectious agent may be transferred directly or indirectly from an infected person or animal to a susceptible host.

periodic or seasonal parasites found on body of their hosts during a certain period of the year, e.g. *Oestrus ovis*, mosquitoes.

periodic parasite a parasite that parasitizes a host for short periods.

periodically forced model a model where some parameters vary over time with a repeating pattern.

periodicity the phenomenon seen in filarial parasites when the parasites are found in the blood in peak numbers at particular times of the 24-hour cycle.

periodontitis a biofilm-induced chronic inflammatory disease, which affects the integrity of the tissues that surround and support the teeth (periodontal ligament, gingiva, and alveolar bone; collectively known as the periodontium) and may exert an adverse impact on systemic health.

peripheral blood blood from the peripheral circulation.

peripheral tolerance specific immunological tolerance occurring outside of the primary lymphoid organs.

peripherally induced T regulatory cells (iTregs) regulatory Foxp3$^+$ CD4 T cell that develops in the periphery after infection or immune insult, exhibits specificity for either self- or non-self-antigens, and exerts suppressive function through both contact-dependent mechanisms and secretion of IL-10 and/or TGF-β.

periplasm the fluid occupying the compartment between the plasma membrane and outer membrane in bacteria. In phage display, proteins of interest are translocated from the bacterial cytoplasm to the periplasm, where they fold and assemble in phage coats.

periportal associated with portal vein (liver).

peristalsis rhythmical, wave-like constrictions of the wall of the intestine that move the contents.

peristome any parts or set of parts around the mouth or oral opening of invertebrates; comparable to lips in higher animals.

peritrabeculae adjacent to the trabeculae of the spleen; rod-shaped structures of fibrous tissue that subdivide the spleen.

peritrophic matrix (pm) a sheath of acellular matrix separating the midgut lumen of some insects, including tsetse, from the gut epithelium. There are two types of insect PMs: type I, in which the PM is secreted directly from gut epithelial cells, sometimes in response to feeding, and type II (e.g. *Glossina*), which continually secretes the PM from specialized cells in the PV, anterior to the midgut.

peritrophins proteins of the pm characterized by containing cysteine-rich CBDs, and usually modified with both *o*-linked and *n*-linked glycans.

perkinsids several genera of parasitic flagellates (e.g. perkinsus and parvilucifera) that diverge near the phylogenetic origin of dinoflagellates. These flagellates share several homologous characteristics with colpodellids and apicomplexans, such as myzocytosis and an apical complex comprising micronemes, rhoptries and an open conoid.

permanent parasites spend most of their life cycle in association with their hosts, e.g. *Entamoeba histolytica*, liver flukes, *Taenia* species; *temporary parasites*, which only visit their hosts occasionally and at intermittent times just for taking their meal, e.g. mosquitoes, bugs.

permeability (*p*) a measure of how easily a solute can cross a biological membrane. for uncharged solutes, the rate of transport across a membrane is proportional to *p*, the transmembrane concentration gradient, and the surface area of the membrane.

permissive vector a vector that supports development of multiple *Leishmania* species.

permutation a strategy for assessing the probability of observing the value of a particular statistic. the probability is computed from a data set in which the data are randomly shuffled and the statistic is recomputed from the shuffled data many times and ultimately compared to the value of the statistic obtained with the non-shuffled data.

persistence (of a population) a population may persist if the reproductive rates are higher than the mortality rates.

persistent infection an infection that is not cleared by the combined actions of the innate and acquired immune response.

personal hygiene the protective measures within the responsibility of the individual that limit the spread of infectious diseases.

personal protective equipment the equipment to be worn when performing duties that may involve possible occupational exposure to blood, splashing or aerosols from cleaning processes, e.g. gloves, goggles, lab coats and masks.

personalized medicine an emerging practice of medicine that uses an individual's genetic profile to guide decisions made in regard to the prevention, diagnosis, and treatment of disease. Knowledge of a patient's genetic profile can help doctors select the proper medication or therapy and administer it using the proper dose or regimen. Personalized medicine is being advanced through data from the human genome project.

pesticide a mixture containing a toxic active ingredient used for killing arthropods or other pests destructive to crops or animals.

petechia (-iae) a small purplish haemorrhagic spot on skin or mucous membrane.

Peyer's patches part of the gut associated lymphoid tissue (galt) and found as distinct lymphoid nodules mainly in the small intestine.

PF1022A one of the anthelmintically active cyclooctadepsipeptide fermentation products of the fungus *Rosellinia* sp.

PfCRT *P. falciparum* chloroquine-resistance transporter. It is a large protein with ten transmembrane domains and is located in the digestive vacuole of the parasite. Point mutations in the coding sequence are associated with resistance to chloroquine.

PfEMP1 *P. falciparum* erythrocyte membrane protein 1. It is encoded by var genes and is a large protein located in the membrane of the infected erythrocytes conferring adhesion to various endothelial receptors (cytoadherence). The protein is a target of natural immune responses and shows antigenic variation.

P-glycoprotein (P-gp) cell transporter presents in tumour cells and in several healthy tissues particularly in organs actively involved in drug pharmacokinetics. P-gp is located in different tissues implicated in the processes of drug absorption, tissue distribution (i.e. blood–brain barrier, enterocytes) and excretion (luminal surface of hepatocytes and duct cells, kidney tubules and enterocytes).

p-granules translationally silent storage particles containing high levels of non-translated mRNA. related to p-bodies, but differ in their inability to degrade mRNA.

pH scale used to specify acidity or alkalinity. The hydrogen ion (H^+) concentration of a sample determines its pH [pH = –log10 (H^+)]; the higher the H^+ concentration, the lower the pH. A solution with a pH value of 7 is neutral; less than 7 is acidic; and greater than 7 is alkaline or basic.

pH sensor amino acid residues that change protonation state according to variations in pH. A change in protein structure may consequently occur, owing to altered hydrogen bonding and electrostatic interactions.

PHA (phytohaemagglutinin) a plant lectin which acts as a T-cell mitogen.

phage a virus for which the natural host is a bacterial cell.

phage antibody library a collection of cloned antibody variable region gene sequences that can be expressed as Fab or scFv fusion proteins with bacteriophage coat proteins. These can be displayed on the surface of the phages. The gene encoding a monoclonal recombinant antibody is enclosed in the phage particle and can be selected from the library by binding of the phage to specific antigen.

phage-display library a phage-display library contains a large repertoire of bacteriophage particles that display foreign proteins or peptides on their surfaces. phage libraries are used to isolate ligands with binding specificity to selected receptors. Phage-display libraries can be classified according to the types of displayed peptides and proteins. Random libraries display peptides, which are derived from degenerate oligonucleotides, whereas natural libraries display natural peptides or protein domains. Genomic libraries are produced from fragments of chromosomal DNA, whereas cDNA libraries encode peptides derived from messenger RNA.

phagocytic cells innate immune cells specialized in the ingestion and degradation of foreign particles such as bacteria and tissue debris, including monocytes, macrophages, and granulocytes (neutrophils).

phagocytosis the process of engulfing material via the extension of pseudopodia by specialized cells called phagocytes. Once inside the cell, the engulfed material gets broken down within the phagolysosome.

phagolysosome a specialized membrane bound compartment formed when a phagosome, carrying a cargo from phagocytosis, fuses with a lysosome. The mammalian life cycle stage of *Leishmania* spp. is an intracellular parasite within the macrophage phagolysosome where apol1 might also have ion pore-forming activity.

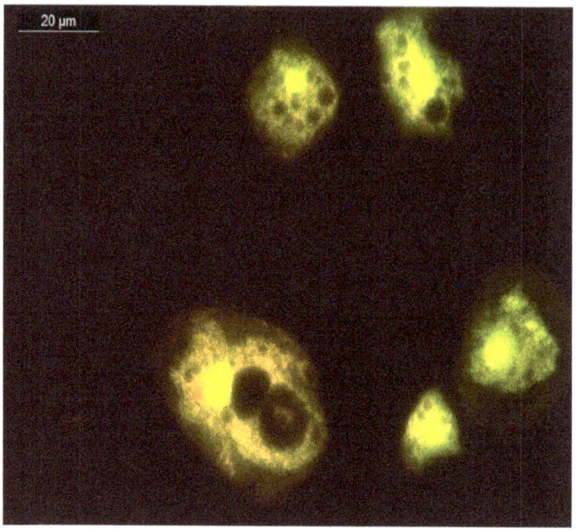

pharmacodynamic properties these include the physiological effects of a drug on the body, on microorganisms or on parasites in or on the body; the mechanisms of drug action; and the relationship between drug concentration and effect. Pharmacodynamics is often summarized as the study of what a drug does to the body, whereas pharmacokinetics is the study of what the body does to a drug.

pharmacogenetics the study of genetically determined variations in drug response, efficacy and frequency of adverse reactions.

pharmacogenomics a branch of pharmacology concerned with using DNA and amino acid sequence data to inform drug development and testing. An important application of pharmacogenomics is correlating individual genetic variation with drug responses.

pharmacokinetic properties characteristics of a drug, including its mechanisms of absorption and distribution, the rate at which a drug action begins and the duration of the effect, the chemical changes of the agent in the body, and the effects and routes of excretion of drug metabolites.

pharyngeal or intestinal lumen the central space through which the tubular pharynx (muscular feeding pump) or intestine conducts food posteriorly.

pharyngeal pumping rhythmical contractions (pumps) of the feeding tube (pharynx) that pass substances from the mouth towards the intestine.

pharyngeal stimulant a drug or similar substance that increases activity (i.e. pumping) of the nematode feeding tube (pharynx) and so triggers uptake through the mouth.

phase diagram a diagram that can be used to graphically display complex relationships between two or three parameters and cell phenotype.

phase Ia and IIa studies clinical trials of malaria vaccines that are performed in malaria-naive volunteers to establish safety (Ia) and efficacy after infection with sporozoites (IIa); the suffix ‘a’ is used to distinguish such trials from those given the suffix ‘b’ that are performed with malaria-exposed volunteers.

phase II clinical trial the second of four phases of evaluating a drug in affected subjects that establishes the efficacy of a drug compared to a placebo.

phase III clinical trials the third of four phases of evaluating a drug in affected subjects that confirms its safety and efficacy.

phased genomic sequences genome sequences for which the haplotype phase (that is, the combination of alleles or genetic markers that coexist on a single chromosome) has been determined.

phased testing pathway a step-wise testing procedure, involving stop/go decisions, for the evaluation of vector control products.

phasmids a pair of chemoreceptors located on the posterior part of certain nematodes of the class Secernentea (i.e. Phasmidia).

phenocopy a phenotype that results from a genetic (or epigenetic) alteration distinct from a reference, recognized alteration. It is a trait that is not caused by inheritance of a gene but appears to be identical to a genetic trait.

phenology the study of recurring biological phenomena.

phenomics collective study of multiple phenotypes (e.g. all phenotypes associated with a particular biological function).

phenotype any observable characteristic or trait of an organism, such as its morphology, development, biochemical or physiological properties, or behaviour.

phenotypic buffering or capacitance a phenomenon where individuals in a population produce the same phenotype despite genetic or environmental variation. Hsp90 is an example of a phenotypic buffer that can mask the effect of mutations in their substrate proteins.

phenotypic plasticity the adaptability of an organism to environmental changes.

phenylpyrazoles fipronil and pyriprole fipronil is a broad-spectrum phenylpyrazole insecticide that is active against fleas and ticks in dogs and cats. Fipronil blocks the passage of chloride ions through the GABA receptor and glutamate-gated chloride channels (GluCl), of the arthropod's nervous system.

pheromone infochemicals used among individuals for communication in animals. It can be secreted or excreted by an organism to trigger responses/behaviour of others of the same species (animals include insects).

phloem vascular tissue that distributes sugars and nutrients throughout a plant.

phorbol myristate acetate (PMA) a mitogenic phorbol ester, which directly stimulates protein kinase C and acts as a tumour promoter.

phosphate backbone the portion of the DNA double helix that provides structural support to the molecule. DNA consists of two strands that wind around each other like a twisted ladder. Each strand has a backbone made of alternating sugar (deoxyribose) and phosphate groups. Attached to each sugar is one of four bases – adenine (A), cytosine (C), guanine (G), or thymine (T). The two strands are held together by bonds between the bases, with adenine forming a bp with thymine, and cytosine forming a bp with guanine.

phosphatidylinositol bisphosphate (PIP2) phospholipid component of membranes; PIP2 serves as a substrate for hydrolysis by phospholipase C.

phosphatidylinositol-specific phospholipase C (PIPLC) enzyme responsible for hydrolysis of phosphatidylinositol 4,5-bisphosphate into the second messenger molecules inositol 1,4,5-trisphosphate and diacylglycerol.

phosphatidylserine phospholipid located in the inner leaflet of the plasma membrane in healthy cells but translocates to the outer leaflet in early apoptotic cells.

phosphomimetic an amino acid substitution, typically by an aspartic acid or glutamic acid, at the site of a phosphorylated residue in a protein, which mimics constitutive phosphorylation.

phosphoproteomic the use of experimental technology such as mass spectrometry to identify and quantify protein or peptide phosphorylation.

photobleaching depletion of a sample's fluorescence due to destruction of fluorophores while in a less stable excitation state. It may present a technical challenge for detection of a fluorophore, but also generates methodological opportunities (e.g. FRAP).

photoisomerization the light-initiated process of change from one isomeric form of a compound, radical, or ion to another. Importantly, in this process the changing in shape occurs without breakage of chemical bonding, an example being conversion of 11-*cis*-retinal to conversion into all-*trans* by photons.

photosensitization contact allergic dermatitis on sun-exposed skin areas involving the interaction between photosensitizing substances and sunlight.

photosynthate organic carbon produced by photosynthesis.

photosynthesis a complex process of light and dark reactions by which plants, algae, and certain types of prokaryotic organisms capture light energy and use it to drive the transfer of electrons from inorganic donors (e.g. water) to carbon dioxide to produce energy-rich carbohydrates, ATP and reducing power.

photosystem large, membrane-bound molecular complex consisting of multiple proteins containing pigment molecules (e.g. chlorophylls) that absorb light at a particular wavelength and transfer the energy from the absorbed photon to a reaction centre initiating a series of electron-transport reactions.

phototransduction machinery specialized membrane domains of sensory cilia in retinal rod and cone cells used for detection of light.

phototroph organism capable of photosynthesis.

phyletic gradualism the view that species evolve by gradual lineage evolution (anagenesis) and splitting (speciation) as a subsidiary process.

phyllosphere the microbial habitat defined by the surface of above ground plant organs (mainly leaves).

phylogenetic clade a subgroup within a phylogenetic tree that all trace back to a common point.

phylogenetic incongruence disparate topologies among gene trees, or between gene trees and organismal trees.

phylogenetic profiling a comparative genomics method for suggesting biological processes associated with a gene of unknown function; phylogenetic profiling is based on the observation that sets of genes that form functional modules tend to be gained or lost from the genome together.

phylogenetic signal the observable consequences with appropriate phylogenetic tools of a situation where a given collection of organisms is composed of distinct entities among which genetic exchange is either absent or severely restricted.

phylogenetic similarity pathogens are more likely to colonize new host species that present conditions similar to those associated with their current hosts; therefore, they spread more readily among closely related hosts.

phylogenetic tree an estimate of the true phylogeny, for example as might be reconstructed from species characters such as molecular data, often shortened to tree.

phylogenetically common found across a wide range of taxa.

phylogenetics study of evolutionary relationships among groups of organisms (e.g. species, populations), based on their DNA sequences.

phylogenomics comparison and analysis of entire genomes, or large portions of genomes, to determine the relationship of the function of genes to their evolution.

phylogeny the evolutionary relationships among a group of organisms.

phylogeography the study of spatial distribution of genetic differences within and between populations of closely related species using a historical framework.

phylotypes sequences or groups of sequences that share a certain level of homology, which enables evolutionary relatedness to be inferred.

physical map a map of the locations of identifiable landmarks on DNA (e.g. restriction-enzyme cutting sites, genes), regardless of inheritance. Distance is measured in bps. Physical map of a chromosome or a genome that shows the physical locations of genes and other DNA sequences of interest. Physical maps are used to help scientists identify and isolate genes by positional cloning.

physicochemical relating to both physical and chemical properties.

physiological homeostasis the ability of an animal to maintain a stable internal environment by physiological processes.

physiology the branch of science that deals with normal functions of living things or their organs.

phytohaemagglutinin (PHA) a lectin derived from the red kidney bean. It acts as a T-cell mitogen.

phytoplankton free-floating, microscopic photosynthetic organisms (e.g. algae, cyanobacteria, dinoflagellates found in the surface layers of marine and freshwater environments).

phytosiderophores chemical compounds released from the roots of certain plants (e.g. grasses) to sequester iron from the environment.

pia mater the innermost layer of the meninges (membranous connective tissue)that surrounds the brain and spinal cord.

pigments chemicals that are coloured. For example, the pigment melanin determines skin colouration.

Ping-Pong amplification cycle a mechanism for PiWi RNA (piRNA) biogenesis hypothesized from observations in *Drosophila* and later in other organisms. This mechanism obviates the need for RNA-dependent RNA polymerase. The mechanism is initiated by mature sense primary piRNAs, derived from transposable elements (TEs) and transcripts from genomic loci termed piRNA clusters, bound to proteins of the PiWi clade which recognize complementary sequences, antisense transcripts from the same piRNA cluster. The antisense piRNAs are cleaved, generating secondary piRNAs with new 5′-ends targeted by another PiWi protein. The 3′-end is then trimmed and then the mature antisense secondary piRNAs are now able to target sense transcripts from the piRNA clusters generating new sense piRNAs.

pinocytosis ingestion of liquid or very small particles by vesicle formation in a cell.

Piot model a framework used to analyse case management, describing the different steps a person has to go through after becoming ill until finally being cured.

piscivorous fish eating vertebrates such as herons, egrets, and raccoons.

PiWi-interacting RNAs (piRNAs) small RNAs of 24–30 nt in length thought to arise from long single-stranded RNA precursors derived from TEs, intergenic regions, and protein-coding genes. piRNAs preserve genome integrity and stability, and play a role in antiviral immunity. piRNAs associate with specific Piwi proteins and together function in silencing TEs. Piwi proteins function in germline development, gametogenesis, germline stem cell maintenance, and meiosis.

PL10 an ATP-dependent DEAD-box RNA helicase, also termed DEAD-box helicase 3 (DDX3), closely related to Vasa. PL10 is not restricted to the germline or somatic stem cells. It is conserved in eukaryotes from yeast: plants to animals. It participates in RNA metabolism, tissue differentiation, embryogenesis, asexual reproduction, cell regeneration, tumorigenesis, apoptosis, co-option in viral pathology, and innate immunity.

placental insufficiency the process of progressive deterioration in placental function and a decrease in transplacental transfer of oxygen and nutrients to the fetus; a common cause of fetal growth restriction.

placental malaria the concentration of *Plasmodium falciparum*-infected erythrocytes in the intervillous space, often accompanied by inflammation.

plaque-forming cell (PFC) antibody-secreting plasma cell detected *in vitro* by its ability to produce a ‘plaque’ of lysed antigen-sensitized erythrocytes in the presence of complement.

plasma the fluid portion of the blood.

plasma cell a fully differentiated B cell able to produce and secrete large numbers of antibodies.

plasma membrane (or cell membrane) **1** the membrane found in all cells that separates the interior of the cell from the outside environment. In bacterial and plant cells, a cell wall is attached to the plasma membrane on its outside surface. The plasma membrane consists of a lipid bilayer that is semipermeable. The plasma membrane regulates the transport of materials entering and exiting the cell. **2** the outermost membrane of the cell which insulates or protects it from the environment. It is made up of phospholipid, cholesterol and membrane protein (often pumps to allow passage of molecules through the barrier).

plasmacytoma is a type of cancer that begins in the plasma cells (sensitized B cell that produce antibodies). Malignant plasma cell tumours grow within bone marrow or soft tissue (extramedullary).

plasmatocytes a class of insect haemocytes important in the cellular immune response.

plasmid a small, circular piece of DNA found in bacteria and yeasts, which is able to replicate independently of the chromosome. They generally carry only a small number of genes, notably some associated with antibiotic resistance. Plasmids may be passed between different bacterial cells. A number of artificially constructed plasmids are used as cloning vectors.

plasmodesmata a thin plasmatic bridge that connects neighbouring plant cells.

plasmodial surface anion channel (PSAC) an ion channel on the host membrane of infected erythrocytes. PSAC is primarily responsible for increased permeability after infection; this channel and the associated *clag* genes are restricted to malaria parasites.

Plasmodium berghei one of the rodent parasites that is currently used to test new compounds in preclinical mouse models.

Plasmodium cynomolgi a parasite that infects monkeys (e.g. rhesus monkeys) and that produces dormant forms in the liver of its host similarly to those produced in humans by *Plasmodium vivax*.

Plasmodium falciparum one of the five human unicellular eukaryotic parasites (with *P. vivax*, *P. ovale*, *P. malariae*, and *P. knowlesi*).

***Plasmodium falciparum* chloroquine resistance transporter (PfCRT)** a transporter on the *P. falciparum* digestive vacuole membrane that when mutated constitutes the primary determinant of chloroquine resistance.

***Plasmodium falciparum* erythrocyte membrane protein 1 (PfEMP1)** predominant family of proteins expressed on the surface of infected erythrocytes. These proteins are encoded by the *var* genes (of which there are 60 different variants) and are actively transported to the surface of the erythrocyte during the middle third of the intraerythrocytic maturation stage of the parasite. Only one variant of PfEMP1 is expressed at any one time. Structurally, PfEMP1 molecules are mainly composed of several DBL domains, cysteine-rich interdomain regions, a transmembrane region, and an intracellular acidic-terminal sequence. This family of proteins is involved in parasite–receptor binding (e.g. to ICAM-1 or to CSA)

***Plasmodium falciparum* multidrug resistance protein 1 (PfMDR1)** integral membrane protein found within the DV membrane that is often, but not always, either mutated or overexpressed in chloroquine resistance (CQR) *P. falciparum*.

***Plasmodium falciparum* parasite rate (PfPR)** a commonly reported index of malaria transmission intensity that describes the percentage of the population found to carry asexual blood-stage parasites.

***Plasmodium falciparum* resistance transporter (PfCRT)** integral membrane protein found within the DV membrane that is mutated in CQR *P. falciparum*.

***Plasmodium vivax* Duffy binding protein (PvDBP)** ligand of the Duffy receptor present on the merozoite surface.

plasmotomy fission of a multinucleated protist into two or more multinucleated daughter cells without direct relationship to nuclear division.

plasticity the ability of cells or tissue and their function to be influenced by an activity and how they respond to distinct environmental conditions.

plastid endosymbiotic organelle found in plants, algae and some protists. It houses photosynthetic machinery in some eukaryotes as well as several biosynthetic pathways.

plastid-encoded polymerase (PEP) a bacterial-like RNA polymerase commonly found in plant chloroplasts that is often encoded on the plastid genome.

plastids a eukaryotic cell compartment in which photosynthesis occurs and any homologous compartment in which photosynthesis has been secondarily lost (e.g. the apicoplasts of apicomplexans). Plastids have their own genomes and can have different suites of light-harvesting pigments (e.g. those with chlorophyll a and b are most often referred to as 'chloroplasts').

plastin chromatic substance within the nucleus that stains only with special stains and is probably the material from which chromosomes are formed.

platyhelminthes phylum Platyhelminthes includes parasitic clades: Cestoda (tapeworms), Trematoda (flukes), and Monogenea (mostly ectoparasites of fish), and the free-living planarian-like worms of order Tricladida. Platyhelminthes consists of the flatworms, which have flattened bodies, which are either leaf-like (trematodes) or resemble ribbon segments (cestodes).

plausibility the programme has an effect above and beyond the effect of external, non-programme influences, by the inclusion of control groups.

pleiotropic epistasis statistical interaction signals shared in multiple traits.

pleiotropy multiple effects of alleles on more than one character. Pleiotropy leads to correlated responses to selection because allele frequency change caused by selection alters the value of all of the traits affected by a pleiotropic allele.

pleomorphic having many forms. The term is used to describe the isolates of *Trypanosoma brucei* that produce both long slender and short stumpy morphotypes during the mammalian bloodstream stage of the life cycle.

plerocercoid the second stage parasitic larva of diphyllobothriid tapeworms with aquatic life cycles, which grows in the body cavity of the second intermediate hosts (usually fish).

plethysmography a non-invasive barometric technique used in lung physiology to determine respiratory parameters in live animals and humans, based on lung volume changes.

pleural effusion excessive fluid accumulation between the two pleural membranes that surround the lungs. Excessive amounts of such fluid impair breathing by limiting the expansion of the lungs during ventilation.

pleurocercoid the second larval type in Pseudophyllidian tapeworms in which after the crustacean is eaten by freshwater fish and the procercoid is liberated, the pleurocercoid is develops in the muscles of the new host and possesses the characteristic scolex and is the only stage which is infective to final host.

ploidy the ploidy of a cell is the number of copies of a complete genome in a cell or nucleus. Haploid means 1 copy, diploid means 2, tetraploid means 4.

pluralism the belief that there are multiple opinions about an issue, each of which contains part of the truth, but none that contain the whole truth.

pluripotency the potential of a cell to develop into more than one type of mature cell, depending on environment.

pluripotent the ability to produce any cell in the body. Usually used when referring to embryonic stem cells.

pluripotent stem cell a cell which has the potential to differentiate into many different cell types.

P-nucleotides palindromic nucleotide sequences generated at the junctions between antibody (and T-cell receptor) variable (V), diversity (D) and joining (J) gene segments during gene rearrangement.

point mutation is when a single bp is altered. Point mutations can have one of three effects. First, the base substitution can be a silent mutation where the altered codon corresponds to the same amino acid. Second, the base substitution can be a missense mutation where the altered codon corresponds to a different amino acid. Or, third, the base substitution can be a nonsense mutation where the altered codon corresponds to a stop signal.

pokeweed mitogen (PWM) a plant lectin which is a T-cell dependent B-cell mitogen.

pol (polymerase) gene a retroviral gene that encodes reverse transcriptase, RNase H and integrase.

polar filament a coiled filament or tubule in the spore of the phyla Myxozoa and Microspora. This filament is extruded when the spore is ingested by a host and serves to inject the sporoplasm into a host cell.

polar head group hydrophilic moiety of polar lipids.

polar lipid lipids having a hydrophobic moiety and a hydrophilic polar head. Polar lipids make up the bulk of biological membranes.

polar plugs the mucoid extremities located at both ends of the lemon-like or barrel-like eggs shed by nematodes belonging to

Trichuris and other members of the family Trichuridae.

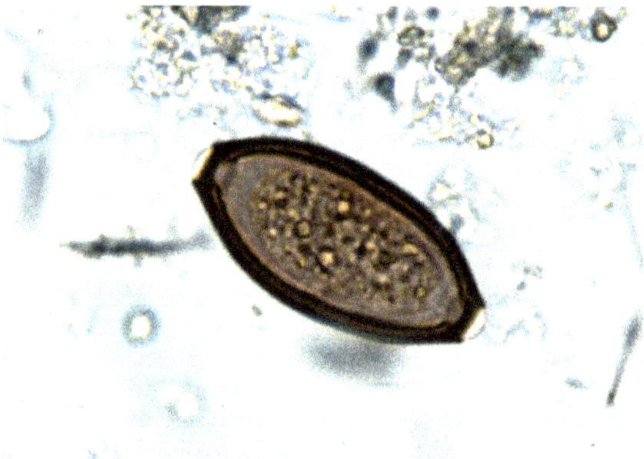

polarized T cell a T cell that is activated and expresses molecules associated with a particular T helper cell subtype.

polo-like kinases (Plks) these are involved in the control of cell-cycle progression during M-phase, activating phosphatases and thus CDKs. Plk activity is regulated in part by the polo-box domain, which is important for both auto-inhibition and subcellular localization of Plks.

polyadenylation sites (PAS) mark the 3′ end of the 3′ UTR of each gene. A poly(A) tail is added to the 3′-most nucleotide of each gene, which is referred to as the PAS.

polyandrous mating system a female mates with multiple males during the breeding season.

polyandry a type of breeding adaptation in which one female mates with multiple males. The mating of a female with several males.

polychromatic erythroblast erythrocyte progenitor.

polyclonal many different clones, or the product of many different clones, e.g. polyclonal antiserum.

polyclonal activator a substance that activates many individual clones of either T or B cells.

polyclonal gammopathies the appearance in serum of a high level of immunoglobulins of many different specificities originating from many different clones.

polyclonal mitogen a substance that induces activation of many individual clones of either T or B cells.

polyembryony monogenean trematodes of the family Gyrodactylidae, such as *Macrogyrodactylus clarii* are viviparous. They are unique in that worms give birth to fully developed adults. Intra-uterine embryos already contain second and often third generation embryos, recognized by their developing marginal hooklets and anchors. The succession of embryos is formed through a process of polyembryony.

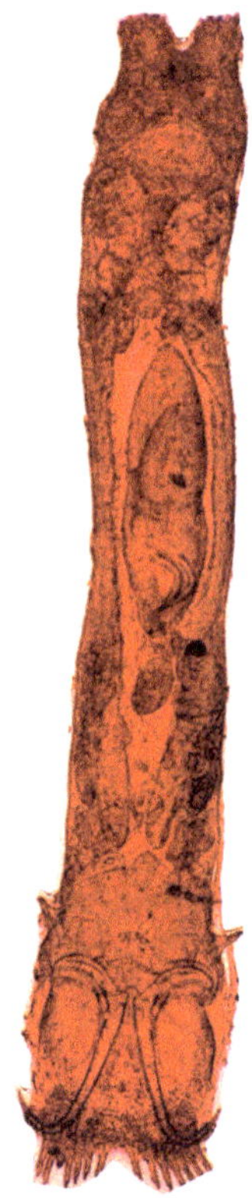

polyethism variation in the allocation of nest-related tasks among individuals in a colony. Polyethism typically refers to the special case of age-dependent changes in an individual's behaviour (that is, age-dependent or temporal polyethism); however, it more generally denotes differences in behaviour associated with caste (that is, temporal as well as morphological or physical polyethism).

polyfunctionality the ability of a single cell to display multiple functions such as cytotoxicity and secretion of multiple cytokines in response to an antigenic stimulus.

polygamy a mating system in which one male has a mating relationship with two or more females (i.e. polygyny) or in which one female has a mating relationship with two or more males (i.e. polyandry).

polygenic architecture a trait genetic architecture under which many genes of small effect contribute to trait variation.

polygenic trait a train whose phenotype is influenced by more than one gene. Traits that display a continuous distribution, such as height or skin colour, are polygenic. The inheritance of polygenic traits does not show the phenotypic ratios characteristic of Mendelian inheritance, though each of the genes contributing to the trait is inherited as described by Gregor Mendel.

polygynandry (or promiscuity) mating system in which any male mates with any female.

poly-Ig receptor a receptor molecule which specifically binds J-chain containing polymeric Ig, i.e. dimeric secretory IgA and pentameric IgM, and transports it across mucosal epithelium.

polymer a large molecule formed by the linking together of smaller subunit molecules.

polymerase chain reaction (PCR) a laboratory technique used to amplify DNA sequences. The method involves using short DNA sequences called primers to select the portion of the genome to be amplified. The temperature of the sample is repeatedly raised and lowered to help a DNA replication enzyme copy the target DNA sequence. The technique can produce a billion copies of the target sequence in just a few hours. The resulting amplified DNA can then be visualized with stains or radioactive labelling, or sized with fluorescent markers in a sequencer.

polymerase, DNA or RNA enzyme that catalyses the synthesis of nucleic acids on pre-existing nucleic acid templates, assembling RNA from ribonucleotides or DNA from deoxyribonucleotides.

polymeric IgG IgG can polymerize into complexes.

polymorphic highly variable in structure or sequence.

polymorphism involves one of two or more variants of a particular DNA sequence. The most common type of polymorphism involves variation at a single bp. Polymorphisms can also be much larger in size and involve long stretches of DNA. Called a single nucleotide polymorphism, scientists are studying how SNPs in the human genome correlate with disease, drug response, and other phenotypes.

polymorphonuclear leucocyte (PMN) white blood cell (WBC), granular cytoplasm and nuclear lobes ranging from 1 to 5 (hence the name PMN). There are three types of granular WBCs, classified according to is ability to be stained by different dies. Neutral staining (neutrophil, PMNs) most frequent, phagocytic. Basophilic staining – basophil. Eosinophilic staining – eosinophil.

polyparasitism the co-occurrence of two or more parasite species in the same organism. This phenomenon is common than single infection especially in regions where different parasites co-exist at high frequencies. The different types of co-existing parasites can result in synergistic or antagonistic interactions in terms of worm burden within a mammalian host.

polypeptide a protein or part of a protein made of a chain of amino acids joined by a peptide bonds.

polyphyletic a polyphyletic group is characterized by one or more character states which represent convergence or reversal traits which appear to be similar but which were not inherited from the last common ancestor. It is neither monophyletic nor paraphyletic.

polyphyly description of a group of organisms that includes organisms, which share convergent or reverted characteristics (identical-by-state characteristics, i.e. characteristics that are similar due to chance and that have not been inherited from a recent common ancestor) or organisms from several distinct common ancestors.

polyploidy cells and organisms that contain more than two paired (homologous) sets of chromosomes. Most eukaryotic species are diploid, meaning they have two sets of chromosomes – one set inherited from each parent. However, polyploidy is found in some organisms and is especially common in plants.

polyprotein a single polypeptide chain that is the primary product of polypeptide synthesis and that is then cleaved into different functional proteins. The genomes of the Flaviviruses, including dengue virus (DENV) and Chikungunya virus (CHIKV), encode a polyprotein that contains three structural and eight non-structural proteins.

polysomic having multiple copies of a given chromosome.

polyunsaturated fat a fat that has more than one double bond in the molecule.

pool genome-wide association studies (pool-GWASs) genotype–phenotype mapping studies in which phenotypically extreme individuals are grouped and sequenced as pools. Causative variants are identified by contrasting the allele frequencies between the pools.

population a group of mutually related organisms; organisms of the same species living together in a given area. A community comprises several different populations.

population attributable fraction (PAF; also known as population attributable risk) for a given disease, risk factor and population, the fraction by which the incidence rate of the disease in the population would be reduced if the risk factor was eliminated.

population bottleneck a severe reduction in population size, which randomly reduces the genetic variability of a population.

population context a collective term for the context in which a single cell displays its activities and behaviours, which can be determined by local and global effects from the population to which the cell belongs. The context is determined not only by the microenvironment of a single cell but also by its physiological state that is a consequence of population effects, such as the cell size.

population differentiation a greater difference in the frequency of a variant between populations than within populations. High population differentiation is a potential sign of positive selection acting on the variant.

population expansion increase in the size of a population over time.

population genetics the study of allele frequency distribution and change under the influence of the four main evolutionary processes: natural selection, genetic drift, mutation, and gene flow. Population genetics also encompasses the factors of recombination, population subdivision, and population structure and attempts to explain such phenomena as adaptation and speciation.

population genomics the application of genomic technologies to understand populations of organisms. In humans, population genomics typically refers to applying technology in the quest to understand how genes contribute to health and well-being.

population replacement strategy (PRS) the use of a genetic drive mechanism, such as cytoplasmic incompatibility induced by *Wolbachia* infection, to replace a natural population with a laboratory-reared one.

population structure population structure describes the fact that populations deviate from the random model (where all individuals are potential partners and mate randomly) and are therefore subdivided. Population structure can be the consequence of non-random mating due, for example, to environmental factors (such as the presence of geographical barriers limiting dispersal or the presence of different conditions leading to differential selection pressure) or the social organization, which can lead, for example, to inbreeding.

population subdivision population structure in which barriers exist to prevent random mixing and mating between members of a species.

portal blood blood coming from the hepatic portal vein that transports blood from the intestines, spleen, pancreas, and gallbladder to the liver.

portal hypertension high blood pressure in the portal vein.

positional cloning a laboratory technique used to locate the position of a disease-associated gene along the chromosome. This approach works even when little or no information is available about the biochemical basis of the disease. Positional cloning is used in conjunction with linkage analysis. It involves the isolation of partially overlapping DNA segments that progress along the chromosome towards a candidate gene.

position-specific iterative (PSI)-BLAST a variation of BLAST in which a profile is constructed from the highest scoring hits from an initial BLAST search. The profile is then used to perform a BLAST search, the results of which are used to refine the profile for use in subsequent iterations.

positive predictive value (PPV) the probability of an individual being diseased or infected given the observed test result. This depends not only on the test characteristics but also on the true prevalence within the population.

positive pressure ventilation mechanical ventilation in which air is delivered into the airways and lungs under positive pressure, usually via an endotracheal tube, producing positive airway pressure during inspiration. This positive pressure may diminish the formation of additional alveolar oedema.

positive selection or directional selection a form of natural selection that increases the frequency of alleles carrying an advantageous mutation. Positive selection reduces nucleotide variation at non-synonymous and synonymous sites linked to the selected site. When positive selection drives a single allele to fixation and nucleotide variation is eliminated, this is referred to as a 'selective sweep'. Positive selection is also commonly referred to as 'directional selection'.

post kala-azar dermal leishmaniasis (PKDL) a chronic, dermal sequel of kala-azar that occurs in some but not all who are treated for kala-azar. Clinically, the disease is characterized by a macular, depigmented eruption found mainly on the face, arms, and upper part of the trunk.

postembryonic developmental rate the rate at which *Caenorhabditis elegans* proceeds through the larval stages (L1, L2, L3 and L4) to the young adult stage and then to adulthood.

posterior probabilities in Bayesian statistics, the posterior probability of a random event is the conditional probability that is assigned after the relevant evidence is taken into account.

posterior probability distribution the probability distribution of a random variable conditional on relevant observed data and prior information. The posterior probability is proportional to the likelihood of the data (conditional on a set of parameter values) multiplied by prior probability of the parameters.

post-genomics the suite of techniques used to explore the functional properties of genes defined by genome sequencing of any organism.

post-nuclear (kinetoplast, kDNA) positioning of a kinetoplast between the nucleus and the posterior end (the end opposite to the protruding flagellum.

post-transcriptional modification of RNA changes to RNA transcripts before formation of mature messenger RNA (mRNA), which is translated into amino acids (e.g. the removal of non-coding introns, thus retaining only the coding exonic regions in the mature mRNA for translation into amino acids).

post-translational modifications (PTMs) chemical modifications of, and ligation of small proteins (e.g. ubiquitin and SUMO) on, substrate proteins that can serve as platforms for the binding of other proteins containing a 'reader' domain for specific PTMs.

postzygotic isolation reproductive isolation that occurs after fertilization, such as hybrid sterility and hybrid inviability.

POT1 protection of telomere 1, a protein that binds G-rich telomeric DNA and is a co-effector of the TRF1 complex by transducing information to the telomere terminus.

poultry cestodes tapeworm of the genera, *Davainea*, *Raillietina*, *Cotugnia*, *Hymenolepis*, *Amoebotaenia*, *Chosnotaenia*, and *Fimbriaria* include species that are pathogenic to chicken, pigeon, ducks, and geese. They inhabit the small intestine and have indirect life cycles (slugs, *Cyclops*, beetles, earthworms and flies as intermediate hosts.). The importance of poultry cestodes depends mainly on age and general conditions of the birds and the worm burden. Although *D. proglottina* is the smallest cestode, it is the most harmful cestode of poultry. This worm penetrates deeply into mucosa, producing marked enteritis often haemorrhagic in heavy infection. *R. tetragonai* and *R. echinobothrida* follow next in order of pathogencity. Young worms penetrate deeply into duodenal mucosa with their anterior ends causing nodular formation visible from peritoneal surface, emaciation, mucoid diarrhoea. Adult worms occur in a posterior location. *C. digonophora* causes catarrhal enteritis. The other tapeworm species are less pathogenic. Young birds are most frequently affected. They show loss of appetite, droopiness, thirst, emaciation, anaemia, and death in heavy infections. In mature chicken, signs include decreased egg production, lowered body weight.

power the probability that the null hypothesis is rejected when it is false. For example, the null hypothesis is that there is no difference between mean intensities pre- and post-intervention. Mathematically, power is $1 - \beta$, where β is the type 2 error, or probability of failing to reject the null hypothesis when it is false. The ability of a study to demonstrate an association or causal relationship between two variables, given that an association exists. For example, 80% power in a clinical trial means that the study has a 80% chance of showing a statistically significant treatment effect if there really was an important difference between outcomes. If the statistical power of a study is low, the study results will be questionable (the study might have been too small to detect any differences).

Praziquantel a pyrazinoisoquinoline derivative that is effective against both adult and juvenile stages of tapeworms. It is also available in combinations with pyrantel pamoate, ivermectin, and moxidectin to kill nematodes. The primary effects of praziquantel are tetanic contractions of the parasitic musculature, and an intense, irreversible, focal vacuolization and disintegration of the worm's integument.

pre-adult immature mobile louse. There are two pre-adult stages in *Lepeophtheirus* spp. and there is some debate whether they occur in *Caligus* spp.

precipitin precipitate of antibody and multivalent antigen due to the formation of high molecular weight complexes.

precipitin test this test depends upon the interaction between a saline extract of the blood meal under examination and animal antiserum. A positive reaction is indicated by the formation of a white precipitate that results from the production of insoluble antigen–antibody complexes.

precision (statistical precision) the degree to which repeated measurements or calculations show the same or similar results. It is also frequently referred to as positive predictive value.

precyst and pseudocyst Precyst refers to an immature cyst stage whereas pseudocyst refers to a cell that may resemble a precyst but may or may not have a role in the life cycle of the organism. Both, in general, lack the thick wall of the cyst stage.

predation enhancement an evolved ability of a parasite to increase predation on its current host after it becomes capable of establishing in its next host, thereby facilitating transmission.

predation suppression an evolved ability of a parasite to decrease predation on its current host until capable of establishing in its next host.

predator animal that kills another animal for food.

prediction the use of a signature to predict the state of an unknown sample, either through cross-validation procedures that use the training samples to evaluate the robustness of the signature, or by predicting an independent data set.

predictive biomarker appears in the absence of any injury with an ability to foretell future injury with some certainty.

predictive value the probability that a person with a positive test is a true positive (i.e. does have the target disease), or that a person with a negative test truly does not have the disease. The predictive value of a screening test is determined by the sensitivity and specificity of the test, and by the prevalence of the condition for which the test is used.

predilection host the host preferred by a parasite.

predilection site the site within a host where the parasite is normally found.

predisposing cause a mechanism that makes a patient more susceptible to the precipitating cause.

pre-eclampsia a common multi-systemic pregnancy disorder, characterized by serious maternal symptoms including hypertension, proteinuria and high risk of fetal growth restriction.

pre-emptive therapy an early, targeted therapy in high-risk patients who already have a laboratory or other test indicating that an asymptomatic patient has become infected. The principle is that delivery of therapy prior to development of symptoms (presymptomatic) aborts impending disease, and the therapy is for a short and defined duration.

pre-loading experiment in which cells are first loaded with a substance and then the subsequent movement of the substance is followed.

premature chromosome condensation (PCC) a method of studying chromosomes in the interphase stage of the cell cycle.

premunition a lengthy repeat allele that is prone to extreme expansion, leading to pathological mutation as seen in the 'triplet repeat diseases'.

pre-nuclear (kinetoplast, kDNA) positioning of a kinetoplast between the nucleus and the anterior end of the parasite cell (the end with the protruding flagellum).

pre-parasitic phase of life cycle development to the infective stage occurring outside the definitive host in the environment or in a second (intermediate) host in the parasite's life cycle.

prepatent infection snail infection by *Schistosoma* parasitic forms during the period of sporocyst development, before shedding infectious cercariae. Depending on environmental conditions and snail susceptibility, the prepatent period might last from 1–2 months up to >1 year in duration. However, only a fraction of infected snails progress from prepatency to become patent shedders of cercariae.

prepatent period (PPP) the span of time from infection until the infection is detectable. The biological incubation period, usually between infection and initial clinical manifestations.

presynaptic receptor a receptor that is localized on the surface of the presynaptic membrane; it is either protein or lipid in nature.

prevalence the proportion of a population with a condition (typically a disease or a risk factor such as smoking or seat-belt use). It is arrived at by comparing the number of people with the condition with the total number of people studied, and is usually expressed as a fraction, a percentage, or as the number of cases per 10,000 or 100,000 people.

preventative measures actions or measures (e.g. vaccination, use of repellents) adopted to reduce the incidence of a disease or infection.

prevention measures taken prior to any infection/infestation of the animal with parasites, to prevent the establishment of an infection/infestation.

prezygotic isolation reproductive isolation that occurs before fertilization. Includes species differences in traits such as sexual behaviour, habitat preference, seasonal breeding (all premating), and gametic compatibility (postmating but prezygotic).

Primaquine antimalarial of the 8-aminoquinoline chemical class that was discovered by the US army nearly 60 years ago. The drug is administered daily for 14 days to clear the liver of patients from *P. vivax* hypnozoites and provide a radical cure of patients by avoiding relapses. The current hypothesis is that a transient metabolite of primaquine formed in the human liver kills hypnozoites and developing liver forms of *P. vivax* extremely rapidly.

primary antibody response the initial response of B cells when first exposed to an infection.

primary hepatocytes parenchymal liver cells that perform all the functions ascribed to the liver.

primary immune response the relatively weak immune response, which occurs upon the first encounter of naive lymphocytes with a given antigen.

primary immunodeficiencies inherited immunodeficiency diseases.

primary lymphoid organs organs in which the maturation of T and B lymphocytes takes place and antigen-specific receptors are first acquired.

primary production synthesis and storage of organic molecules (biomass), starting with fixation of CO_2 by photosynthesis, in plants and microorganisms.

primary research individual studies such as randomized controlled trials, cohort studies, case–control studies, cross-sectional studies, etc.

primary reservoir host a host that is responsible for maintaining the parasite indefinitely in nature. In these hosts infection is normally without clinical signs.

primary responses the immune response resulting from an individual's first encounter with an antigen.

primary sex ratio the ratio between the number of males and females at the time of conception.

prime the process of giving an initial sensitization to antigen.

primer short, pre-existing single-stranded polynucleotide chain to which new deoxyribonucleotides can be added by DNA polymerase (to 'prime' PCR amplification). The primer anneals to a nucleic acid template (DNA of the organism of interest) and promotes copying of the template, starting from the primer site. To amplify microsatellites one uses a forward and reverse primer pair.

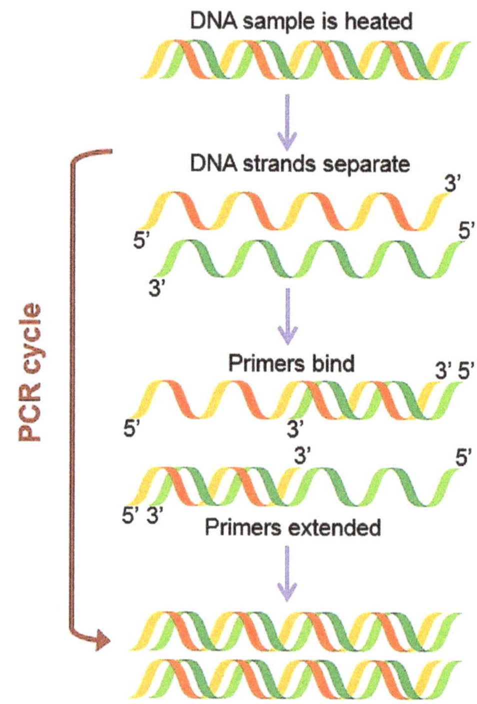

primitive gut the embryonic structure that gives rise to the gut (intestine).

primordial germ cells the precursors of reproductive cells within the embryo. They are detectable in an embryo after four weeks of development and will develop into either sperm or eggs.

prior probability distribution the probability distribution of a random variable that captures one's uncertainty before (prior to) observing relevant data. An uninformative or vague prior expresses a high degree of prior uncertainty. This results in a posterior distribution which is dominated by the likelihood of observed data. Conversely, an informative prior will dominate the posterior if the data holds little information on the variable of interest.

private sector facility a treatment source that is not funded or controlled through the government.

privately inherited pertaining to a genetic variant: confined to a single individual, family or population.

privileged sites some (rare) locations within the body where the immune response can not see (e.g. foreign grafts are not rejected). An example is the cornea of the eye.

proactive infection detection (proactive ID) investigation tactics focused on populations defined by high-risk geography (hotspot) or high risk demography (hotpop). Proactive ID may be preceded by fever screening of all patients with a recent history of malaria symptoms or by testing the target population without prior screening.

probability the program has an effect, with a small and defined probability that the difference between program and control areas were due to chance or confounding.

proband the index case being studied usually based on it having a particular phenotype or disorder.

probe single-stranded DNA or RNA molecules of specific base sequence, labelled either radioactively, immunologically, or by other means, that are used to detect the complementary base sequence by hybridization. Some genetic markers (e.g. minisatellites) depend on probe-based techniques. *See* **coefficient of relatedness**.

probiotics viable non-pathogenic microorganisms that confer health benefits by augmenting the activities of commensal microorganisms in the gut. Examples of these activities include biosynthesis of vitamin K, fermentation of indigestible dietary fibre, and competition with pathogenic microorganism for nutrients, among others. Probiotics can modulate the host immune response directly through interaction with intestinal epithelial cells (IEC) or immune cells in the gut, or indirectly through the modulation of the intestinal microbiota. The immunomodulatory effects of probiotics seem to be microbe species- and strain-specific. The most important benefit of probiotics is likely the maintenance of gut homeostasis (sustaining a balanced and beneficial flora). Currently, the most common probiotics used are *Lactobacillus* and *Bifidobacterium* species.

problem-oriented medical diagnosis a system of diagnosis that identifies and focuses on a patient's specific health problems, by considering potential causes (differential diagnoses) of each clinicopathological abnormality (problem) detected from the history, physical examination, and from laboratory and ancillary test results.

proboscis in some insects, a proboscis is a filamentous structure that projects outwards from the mouthparts, functioning as a piercing and sucking device for obtaining liquid food. In acanthocephalans, a proboscis is an anterior protrusible organ, typically studded with hooks.

procercoid the first stage parasitic larva or metacestode of pseudophyllidian tapeworms with aquatic life cycles that usually develops from the onchosphere in the body cavity of copepod hosts.

processed food any food product that has undergone physical or chemical treatment resulting in a substantial change in the original state of the food.

procyclic form (PF) trypanosome trypomastigote form in which the kinetoplast is posterior to the nucleus and which colonizes and proliferates in the midgut of infected tsetse flies.

procyclins GPI-anchored acidic glycoproteins expressed abundantly on the surface of procyclic trypanosomes of *Trypanosoma brucei* and other insect stages of *T. congolense*.

prodromal biomarker represents a symptom of the initial stage at onset of an injury before any observation of certain injury.

pro-drug a compound that is administered in an inactive/less than fully active form that requires conversion to its active form through a normal metabolic process. Flagyl® (metronidazole) is an example.

production parameters indicators of the level of performance of a production animal. Examples are milk production per day (kg), daily weight gain (kg), or number of inseminations per conception.

productive infection infection that comprises all steps of the viral life cycle, including viral protein and particle production.

productive invasion infection of a hepatic cell by *Plasmodium* sporozoites with formation of a parasitophorous vacuole inside which the parasite replicates and grows.

professional antigen-presenting cells dendritic cells, macrophages, and B cells; defined by their ability to take up antigens and present them to T lymphocytes in the groove of a MHC class II molecule.

profile model a bioinformatic model for identifying a protein functional domain; the model represents the frequency of amino acids observed at each position in a multiple alignment of protein sequences taken from a range of organisms.

progenesis precocious development of the reproductive system, leading to early sexual maturity in juvenile stages.

proglottid a body segment of a tapeworm containing a complete set of reproductive organs. Proglottids may be immature, mature (male and/or female sex organs present) or gravid (full of eggs).

prognosis the possible outcomes of a disease or condition and the likelihood that each one will occur.

prognostic biomarker predicts the course or outcome (e.g. end, stabilization or progression) of an illness.

prognostic factor a factor or indicator (such as age or gender) related to an individual's probability of developing a disease or other outcome. Compare with risk factors. Neither prognostic nor risk factors necessarily imply a cause and effect relationship.

programmed cell death (PCD) the suicide of a cell, usually in a multicellular organism. It is a genetically determined, regulated process that follows intracellular or extracellular stimuli. Types of PCD are apoptosis and autophagic cell death.

programmed death 1 (PD-1) a surface membrane protein that belongs to the immunoglobulin (Ig) superfamily, is expressed by a wide variety of cells, including T cells, B cells, and natural killer cells. PD-1 interaction with its ligand PD-L1 negatively regulates the immune response. In the context of exhaustion, progressive expression of PD-1 correlates with exacerbation of CD8 dysfunction (PD-1 intermediate < PD-1 intermediate/high < PD-1 high).

programmed death ligand 1 (PD-L1) one of the ligands for PD-1 receptor, is constitutively expressed on antigen-presenting cells, T cells, and a wide range of non-haematopoietic cell types. PD-L1 expression is up-regulated after activation of antigen-presenting and T cells. The PD-1/PD-L1 pathway controls the balance between stimulatory and inhibitory signals needed for optimal immunity.

prohaptor the anterior holdfast of members of the trematode class Monogenea.

proinflammatory cytokines cytokines produced predominantly by activated immune cells; responsible for amplification of inflammatory reactions.

proinflammatory mediators signalling molecules, such as cytokines and chemokines, released by cells to recruit immune cells and promote inflammation.

prokaryote single-celled organism lacking a membrane-bound, structurally discrete nucleus and other subcellular compartments. Bacteria and archaea are prokaryotes.

promastigote one of the morphological stages in the development of certain protozoa (*Leishmania*), characterized by a free anterior flagellum and the kinetoplast and axoneme at the anterior end of the body. There is no undulating membrane. Also termed: 'leptomonad' stage.

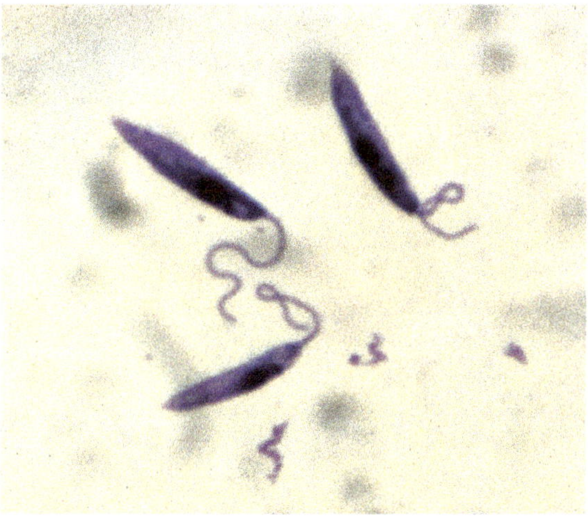

promastigote surface antigen 2 (PSA2) a surface glycoprotein of *Leishmania* expressed on both promastigotes and amastigotes. It has been proposed to be involved in host–parasite physical interactions. It plays a role in parasite attachment through complement receptor 3 (CR3) and is associated with the resistance against complement-mediated lysis. PSA2 is an immunologically reactive protein that induces high levels of Th1 cytokines and is therefore considered as a vaccine candidate.

promiscuity mating with multiple partners.

promoter a sequence of DNA needed to turn a gene on or off. The process of transcription is initiated at the promoter. Usually found near the beginning of a gene, the promoter has a binding site for the enzyme used to make a mRNA molecule.

pronucleus the nucleus of either the sperm or egg during fertilization.

propagative a biological relationship in which the parasite multiplies in the host but does not undergo morphological change.

propagative stages juveniles belonging to the propagative part of the life cycle of nematodes in which the nematode reproduces. In the context of PWN, the propagative stages multiply inside the pine tree host, feeding on it, and leading to its demise.

propagative transmission transfer, as of an infection from one patient to another, in which the agent multiplies in the transmission vehicle.

proper pairs paired-end reads where both pairs can be mapped to the same chromosomes within a distance pre-specified by the insert size chosen during library preparation.

prophenoloxidase activating system melanization is controlled by phenoloxidase, a key enzyme that catalyses the oxidation of phenols to quinones leading to the formation of melanin. Phenoloxidase exists as an inactive precursor called prophenoloxidase and, following either physical injury or microbial challenge, it is released by haemocytes and is activated via limited proteolysis through the action of a serine protease cascade.

prophylaxis prevention/protection; measures that are carried out to prevent the transmission of the parasitic agent or the occurrence of disease.

proportion gametocytaemic the proportion of individuals with asexual parasites who also have sexual-stage gametocytes.

propulsive parasite motility a sustained, forwards movement of a parasite. Propulsive motility is distinguished from general writhing of the parasite, which is generated by unregulated beating of the flagellum.

prospective study a study design where one or more groups (cohorts) of individuals who have not yet experienced the outcome event in question are followed forward in time and monitored for the number of such events which occur.

prostaglandins acidic lipids derived from arachidonic acid, which are able to increase vascular permeability, mediate fever, and can both stimulate and inhibit immunological responses.

protease activated receptors (PARs) receptors activated by cleavage of an extracellular domain by serine proteases. Diverse biological functions but are widely expressed on immune system cells and modulate inflammatory responses.

protective immunity implies that memory B and T cells are produced in response to a vaccine or natural infection, which are appropriate to defend against a future infection from an actual wild-type pathogen.

protein a large molecule composed of one or more chains of amino acids in a specific order; the order is determined by the base sequence of nucleotides in the gene coding for the protein. Proteins are required for the structure, function, and regulation of the body's cells, tissues, and organs, and each protein has unique functions. Examples are hormones, enzymes, and antibodies.

protein complex aggregate structure consisting of multiple protein molecules.

protein expression subcomponent of gene expression. It consists of the stages after DNA has been transcribed to mRNA. The mRNA is then translated into polypeptide chains, which are ultimately folded into proteins.

protein interaction network (PIN) a schematic mapping proteins (nodes) to other proteins, where the lines connecting two proteins (edges) represent a physical binding of the two proteins or a predicted interaction between them.

protein kinase Cδ (PKCδ) a member of the PKC family which is thought to play a crucial role in cell proliferation, differentiation and apoptosis.

protein kinases (PKs) enzymes exerting regulatory effects on growth and differentiation (including malignant transformation) by phosphorylating proteins. PKs catalyse the transfer of phosphate from ATP to hydroxyl groups.

protein phosphatase with EF-Hands (PPEF) function unknown in mammals but may correlate with stress responses and proliferation signals; in trypanosomatids, dually acylated proteins expressed in the endomembrane system but do not bind calcium.

protein tyrosine kinases enzymes able to phosphorylate proteins on tyrosine residues, and often act in a cascade-like fashion in the signal transduction systems of cells.

protein tyrosine phosphatases a group of enzymes that remove phosphate groups from tyrosine residues and participate in regulation of signal transduction.

protein–protein interaction (PPI) chemical inhibitors that bind to the protein–protein interaction surface of two proteins and thereby block their interaction.

proteoglycan a protein bearing covalently attached glycosaminoglycan chains.

proteoliposomes mixtures of proteins and lipids that are induced to form artificial thin-walled vesicles comprising both of these components.

proteome a collection of proteins expressed by a cell at a particular time and under specific conditions.

proteomics large-scale analysis of the proteome to identify which proteins are expressed by an organism under certain conditions. Proteomics provides insights into protein function, modification, regulation, and interaction.

proteorhodopsin light-dependent proton pumps in marine bacteria.

protocells membrane-enclosed compartments that may not contain all the components of present-day cells but that were presumably capable of some rudimentary means of growth and division. They can also refer to artificial cell-like structures created in the laboratory.

protofilaments polymers of α- and β-tubulin heterodimers arranged end-to-end to form filaments. Thirteen protofilaments interacting along the long axes form the canonical microtubule. The arrangement of the tubulin dimers gives the protofilaments polarity, and this is conferred to the microtubule because all protofilaments are orientated in the same direction.

protonephridia a primitive excretory system consisting of flame cells and tubules for the elimination of liquid wastes, in Platyhelminthes.

protonophore reagent that eliminates proton gradients. The special case of an ionophore (q.v.) that is specific for protons.

protoscolex (protoscolices) a holdfast of tapeworms of the order Cyclophyllidea which forms from a germinal epithelium in a coenurus, hydatid, or alveolar cyst.

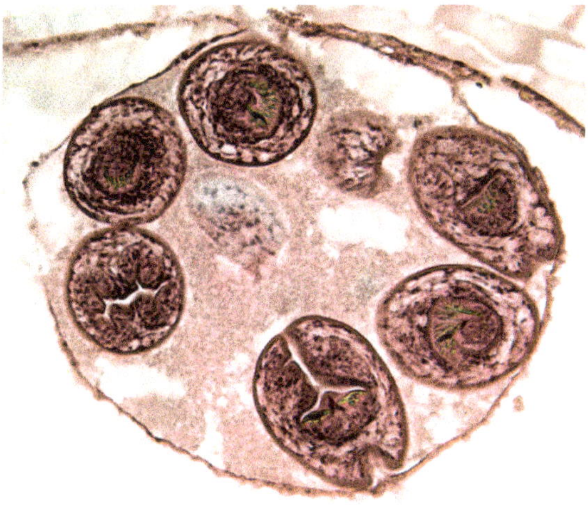

protospacer adjacent motif (PAM) a short, conserved DNA sequence (usually 2–5 bp) that is adjacent to the binding site of crRNA.

Protostrongylus rufescens long slender brown worms. Adult worms are found in the terminal bronchioles. It has an indirect life cycle, requires a snail or slug intermediate host. No age related immunity; adults continue to collect foci of infection throughout life. Affected hosts show weight loss, loss of condition, respiratory signs, and diarrhoea. Worms are found in granulomatous foci in the parenchyma and terminal bronchioles.

protozoa simple, single-cell microorganisms that range in size from 2 to 100 µm. Their protoplasm is enclosed by a cell membrane and contains numerous organelles, including a membrane-bound nucleus, an endoplasmic reticulum, foodstorage granules, and contractile and digestive vacuoles. Organs of motility vary from simple cytoplasmic extrusions or pseudopods to more complex structures such as flagella or cilia (taxonomic importance). Protozoa comprises seven major phyla, four of which are the concern of veterinary medical parasitology. Most are free-living, but some lead commensalistic, mutualistic, saprophytic or parasitic existences.

protozoal pertaining to or caused by protozoa.

protozoal encephalomyelitis probably the same disease as equine protozoal myeloencephalitis caused by *Sarcocystis neurona*. Protozoal hepatitis (parasitology) a disease caused usually by *Toxoplasma*, *Neospora*, and *Leishmania*.

protozoan (plural **protozoa**) an organism belonging to the kingdom Protozoa.

protozoology the scientific study of protozoa.

provision measure of the quantity of an intervention or service delivered.

provisional or temporary host may become infected but throw off the infection after a short time.

proximal factors mechanistic explanations that answer questions regarding 'how' a phenomenon takes place.

proximity ligation assays (PLAs) a method that can detect proteins and protein–protein interactions with high specificity. The protein of interest is recognized by two primary antibodies raised in different species, followed by binding of species-specific secondary antibodies conjugated to a short DNA oligonucleotide. Upon specific binding of both primary antibodies to the target (two epitopes either on the same protein or on two distinct interacting proteins), the two oligonucleotides are brought together, which aids the formation of a circle from an extra pair of oligonucleotides added to the mixture by the use of *in situ* ligation. Fluorescent detection is then achieved using rolling circle amplification.

PRR (parasite reduction ratio) the fractional reduction in parasite numbers per asexual cycle. Values typically vary between 10 and 10,000 per cycle.

PRR lectins lectins are a diverse set of proteins binding with varying affinities to molecular patterns formed by sugar residues and generally involved in biological recognition phenomena, including functioning as PRRs in the recognition of pathogens. Those implicated in the recognition of helminth glycans include C-type lectins and galectin-3.

pruritus an unpleasant sensation that induces someone to scratch or rub the skin.

pseudocoelom a body cavity of a metazoan that is not completely lined with mesoderm.

pseudogene a DNA sequence that resembles a gene but has been mutated into an inactive form over the course of evolution. A pseudogene shares an evolutionary history with a functional gene and can provide insight into their shared ancestry.

pseudogenization the loss of function for a given gene/sequence. Pseudogenes are not expressed and are, therefore, nonfunctional relatives of functional genes. Gene duplication is a major contributor to pseudogenization.

pseudokinase a protein containing a kinase-like fold that lacks one or more residues crucial for catalysis, rendering it enzymatically inactive.

pseudoknot a higher-order RNA structure that involves base pairing between the loop of a stem–loop structure and an upstream or downstream RNA sequence that is located outside of the stem–loop.

pseudolinkage artificial 'manufactured' linkage created by a translocation or inversion.

Pseudomonas aeruginosa widely distributed Gram-negative bacterium that show a predilection for acute and chronic infection of the lungs of individuals with cystic fibrosis.

pseudophyllidian refers to an order of tapeworms in which the scolex has a single terminal or two, opposite, lateral organs of attachment, the bothria, e.g. in *Diphyllobotrium latum*.

pseudopod protrusions of the macrophage membrane that is the result of extensions of actin filaments near the edges of the cell.

pseudopodium (ia) a clear projection of the ectoplasm of protozoa, especially amebae, that is usually associated with movement and/or food gathering.

***Psorergates ovis* (itch mites)** the entire life cycle of this mite, eggs, larvae, two nymphal stages, and adults, takes place entirely on the host. In sheep the cycle takes 4–5 weeks. All stages occur in the superficial layers of the skin. Clinically, the disease resembles louse infestation, but may be distinguished on the smaller proportion of the flock affected (10–15%), the less severe irritation and tendency of the sheep to bite those areas it can reach. Hence lesions are confined to parts of the flank and the hindquarters and the wool tufts have a chewed appearance. Only the adults are mobile on the skin surface and they affect spread of the disease by direct contact. Mite feeding activity, in addition to excreta causes skin irritation leading to rubbing and biting of the affected parts (the sides, flanks, and thighs) and raggedness, sometimes shedding, of the fleece.

psoroptic mange a serious disease in cattle and sheep but less so in goats. The causative mites are species of *Psoroptes*. The species present in ruminants are in cattle *P. ovis* and (less common) *P. natalensis*, in sheep *P. ovis* and *P. cuniculi* and in goats *P. cuniculi*. The presence of mites on the skin causes a local reaction with vesicle formation. The exudate from the vesicles coagulates and dries on the skin surface, resulting in the formation of a crust or scab of varying thickness. There is marked pruritus, and the host's scratching results in alopecia, erosions and crust development. In goats, *P. cuniculi* is known as the 'earcanker' mite because of its predilection for the ear (causing otoacariasis).

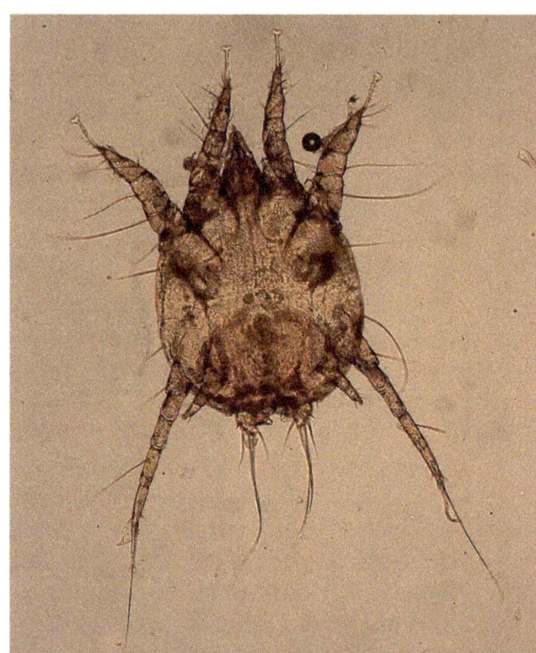

publication bias a bias that arises owing to only certain types of results (e.g. those that successfully reject the null hypothesis) being much more likely to be published than others, leading to a disproportionate representation in the literature.

pulicide an agent that kills members of the family Pulicidae, specifically fleas, used for flea control.

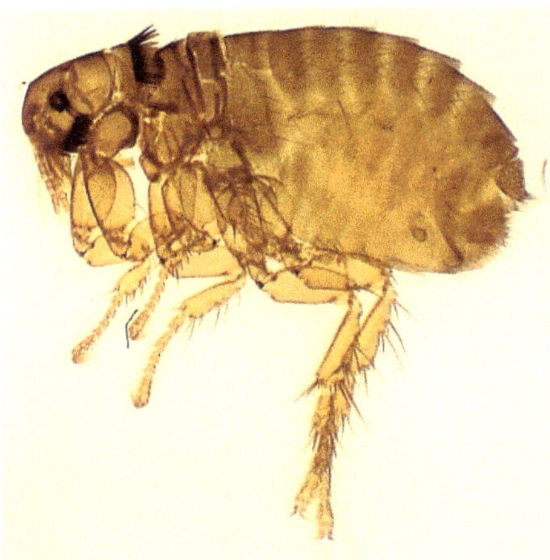

pulmonary capillary wedge pressure provides an indirect measure of the left atrial pressure and is measured by wedging a pulmonary catheter with an inflated balloon into a small pulmonary arterial branch.

pulmonate gastropods that belong to the subclass Pulmonata that include terrestrial snails and slugs as well as some freshwater and marine snails that breathe air using lung-like sacs.

punctuated equilibria the view that speciation occurs relatively rapidly, and that virtually all evolution happens at speciation (cladogenesis), with long intervening phases of stasis of lineages.

punctuation a rapid speciation event, forming part of the model of evolution by punctuated equilibria.

pupa (plural **pupae**) the second stage in the development of a holometabolous insect, between the larva and the imago, that is usually an inactive stage such as a cocoon.

pupal chamber a hollowed out space in the xylem of the pine tree that is created by the chewing action of the *Monochamus* beetle larvae and that serves to protect the beetle as it pupates.

puparium the hard pupal case of the insect pupa.

pupate to proceed to the stage of pupa in an insect life cycle.

purifying (negative) selection process favouring the pre-mutation genotype over any new mutations. Loci under purifying selection rarely vary in the population. It is a selection regime that results in the removal of deleterious substitutions from the population.

purine a nitrogen-containing, double-ring, basic compound that occurs in nucleic acids. The purines in DNA and RNA are adenine and guanine. *See* **bp**.

purity of repetition the degree to which all motifs within an SSR are identical. In a 'pure' or 'perfect' repeat, none of the motif copies displays any variation. In contrast, an 'imperfect' repeat has some substitutions in the sequence of one or more of the repeating motifs. Imperfect repeats are more stable (less prone to slippage mutations) than pure repeats.

putative vector a vector for which a definitive proof of vector capacity is lacking. Usually, only epidemiological or molecular evidence are available.

***p*-value** the possibility that any particular outcome would have occurred by chance. Statistical significance is usually $p < 0.05$.

pycnocline a zone of rapid change in density in a water body, typically attributable to temperature and/or salinity.

Pyrantel (PYR) belongs to the tetrahydropyrimidine group of anthelmintics. PYR is available as tartrate and pamoate (synonym embonate) salts. Different salts of PYR have different pharmacokinetic properties. The pamoate salt is almost insoluble in water, poorly absorbed from the gastrointestinal tract, and most passes unchanged in the faeces. Reduced systemic absorption of the pamoate form potentially increases availability in the lumen of the intestine. The tartrate salt of PYR is soluble in water and absorbed rapidly and extensively from the intestine of monogastric animals.

pyrethrins are naturally occurring botanical compounds derived from members of the Chrysanthemum flower family. While they have a quick knock-down effect against insects, they are unstable in the environment, so may not last long enough to kill the insects. Pyrethroids are synthetic analogues of pyrethrins, designed to be more stable in the environment, thus provide longer-lasting control. The pyrethrins and pyrethroids have efficacy against flies, fleas, lice, and ticks. Pyrethroids have repellent activity. Dog preparations should not be used in cats. Occasionally products containing low concentrations of permethrin have been licensed for cats. These compounds act on voltage-sensitive ion channels through which sodium is pumped to cause excitation of neurons. They prevent the sodium channels from closing, resulting in continual nerve impulse transmission, tremors, and eventually, death. The pyrethroids are classified as type 1 or type 2 based on their neurophysiological effects. Type 1 pyrethroids (permethrin, resmethrin) prolong opening of individual sodium channels long enough to cause repetitive discharges, leading to tremors. The type 2 pyrethroids (fenvalerate, cypermethrin) cause a more prolonged opening of sodium channels, leading to eventual suppression of neurotransmission. They also increase release of the inhibitory peripheral neurotransmitter GABA.

pyrethroids toxicity pyrethroids are applied on dogs in the form of sprays, shampoos, lotions, collars, or spot-ons. They are also used in the environment, sometimes combined with insect growth regulators, in the form of sprays, diffusers, and solutions. Pyrethroids are mainly used to prevent flea and tick infestation, but also to repel flying insects, especially sand flies (*Phlebotomus*), mosquitoes (*Aedes*, *Culex*), and biting flies (*Stomoxys*, *Haematobia*). Pyrethroids have relatively low toxicity to mammals but are highly toxic to poikilothermic vertebrates (fish). Cats are physiologically deficient in glucuronoconjugation enzymes, which explain their hypersensitivity to pyrethroids, especially to permethrin.

pyriform apparatus the inner membrane of the eggs of certain tapeworms (e.g. *Monezia*), which often is pear-shaped and bears hooks.

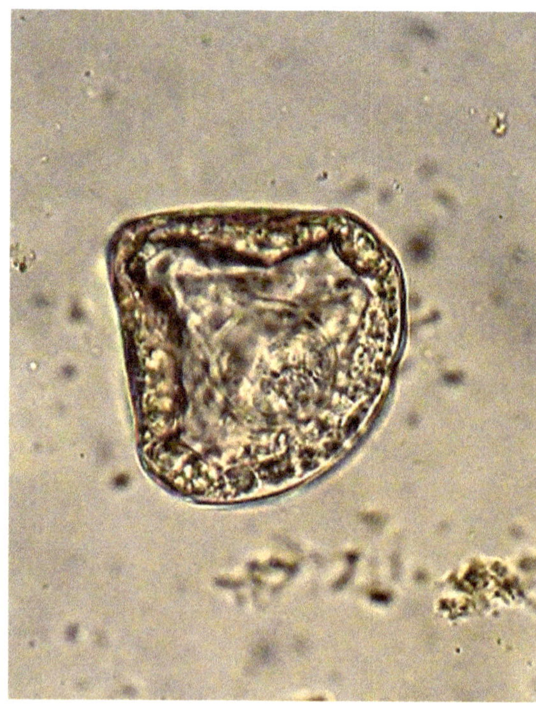

pyrimidine a nitrogen-containing, single-ring, basic compound that occurs in nucleic acids. The pyrimidines in DNA are cytosine and thymine; in RNA, cytosine and uracil. *See* **bp**.

pyrogens proteins produced by phagocytes that induce fever.

pyroptosis an inflammatory form of programmed cell death, activated following cellular insults such as bacterial infection or exposure to toxins, and mediated by caspase-1 and/or caspase-11.

pyroptosome a large supramolecular complex composed of Pycard dimers that mediates inflammatory programmed cell death (pyroptosis) through caspase-1 activation.

pyrosequencing a sequencing-by-synthesis method based on the detection of phyrophosphate released upon nucleotide incorporation.

pyruvate dehydrogenase an enzyme that catalyses the conversion of pyruvate into acetyl-CoA.

qualicode cysticercose IgG ELISA kit using crude cysticercus antigens to detect IgG in serum.

quality-assured artemisinin-based combination therapies (QAACTs) ACTs that comply with the Affordable Medicines Facility – malaria (AMFm) and Global Fund to Fight AIDS, Tuberculosis, and Malaria's Quality Assurance Policy.

quantitative trait a phenotype that differs in degree among individuals.

quantitative trait loci analysis (QTL analysis) a method of comparing the inheritance of genetic and phenotypic traits such that the log of difference probability scores can be assigned across the genome to define genetic regions that track with the inheritance of a particular phenotype.

quantitative trait locus (QTL) a stretch of DNA/chromosome region containing or linked to the genes that underlie a quantifiable phenotypic trait, such as body weight or blood pressure in animals. These traits are typically affected by more than one gene, and also by the environment. It has a statistically significant effect on a phenotypic difference between two lines, populations, or species. They are identified on the basis of statistical association between genetic markers and phenotypes that can be measured. QTL studies typically involve the crossing of strains that differ markedly in a particular phenotype of interest to produce a hybrid generation, with subsequent crossing back into one of the strains; the more generations of crossing are made, the more that recombination will occur and the shorter the resulting QTL region will be.

quantitative trait locus mapping use of a genetic linkage map to locate chromosomal segments that influence phenotypic traits of interest.

quarantine the restricted freedom of movement imposed upon apparently healthy humans or animals exposed to infectious disease to contain the spread of a communicable disease. The length of the quarantine is slightly longer than the longest known incubation period of the disease in question.

queen a morphologically and/or behaviourally distinct reproductive caste in eusocial insects that often shows specialization in both reproduction and dispersal abilities. Depending on the species, a colony may contain one queen (monogyny) or multiple queens (polygyny). Queens, together with males (referred to as drones in some societies), constitute the 'germline' of a eusocial insect colony.

quinine (QN) a natural product, antimalarial drug derived from the cinchona tree that was the preferred antimalarial treatment until the 1940s.

quinolones decoquinate, buquinolate, and nequinate are coccidiostats. These agents permit the coccidial sporozoites to enter the host, but interfere with further development. Quinolones act by inhibiting electron transport system (respiration) within coccidial mitochondria. Also, they may block DNA synthesis by inhibiting DNA gyrase.

quinones lipid-soluble electron carriers, which are often associated with chemiosmotic respiratory chains.

quorum sensing a regulatory mechanism by which individuals sense signals of the entire population and respond in a specific manner. It is the main mechanism that bacteria use to communicate and coordinate behaviour and regulate their gene expression according to the density of the population via signalling molecules. The concentration of these molecules is an indicator of local population density.

R_0 the basic reproduction number of an infectious agent, which is the mean number of secondary infections caused by a single infected individual in a susceptible population. An infectious agent can persist only if $R_0 > 1$.

R1 inhibitory peptide a 20-residue peptide capable of inhibiting erythrocyte invasion by *P. falciparum* merozoites by binding to AMA1 and preventing tight junction (TJ) formation.

Rab a family of coat proteins. Mammalian cells have at least 60 different Rab proteins.

RAC1 a small G protein that indirectly regulates ARP2/3 function and promotes the formation of lamellipodia.

Rac-dependent pathway a signalling pathway that controls cytoskeleton and transcription activity. Rac activation is responsible for stimulating cell spreading and migration.

RAD sequencing (RAD-seq) a sequencing method in which DNA is cut with restriction enzymes prior to sequencing, which exclusively targets the DNA associated with restriction sites.

RAD51 the eukaryotic homologue of bacterial RecA and participates in DNA damage-response pathway, homologous recombination and DSB repair.

radiation attenuated sporozoites (RAS) *Plasmodium* sporozoites that have been irradiated with a radiation source (such as gamma rays or X-rays) such that they can invade the host's hepatocytes but can not fully develop.

radioallergosorbent test (RAST) a solid-phase radioimmunoassay for detecting IgE specific for a particular allergen.

radioimmunoassay (RIA) immunological techniques that require the use of an isotope-labelled reagent. These techniques measure antigen–antibody interactions, and determine the level of important biological substances in mixed samples. RIA takes advantage of the specificity of the antigen–antibody interaction and the sensitivity that derives from measurement of radioactively labelled materials.

radioimmunoconjugate a biochemical conjugate consisting of an immune targeting molecule such as an antibody or antibody fragment together with a cytotoxic radionuclide.

radioimmunodiffusion a type of immunodiffusion in which a radioactive antibody is incorporated in order to increase the sensitivity of the method.

radioimmunosorbent assay a solid-phase assay that can detect extremely low levels of IgE.

radioisotope (unstable) isotope of an element that releases radiation as it decays to a stable form.

raft-like structures (RLS) membrane microdomains in which structure and composition are similar to that of lipid rafts.

random amplified polymorphic DNA (RAPD) uses a set of single and short random primers for amplifying genomic DNA and generating polymorphic banding patterns. The polymorphic DNA fragments are separated on agarose gels and the resulting banding patterns are analysed as described for DNA fingerprinting.

random genetic drift random changes in a gene population over time. These fluctuations can lead to fixation or extinction of an allele without regard to their adaptive value.

random mating a fundamental simplifying assumption for many population genetics models. Non-random mating may be assortative (birds of a feather), disassortative (opposites attract) or skewed (hotshots). For example, random mating is required for Hardy–Weinberg equilibrium.

randomized control trial (RCT) a study in which a number of individuals are randomly assigned to two (or more) groups to evaluate a specific intervention, treatment, device, or strategy with one group (experimental) randomly assigned to the intervention and the other (control group) receiving an alternative intervention or placebo.

RANTES C–C chemokine ligand type 5 (CCL5) plays a chemotactic role in recruiting leucocytes to inflammatory sites. It is released by activated T cells and is a ligand for CCR5.

Rap1 (repressor activator protein 1) a multifunctional protein first described as a transcriptional factor, which is involved in telomere length homeostasis, and telomere-position effect in yeasts and humans.

rapid diagnostic test (RDT) immunoassays for antibody or antigen detection using a visual system, which provides results in minutes. RDTs are simple to operate and read, sensitive, specific, stable at high temperatures, and require only a short time to perform. These tests generally involve interaction of a fixed reagent of either target antigen or antibody linked to a visible detector that reacts with a patient's sample. The most common formats are dot, immunochromatographic strip, dipstick, and agglutination tests.

rare alleles polymorphisms that occupy the lower bounds of the site frequency spectrum, often as singletons.

rare-cutter enzyme *See* **restriction-enzyme cutting site**.

rate ratio (RR) the ratio of the rate of occurrence of an event given a set of covariates compared to the rate of occurrence in their absence. For example, the average number of hookworm eggs counted (by Kato–Katz) 1 week after treatment with albendazole divided by the average number of hookworm eggs counted before treatment.

rays finger-like hypodermic structures which support the copulatory bursa.

Rbp38 RNA-binding protein 38, a trypanosomatid protein that binds single- and double-stranded kinetoplast RNAs. It is involved in *T. brucei* kinetoplast DNA replication and has also been described as a telomere- and GT-rich DNA-binding protein.

Rdl (resistant to dieldrin) an insecticide-resistance allele, not to be confused with RIDL.

reactive infection detection (reactive ID) the detection of malaria infections in community or occupation-based population groups in close proximity to an index case. Reactive ID involves testing (and subsequently treating) co-workers, household members, and neighbours of an index case. Reactive ID may be preceded by fever screening of all patients with a recent history of malaria symptoms or by testing the target population without prior screening.

reactive oxygen species (ROS) oxygen-containing species that have unpaired electrons, making them highly reactive towards transition metals such as iron and copper.

reader enzyme a protein that recognizes a certain histone mark and binds to it.

reads the fragments of DNA sequences generated that represent data from a unique fragment of the sequencing library. A typical next-generation sequencing run generates millions of reads per sample.

reagin the name allergists use for IgE class antibodies.

really interesting new gene (RING) finger domain a type of zinc finger defined by a consensus sequence of approximately 40–60 amino acids that includes cysteine and histidine as the zinc-binding residues.

reassortant a term that usually refers to a virus that contains two or more segments of nucleic acid (segmented genome) that are derived from different viruses.

recall bias systematic error due to the differences in accuracy or completeness of recall to memory of past events or experiences.

recall response occurs after a second encounter with a pathogen or antigen. The ensuing immune response is usually faster and stronger than a primary response and involves memory cells.

receptor a structure on cell membranes that binds specifically to ligands in the surrounding medium.

recessive gene a gene expressed only if there are 2 identical copies.

reciprocal chemical genetics a chemical-compound screen is compared with a genetic screen to find novel inhibitors and their potential targets that control the process of interest.

reciprocal translocation when a pair of chromosomes exchange exactly the same length and area of DNA resulting in a shuffling of genes.

recombinant clone clone containing recombinant DNA molecules. *See* **recombinant DNA technology**.

recombinant DNA (rDNA) DNA formed by combining segments of DNA from different genes or different types of organisms.

recombinant DNA molecules a combination of DNA molecules of different origins joined using recombinant DNA technologies.

recombinant DNA technology a procedure used to join together DNA segments in a cell-free system (an environment outside a cell or organism). Under appropriate conditions, a recombinant DNA molecule can enter a cell and replicate there, either autonomously or after it has become integrated into a cellular chromosome.

recombinant vaccine a vaccine containing antigen prepared by recombinant DNA techniques.

recombination the production of new combinations of alleles, or genotypes, by independent assortment of different chromosomes, by crossing-over, or by gene conversion. Homologous chromosomes undergo crossing-over and gene conversion with high frequency during meiosis, but these forms of recombination can also occur with low frequency in any cell. *See* **crossing-over**, **mutation**.

recombination signal sequence (RSS) conserved heptamer (7-nucleotide) – nonamer (9-nucleotide) sequences, separated by a 12 or 23 base spacer, which occur 3′ of variable gene segments, 5′ and 3′ of diversity gene segments, and 5′ of joining gene segments, in both immunoglobulin and T-cell receptor genes. RSSs function as recognition sequences for the recombinase enzymes that mediate the gene rearrangement process involved in the generation of lymphocyte antigen receptor diversity.

recrudescence **1** the interconversion between bradyzoite and tachyzoite stages has a pivotal role in establishing chronic infection and recrudescence. **2** disease reactivation, reoccurrence of a disease after treatment. This can be caused by parasites that were not completely eliminated during the treatment.

recurrent sweeps a model in which sweeps are considered to occur at a given rate, rather than at a fixed time in the past.

recycling endosomeslate vesicular compartments that are involved in recycling membrane proteins and de novo synthesized cargo from the Golgi complex to the cell surface.

red blood cell ghosts (RBC ghosts, erythrocyte ghosts) haemoglobin depleted RBC, which consists basically of only the RBC membrane.

red pulp non-lymphoid area of the spleen, rich in blood vessels and haematopoietic cells.

redfield ratio optimal ratio of carbon, nitrogen, and phosphorus for phytoplankton growth (106C:16N:1P) based on molecular concentrations.

redia (plural **rediae**) an intermediate larval stage of trematodes that develops asexually within a sporocyst inside the first intermediate host, usually a snail.

Redox clines the interfaces between two spatially distinct areas that differ in chemical composition and redox potential; they are typically used to describe a transition from oxic to anoxic conditions.

reduction electron-transfer reactions in which a substance gains one or more electrons.

reduction division meiosis; division of the nuclear material to give rise to gametes having the haploid number of chromosomes.

reference species the species represented by the microarray.

reference test the reference test is usually an imperfect gold standard. It will be the best available test to define the infection or disease status of an individual. The sensitivity and specificity of the test are usually well defined but often fall short of a perfect gold standard.

referral bias the sequence of referrals that may lead patients from primary to tertiary centres raises the proportion of more severe or unusual cases, thus increasing the likelihood of adverse or unfavourable outcomes. Physicians and medical centres may attract individuals with specific disorders or exposures.

reflection interference contrast microscopy (RICM) microscopy method that uses polarized monochromatic light to illuminate an object. The reflected light interferes and generates patterns that allow distinguishing how close a part of the object is to the substrate.

refractoriness an intrinsic ability of anopheline species or populations to not replicate and not transmit different plasmodium species (or strains).

refractory host a host difficult to be parasitized.

refugia a subpopulation of parasites not exposed to an anthelmintic at the time of treatment. This includes pre-parasitic stages in the environment, parasites of herd members that are left untreated when a selective treatment is applied, and parasitic stages within the animal that are unaffected by treatment due to physical, physiologic, or pharmacokinetic factors. Parasite subpopulation left in refugia represents a reservoir of unselected genes which includes those alleles imparting anthelmintic susceptibility. *Refugia* ensure that susceptible worms are available to mate with resistant worms that have survived treatment. When the number of parasites in *refugia* is increased, the rate of resistance development in the parasite population will be reduced correspondingly.

regenerative medicine new medical advances in which damaged body parts or tissues are replaced or the body is encouraged to heal itself. From our understanding of our genes and how they work to control the growth, building and repair of our body, scientists are using regenerative medicine to study how to create new tissues for transplant, transplant stem cells into the body or how to induce the body to regenerate from the body's own cells.

regulatory elements segments of the genome (e.g. regulatory regions, genes that encode regulatory proteins, or small RNAs) involved in controlling gene expression.

regulatory phenotype a refractory phenotype in which immune cells (including macrophages, T cells and B cells) do not respond to antigenic stimuli and alter the cellular milieu to induce regulatory phenotypes in other nearby cells.

regulatory region or sequence a DNA segment to which a regulatory protein binds to control expression of a gene or group of genes that are expressed together.

regulatory T (Treg) cells T helper cells specialized to promote immune quiescence by down-regulating effector T helper cells (Th1, Th2, and Th17).

reinfection a second infection by the same agent.

reinforcement the adaptive strengthening of prezygotic isolating mechanisms in a zone of secondary contact between two distinct taxa that have previously evolved some postzygotic isolation. In this process, natural selection reduces gene flow between taxa because hybridization is deleterious.

rejection an immune response with both cellular and humoral components directed against transplants and grafts.

relapse the recurrence of asexual parasitaemia in *P. vivax* malaria derived from persisting liver stages. Relapse occurs when the blood stage infection has been eliminated, but hypnozoites persist in the liver and mature to form hepatic schizonts. After variable intervals of weeks to months, the hepatic schizonts burst and liberate merozoites into the bloodstream.

relatedness a genetic correlation between individual loci or organisms.

relational database a database in which data are stored in tables and the relationships among the various data elements are also stored in tables. Microsoft

Access® (Microsoft, Redmond, WA, USA) is a common relational database.

relations the edges that link together terms in a given ontology. Open Biological Ontologies (OBO) Relations (OBO-REL) lists all relations that are accepted for use in an OBO Foundry-hosted ontology.

relative risk (RR) the ratio of the probability of developing, in a specified period of time, an outcome among those receiving the treatment of interest or exposed to a risk factor, compared with the probability of developing the outcome if the risk factor or intervention is not present.

relative risk reduction (RRR) the extent to which a treatment reduces a risk, in comparison with patients not receiving the treatment of interest.

release of insects carrying a dominant lethal (RIDL) refers to an insect control method based on the release of transgenic individuals carrying a dominant, repressible, female-specific lethal gene.

repeated measures measurements or observations made repeatedly on the same unit, e.g. multiple hookworm counts measured from the same individual host.

repellent compound that makes a host unattractive to a parasite and thus can prevent attack or establishment.

replication kinetics surface molecule organization for prolonged antigen exposure.

replication origin locus in the genome where DNA synthesis begins during replication. In eukaryotes, the multiple origins used are not normally defined by conserved sequences; the exception to this is in *Saccharomyces cerevisiae*, where origins are composed of related autonomous replicating sequences (ARSs).

reportable disease a disease that, by law, must be reported to a health authority. In general, such diseases are of special concern to the health of the human or animal population; examples include malaria, foot and mouth disease, and tuberculosis.

repressor a protein that turns off the expression of one or more genes. The repressor protein works by binding to the gene's promoter region, preventing the production of mRNA.

reproducibility (repeatability, reliability) the results of a test or measure are identical or closely similar each time it is conducted.

reproductive character displacement the evolution of increased differences between two established species in sympatry, which might result from the production of hybrids that are less fit or from efforts wasted on unsuccessful interspecific mating interactions.

reproductive cloning making a full living copy of an organism. Currently illegal in Australia and many other countries around the world.

reproductive isolation absence (or severe restriction) of gene flow between populations whose members are in contact with one another.

reservoir host one in which the parasites lives and is available for transmission to another host. In parasitology, the term usually refers to a host which harbours a stage of the parasite found in the typical host. The dog is a host of *Trypanosoma cruzi* and functions as a reservoir host for man.

reservoir of infectious agents any person, animal, or substance in which an infectious agent normally lives and multiplies in such a manner that it can be transmitted to a susceptible host.

resilience the ability of the host to maintain an undepressed production level under parasite challenge. Djallonke sheep are able to gain weight and reproduce even in the presence of a persistent parasitaemia. Therefore, 'resilience' is the most appropriate term to describe the phenomenon of trypanotolerance in sheep.

resistance gene (R-gene) a plant gene encoding a specificity determinant for activation of the plant defence response to a pathogen.

resistance to ectoparasiticides some ectoparasites can develop resistance to the lethal effects of ectoparasiticides following continuous exposure to them. Ectoparasiticides may lose their efficacy due to (i) diminished penetration into target organism, (ii) increased detoxication (metabolism) of ectoparasiticides, or

(iii) decreased sensitivity of the target site (i.e. receptor).

resistance to parasites or pathogens the inherent ability of organisms to prevent establishment or development of a given species or strain of compatible parasite, by opposition to susceptible organisms from the same species. Resistance is heritable and could result from the genetic (or epigenetic) modification of any constitutive or inducible trait (not necessarily a defence trait).

resolution degree of molecular detail on a physical map of DNA, ranging from low to high.

resorcinol a dihydroxybenzene that can be used to stimulate pumping of the feeding tube/pharynx in plant nematodes.

respiration a series of biochemical redox reactions in which the energy released from the oxidation of organic or inorganic compounds is used to generate cellular energy in the form of ATP.

respiratory burst (oxidative burst) the rapid release of reactive oxygen species (ROS) such as hydrogen peroxide or superoxide by phagocytic cells that have a dilaterious effect on phagocytosed microbes.

respiratory frequency the number of breaths taken per minute.

resting memory CD4$^+$ T cells antigen-specific CD4$^+$ T cells that have reverted to the G0 state of the cell cycle from a previously activated state but retain the capability to rapidly respond to a second antigenic exposure.

restrained recombination genetic recombination among loci does not occur at random. In predominant clonal evolution (PCE), restrained recombination is strong.

restricted maximum likelihood (REML) estimation an alternative to ML estimation for models that include random effects. In REML estimation, the dispersion of the random effects is estimated having averaged over some of the uncertainty in the fixed effects. By contrast, in ML estimation, the fixed effects estimates are treated as precisely correct.

restriction enzyme (RE) an enzyme (normally derived from bacteria) that cuts strands of DNA at particular points along its length into shorter fragments.

restriction fragment length polymorphism (RFLP) a type of polymorphism that results from variation in the DNA sequence recognized by restriction enzymes. These are bacterial enzymes used by scientists to cut DNA molecules at known locations. RFLPs (pronounced 'rif lips') are used as markers on genetic maps. Typically, gel electrophoresis is used to visualize RFLPs. *See* **marker**, **polymorphism**.

restriction-enzyme cutting site a specific nucleotide sequence of DNA at which a particular restriction enzyme cuts the DNA. Some sites occur frequently in DNA (e.g. every several hundred bps); others much less frequently (rare-cutter, e.g. every 10,000 bps).

restriction-site-associated DNA markers sequence polymorphisms in close proximity to a restriction enzyme recognition site.

reticulocyte young RBCs characterized by the presence of RNA that could be seen using a specific staining (methylene blue or brilliant cresyl blue). *Plasmodium vivax* merozoites primarily invade due to the presence in the merozoiyes of a reticulocyte binding protein complex.

reticuloendothelial system (RES) a network of phagocytic cells; macrophages are the most important cells of this system.

retinoic acid-inducible gene 1 like receptors (RLRs) cytoplasmic RNA helicase enzymes, which recognize viruses by binding their dsRNA. RIG-1, MDA5, and LGP2 are the currently described members of this pattern recognition receptors (PRRs) family.

retroactivity alteration in the dynamic behaviour of a gene circuit owing to the connectivity of its output gene with downstream circuits.

retroinfection a process of infection in which the parasite leaves the body of the host and then immediately penetrates the skin; an example is *Strongyloides*.

retrotransposon a transposon that was created by reverse transcription of an RNA molecule.

retroviral infection the presence retroviruses (RNA viruses)that use their reverse transcriptase to make DNA to insert their genetic material into the chromosomes of the host's cells. The virus is then propagated by the host cell. *See* **retrovirus**.

retrovirus a virus that uses RNA as its genetic material template upon which to synthesize DNA using an enzyme known as reverse transcriptase. When a retrovirus infects a cell, it makes a DNA copy of its genome that is inserted into the DNA of the host cell. There are a variety of different retroviruses that cause human diseases such as some forms of cancer and HIV. *See* **reverse transcriptase**.

reverse genetics an approach to discovering the function of a gene that proceeds in the opposite direction to classical genetics by seeking the possible phenotypes that might derive from a specific genetic sequence.

reverse line blot a PCR-based assay used to detect different pathogens simultaneously by species-specific diagnostic probes on a panel of PCR amplicons obtained from different biological samples.

reverse line-blot hybridization a method consists of a single-run PCR amplification of a conserved gene by using non-degenerated primers, followed by a reverse line-blot hybridization assay. Amine-conjugated oligonucleotide probes are covalently coupled to a carboxylated nylon membrane and then biotin-labelled PCR products are hybridized to the probes. Hybridized PCR products are visualized colorimetrically by incubating the membrane with streptavidin-labelled peroxidase and a chemiluminescent substrate.

reverse transcriptase an enzyme that transcribes RNA to DNA and is used by retroviruses such as HIV to form a complementary DNA sequence (cDNA) from their RNA. The resulting DNA is then inserted into the chromosome of the host cell.

reverse transcription polymerase chain reaction a reaction in which single-stranded RNA is first converted to its double-stranded DNA complement (cDNA), which is then PCR amplified.

reverse vaccinology (RV) a vaccine design approach in which the entire genome of pathogens is screened to identify antigens with characteristics desirable as vaccine targets, such as cell surface expression.

reverse-engineer examination of the constituent components of a system in order to understand their integrated function. In systems biology, this may involve making perturbations to a cellular network and then constructing a model that describes the relationship between the behaviour of the molecular components and that of the entire system.

Reynolds number (Re) a dimensionless number that describes the relative contribution of inertial and viscous forces to cell movement. $Re = F_i/F_v$ where F= force, i = inertial and v = viscous.

rhabditoid juvenile or rhabditiform larva the first, feeding stage of a juvenile nematode that emerges from the egg in which the oesophagus is functional, is usually muscular, and has an enlarged, bulbous posterior end.

Rhipicephalus sanguineus the brown dog tick or kennel tick; a putative vector of *Ehrlichia canis*, a Gram-negative, intracellulear bacterium that causes canine ehrlichiosis. Other pathogens are *Babesia vogeli*, a other protozoan parasite that causes canine babesiosis, *Rickettsia rickettsii*, the causative agent of Rocky Mountain spotted fever (RMSF), and other *Rickettsia* species such as *R. belli*, *R. rhipicephali* and *R. montana*. This tick is almost exclusively a parasite of dogs and can complete all its life stages inside the household indoor.

rhopH complex a high molecular weight protein complex comprising RhopH1, RhopH2 and RhopH3, constituting Rhoptries, which are part of the apical complex of Apicomplexan protists.

rhoptries club-shaped or sac-shaped elongated electron-dense secretory organelles within the apical complex that provide some proteins involved in the penetration of host cells (along with micronemes) and are the primary contributor to parasitoforous vacuole (PV) formation. They are largest of the apical organelles in apicomplexan parasites. The contents of these club-shaped organelles are discharged during invasion and are believed to play a role in the formation of the PV in several member of the phylum Apicomplexa such as *Plasmodium* and *Toxoplasma*.

rhoptries and micronemes secretory organelles containing proteins required for parasite motility, adhesion, and invasion of host cells, and for the establishment of the parasites within the host cell.

rhoptry-associated membrane antigen (RAMA) an essential and abundant glycophosphatidylinositol (GPI)-anchored protein that accumulates in the endoplasmic reticulum (ER) and Golgi compartments in lipid rafts prior to rhoptry development, and which is implicated in rhoptry biogenesis.

rhoptry-associated protein (RAP) complex a low molecular weight protein complex localized to the rhoptry bulb, and composed of heterodimers of RAP1 and RAP2, or RAP1 and RAP3.

ribonucleic acid (RNA) a molecule similar to DNA. Unlike DNA, RNA is single-stranded, except in certain double-stranded RNA (dsRNA) viruses. An RNA strand has a backbone made of alternating sugar (ribose) and phosphate groups. Attached to each sugar is one of four bases – adenine (A), uracil (U), cytosine (C), or guanine (G). Different types of RNA exist in the cell: mRNA, rRNA, and transfer RNA (tRNA). In dsRNA viruses the strands are (–) and (+) strands where the + strand is typically the mRNA.

ribonucleotide *See* **nucleotide**.

riboregulator an RNA molecule that responds to a signal nucleic acid (DNA or RNA) by Watson–Crick base pairing.

ribose the five-carbon sugar that serves as a component of RNA. *See* **ribonucleic acid**, **deoxyribose**.

ribosomal frameshifting a translational recoding event in which a proportion of translating ribosomes change their reading frame at a defined point in the mRNA, resulting in the addition of an alternative carboxy-terminal extension.

ribosomal RNA (rRNA) specialized RNA found in the catalytic core of the ribosome, a molecular machine that serves as a substrate for the synthesis of proteins in all living organisms.

ribosome a cellular particle made of rRNA and protein that serves as the site for protein synthesis in the cell. Typically their size is measured in terms of Svedburg (S) units, which are measures of their behaviour in a centrifugal field. Eubacterial ribosomes are 70S (50S + 30S subunits) and eukaryotic ribosomes are 80S (60S + 40S subunits) referred to as large and small subunits, respectively. The ribosome reads the sequence of the m RNA (mRNA) and, using the genetic code, translates the sequence of mRNA bases into a sequence of amino acids.

riboswitches regulatory RNA elements of mRNAs that bind to small molecules and regulate expression of the mRNA accordingly.

ribozymes RNA molecules that are capable of catalysing specific biochemical reactions.

ring enhancement or ring-enhancing lesion an abnormal radiological sign on magnetic resonance imagining (MRI) or computerized tomography (CT or CAT scan) images obtained using radiocontrast. For example, there is an area of decreased density surrounded by a bright rim from concentration of the enhancing contrast dye. This can be a finding in numerous disease states. In the brain, it can occur with an early brain abscess as well as in CNS parasitic protozoa infections such as *Toxoplasma gondii* or parasitic or pathogenic free-living amoebae.

ring-infected surface antigen (RESA) also known as Pf155 due to its molecular weight of 155 kDa; a protein that is released from *Plasmodium* merozoites and associates with the membrane of the erythrocyte. Recombinant fragments of this protein have been used in Phase II trials of malaria vaccines.

rings, trophozoites, schizonts successive stages of the *Plasmodium* replicative asexual cycle in the host red blood cell. Each cycle consists of invasion, replication, and egress, and requires ≈48 h, resulting in the production of ≈8–32 infectious merozoites per red blood cell.

ring-stage assay (RSA) an assay developed by the group of D. Ménard in Cambodia to synchronize *P. falciparum* cultures and study specifically the action of endoperoxides on both artemisinin-sensitive and -resistant clinical isolates. So far this assay is the only one that has led to the characterization of the artemisinin resistance phenotype *in vitro*.

risk the probability that an individual will be affected by a particular genetic disorder. Both genes and environment influence risk. An individual's risk may be higher because they inherit genes that cause or increase susceptibility to a disorder. Other individuals may be at higher risk because they live or work in an environment that promotes the development of the disorder.

risk factor patient characteristics or factors associated with an increased probability of developing a condition or disease initially. Compare with prognostic factors. Neither risk nor prognostic factors necessarily imply a cause and effect relationship.

risk ratio (RR) the ratio of the probability of an event occurring given a set of covariates compared to the probability of the event occurring in their absence. For example, the probability of observing worm eggs 1 week after treatment with an antiparasitic drug divided by the probability of observing worm eggs before treatment.

R-loop a three-stranded nucleic acid structure in which RNA is hybridized to the complementary strand of dsDNA and the other DNA strand is exposed in a ss loop.

RNA editing a special case of post-transcriptional modification of RNA involving one or more changes in the base composition of an mRNA molecule, such that the information in the mRNA differs from that of the DNA from which it was transcribed, often resulting in translation into a different amino acid.

RNA flare a technique to quantify RNA expression. It is based on gold particles carrying a sequence complementary to the RNA of interest that are coated with small RNA strands. On binding of the target RNA to the probe, the small RNA strands are released and become fluorescent.

RNA fluorescence *in situ* hybridization (FISH) a technique that allows the localization and quantification of RNA in the cell. RNA is stained in cells using fluorescent probes that are complementary to the RNA sequence of interest.

RNA induced silencing complex (RISC) a multiprotein complex that uses a bound miRNA to recognize complementary mRNA molecules, and which results in translation inhibition and/or degradation.

RNA interference (RNAi) a process in which double-stranded RNA triggers the degradation of a homologous mRNA.

RNA polymerase the enzyme that catalyses the synthesis of a complementary strand of RNA from either a DNA strand or an RNA strand in some viruses.

RNA polymerase III the enzyme responsible for transcribing many non-coding RNA molecules. These include the tRNA and 5S rRNA molecules essential for translation, and also other structural, catalytic and regulatory RNA molecules.

RNA processing the suite of cellular activities that leads to maturation and proper localization of a nascent transcript, and to the various steps in its eventual degradation.

RNA profiling comparative measure of RNA levels over the complete genome through hybridization of cDNA to microarrays containing features representing all transcribed sequences. The resulting data provide the expression profile.

RNA pulse labelling a method used to measure RNA decay. Cells are incubated with radioactive RNA precursors. RNA is then isolated and fractionated into newly synthesized (radioactive) and pre-existing (non-radioactive) fractions, which are sequenced and quantified. The mRNA half-lives are calculated based on the ratio of newly synthesized RNA to pre-existing RNA.

RNA recognition motif (RRM) an RNA-binding domain comprising about 90 amino acids that mediates binding to single-stranded RNA.

RNA regulatory elements (RREs) sequence or structural elements, located predominantly in the 3′-UTR of messenger RNA, that mediate(s) binding to RBPs.

RNA regulon model a model in which *trans*-acting factors co-ordinate the regulation of multiple functionally related transcripts.

RNA riboswitches RNA molecules that respond to environmental conditions by changing secondary structure and, in some cases, by modulating catalytic function, thereby affecting gene expression.

RNA-binding protein immunoprecipitation and high-throughput sequencing (RIP–seq) a method in which the RBP is immunoprecipitated to isolate associated RNAs. Isolated RNAs are reverse transcribed to cDNA and sequenced to identify target genes.

RNA-binding protein purification and identification (RaPID) a method to identify ribonucleoprotein components associated to RNA-aptamer tagged mRNA *in vivo* by mass spectrometry; also allows detection of different RNA species captured in the same ribonucleoprotein complex by RT-PCR.

RNA-binding proteins (RBPs) proteins that bind to double- or single-stranded RNA via one or more RNA-binding domains. These proteins are involved in a wide range of processes from alternative splicing to RNA degradation.

RNA-dependent RNA polymerase (RdRP) an enzyme that catalyses the copying of RNA from an RNA intermediate. RdRP genes appear to have been lost in many animal species, including Drosophila and mosquitoes, but are encoded by many viruses and are essential for RNA virus replication.

RNA sequencing (RNA–seq) the use of high-throughput sequencing technologies to sequence complementary DNA (cDNA) and obtain information about a sample's RNA content.

RNP (ribonucleoprotein) a protein–mRNA complex that mediates all the steps of retrotransposition, ensuring the fidelity of the reaction.

Robertsonian translocation a chromosome rearrangement that results from the fusion of the long arms of two non-homologous acrocentric chromosomes.

Roll Back Malaria (RBM) the global framework for co-ordinated action against malaria launched in 1998 under the auspices of the World Health Organization (WHO).

rootlet a complex of bundles of microtubules originating from around the basal-body region of the flagellum.

rosette name for the structure of the *Diphyllobothrium latum* uterus.

rostellum a prominence on the anterior end of the scolex of certain tapeworms of the order cyclophyllidea and is usually fitted with rows of hooks.

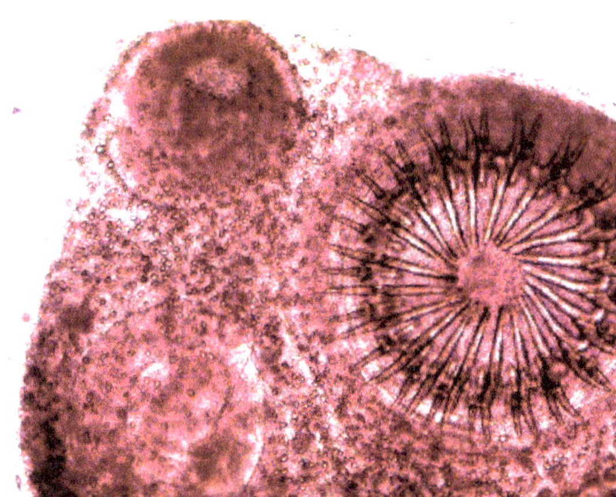

rotenone a botanical insecticide, an alkaloid derived from the root of the derris plant. Rotenone inhibits the respiratory system of the parasite by blocking NADH oxidation and subsequent ATP generation. As a result, oxidation of lactate, glutamate, and other substances is reduced and nerve conduction is inhibited. Although it is more toxic than the pyrethroids or pyrethrins, it is still used in dogs and cats. Signs of toxicosis include gastrointestinal upset, respiratory stimulation, seizures, coma, and death. Do not use in reptiles.

royal jelly a nutrient-rich secretion produced by mandibular and hypopharyngeal glands in honeybee nurses, who feed it to larvae to induce their development into gynes (i.e. virgin queens).

Rpa-1 replication protein A subunit 1, involved in DNA replication and recombination, and in telomere maintenance. This protein contributes to chromosome stability during DNA metabolism.

rRNAs (ribosomal RNAs) non-coding RNAs that form the nucleic acid component of the ribosome; in bacteria, there are three rRNA species: the 16S rRNA, the 23S rRNA and the 5S rRNA.

A B C D E F G H I J K L
M N O P Q R **S** T U V W
X Y Z

16S ribosomal RNA gene PCR analysis a method in which primers are designed for highly conserved regions of the 16S rRNA gene to enable PCR from most bacteria, and variable regions of the sequence can be used for taxonomic identification.

16S rRNA gene the small subunit of the bacterial and archaeal ribosome. The DNA sequence of this gene is the most commonly used marker for taxonomic classification of microbial communities. Other markers include hsp65 and rpoB, which encode a highly conserved heat shock protein and the β-subunit of RNA polymerase, respectively.

16S sequencing DNA sequencing of the variable region of bacterial rRNA genes, which are typically conserved within a bacterial species, and frequently used to taxonomically classify bacteria and determine the composition of bacterial communities.

safety biomarker biomarkers typically used to monitor organ safety and diagnose or predict onset or reversibility of injury.

salicylanilides (such as closantel or rafoxanide) are mainly active against mature liver flukes. The mode of action of this drug is currently not clear, and inhibition of metabolic as well as neuromuscular pathways are possible. These drugs act against oxidative phosphorylation. This could lead, in a manner similar to that of the benzimidazoles, to a decrease in energy stores and finally to starvation of the parasites. However, more direct effects, like the rapid spastic paralysis salicylanilides induce in trematodes, may point at a different mode of action, possibly involving calcium ion influxes through pores created by the drug in the tegument of the worms. These drugs collapse proton gradients across mitochondrial membranes as the primary mechanism of action.

salinity gradient method a simple device based on a salinity gradient that is used for the concentration of parasite eggs, which are then detected by microscopy.

Salivaria the anterior station group or Group B trypanosomes which transmit the protozoa with their mouthparts.

Salivarian said of trypanosomes which belong to the subgenera of *Duttonella*, *Trypanozoon*, *Pycnomonas* and *Nannomonas*. These trypanosomes are passed to the recipient in the saliva of the tsetse fly (*Glossina* spp.) during the act of feeding (biting).

salivary glands (SGs) saliva-producing glands of the mosquito. *Plasmodium* sporozoites invade the salivary glands in order to be transmitted to another vertebrate host in the mosquito saliva.

sample size the size of the sample. Larger samples usually mean more accurate results. Sample size usually depends on the purpose of the study, the population size from which the sample will be drawn, the desired level of precision and acceptable the level of confidence or risk, and the degree of variability in the attributes being measured.

sample size calculation the number of individuals that need to be enrolled in a study to have sufficient statistical likelihood of detecting a specified association if it is real.

sample statistic a quantity calculated from a sample of data using simple mathematical functions which are independent of the sample's distribution.

sampling distribution the hypothetical expected distribution of a quantity estimated from a random sample of observations.

sandwich estimator a standard error (SE) of an estimated quantity that is robust to misspecifications in the variance-covariance of the error distribution in a statistical model. Sandwich estimators are typically used with marginal models so that SEs (and confidence intervals) are invariant to inaccuracies in the specification of the repeated measures correlation structure. In this context, sandwich estimators are based on the empirically observed variation among unit-level statistics rather than on the model-derived variance-covariance matrix which depends on the assumed correlation structure.

Sanger sequencing a widely used method of determining the order of bases in DNA. *See* **sequencing**.

sapronotic agent a saprophage that can establish an infection, multiplying in or on a host, causing disease, e.g. *Naegleria fowleri*.

saprophage a free-living organism that obtains its nutrition by consuming dissolved organic matter deriving from the decomposition of dead organisms or ejecta.

Sarcomastigophora protozoa consists of the amebae (subphylum Sarcodina) and the flagellates (subphylum Mastigophora). Locomotion of amebae is accomplished by the extrusion of pseudopodia ('false feet'), whereas flagellates move by the lashing of their whip-like flagella. The number and position of flagella vary in different species.

Sarcoptes scabiei a parasitic mite responsible for the highly infectious skin disease known as mange or scabies in animals and humans. This ectoparasite can cause severe skin disorders in over 40 mammalian species, including humans, domestic animals and wildlife. Commonly infected species include pigs, dogs, foxes, wombats and humans. *Sarcoptes scabiei* occurs commonly in cattle and small ruminants and causes sarcoptic mange. There is evidence that *S. scabiei* is the only species of the genus *Sarcoptes*. Therefore, it is appropriate to declare in studies with *Sarcoptes* mites exactly the host species or use terms such as *S. scabiei* var. bovis, *S. scabiei* var. ovis and *S. scabiei* var. caprae. These mites burrow into the skin, preferring regions without wool or the less well-haired parts of the body such as head, axillae, groin and tail. Clinical signs are similar in all species and consist of a papillar eruption, the skin becomes thickened, and there is marked loss of hair and severe pruritus.

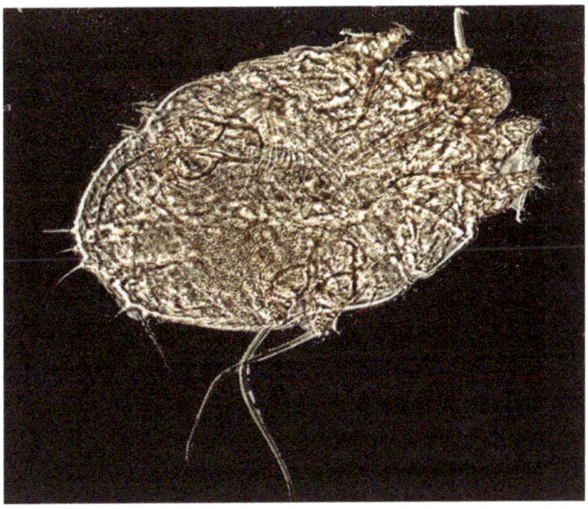

SAS6-like (SAS6L) a novel protein with significant homology to the basal-body protein spindle assembly defective 6 (SAS-6). SAS6L has been localized to the apical complex of *T. gondii* and to the basal-body apparatus of *T. brucei*.

satellite a chromosomal segment that branches off from the rest of the chromosome but is still connected by a thin filament or stalk.

saturated and reduced models there are nine joint genotypes for a pair of SNPs each with three genotypes (e.g. AA, Aa and aa). These can be modelled in full using nine parameters: one as the baseline (for example, aa/aa), two for each SNP (for example, AA/aa and Aa/aa) and four for interactions (e.g. AA/Aa, AA/AA, Aa/Aa, Aa/AA). The saturated model fits all the nine parameters, whereas the reduced model fits the first five parameters and excludes the four interaction parameters.

saturated fat a fat in which all of the fatty acids have only single bonds in their molecule. Typical of most animal fats.

saturated screens genetic screens that have reached the point at which no new gene mutations can be found.

scabies human scabies is caused by an infestation of the skin by the human itch mite (*Sarcoptes scabiei* var. *hominis*). The microscopic scabies mite burrows into the upper layer of the skin where it lives and lays its eggs. The most common symptoms of scabies are intense itching and a pimple-like skin rash. The scabies mite usually is spread by direct, prolonged, skin-to-skin contact with a person who has scabies. Scabies occurs worldwide and affects people of all races and social classes. Scabies can spread rapidly under crowded conditions where close body contact is frequent. Institutions such as nursing homes, extended-care facilities, and prisons are often sites of scabies outbreaks.

scaffold (in genomic mapping) a series of contigs that are in the right order but not necessarily connected in one continuous stretch of sequence.

scavenger receptors surface receptors, for example on phagocytic cells, that recognize cells or molecules that require clearance from the body.

scenopoetic niche a set of ecological variables (a niche) that change very slowly.

scFv a single chain molecule composed of the variable regions of an antibody heavy and light chain joined together by a flexible linker.

Schistosoma a trematode genus commonly referred to as blood-flukes, are sexually dimorphic, parasitic flatworms responsible for significant human infections termed schistosomiasis, considered by the World Health Organization as the second most devastating parasitic disease, after malaria, Hundreds of millions of people are infected worldwide, Three major species and two minor ones infect humans worldwide primarily in the tropics and subtropics. Adults parasitize blood capillaries of mesenteries or plexus of the urinary bladder, depending on the infecting species. Eggs are released in capillaries reaching either the urinary bladder or the intestine (depending upon the infecting species), and these are then excreted in urine or faeces into fresh water. Miracidia penetrate the snail intermediate host going through sporocyst and daughter sporocyst stages producing bifurcate cercariae that then penetrate the human host directly via the skin. The cercariae lose their tails and metamorphose to schistosomules, travel to the liver and descend via the venous flow to their various blood vessels.

***Schistosoma haematobium* adult microsomal antigen (HAMA)** proteins purified by differential centrifugation, urea extraction, molecular sieving, and ion-exchange chromatography that are used as the target antigen in immunological assays for antibody detection.

schistosomal myeloradiculopathy (SMR) known also as spinal cord schistosomiasis, SMR is the schistosomiasis of the brain and spinal cord caused by infection with the digenean trematodes of the genus *Schistosoma*.

schistosomiasis (or schistosomosis) a parasitic disease that mainly occurs in tropical and subtropical regions of the world and is caused by blood flukes of the genus *Schistosoma* (subclass Digenea). In Africa, two forms of the disease, urogenital and intestinal, are dominant being caused by *Schistosoma haematobium* and *Schistosoma mansoni*, respectively. As part of its intricate life cycle, the schistosome needs to develop in freshwater snails. The cercaria, a short-lived larval stage emerging from the snail, is able to penetrate human skin directly.

schistosomiasis elimination reduction to zero of the incidence of schistosomiasis (both disease or infection) in a defined geographical area, as a result of deliberate efforts. Continued intervention or surveillance measures are required.

schistosomiasis eradication permanent reduction to zero of the worldwide incidence of infection caused by *Schistosoma* spp. as a result of deliberate efforts. Continued measures would no longer needed.

Schistosomiasis japonica helminth disease caused by the human blood fluke *Schistosoma japonicum*, primarily endemic in China and the Philippines.

Schistosomiasis mansoni helminth disease caused by the human blood fluke *Schistosoma mansoni*, found mainly in regions of South America and the Caribbean, Africa, and the Middle East.

schistosomula an immature schistosome following penetration of the cercariae into skin and blood vessels of its bird or mammal host. *See* **metacercaria**.

schizogony life cycle event where the nucleus of the original invading protozoan undergoes repeated asexual divisions to form many nuclei within the cell forming a schizont. Then each nucleus becomes surrounded by cytoplasm and a cell membrane forms around each new organism. These new stages within the schizont are known as merozoites. *See* **asexual reproduction**.

schizont asexual reproductive stage in the development of the *Eimeria* spp. and in many other coccidians, e.g. *Toxoplasma* spp. in the cat, *Cystoisospora*, *Hammondia*, *Frenkelia*, *Isospora* spp. and *Plasmodium* spp., following the trophozoite whose nucleus divides into many smaller nuclei. This process of division is known as schizogony.

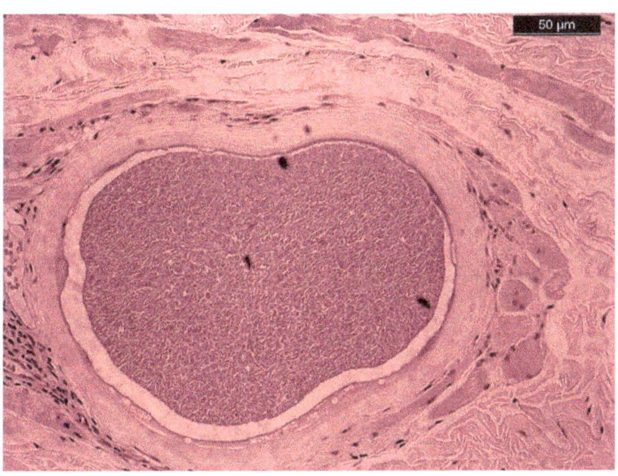

schizophrenia a heterogeneous neuropsychiatric illness that typically arises in late adolescence or early adulthood, with a prevalence of ≈1% worldwide. Symptoms fall into three categories: positive (hallucinations or delusions), negative (affective flattening, alogia, avolition) and cognitive (deficits in memory or executive function, anosognosia).

school exclusion exclusion from school or children's services centre under the Public Health and Wellbeing Regulations 2009

SCID (severe combined immunodeficiency) immunodeficiency affecting both T and B lymphocytes.

scleral conjunctiva conjunctiva (membrane) that covers the sclera (white part of the eye).

scolex (plural scolices) the anterior end ('head') of the adult cestode, it is the holdfast organ that anchors the worm in the GI tract.

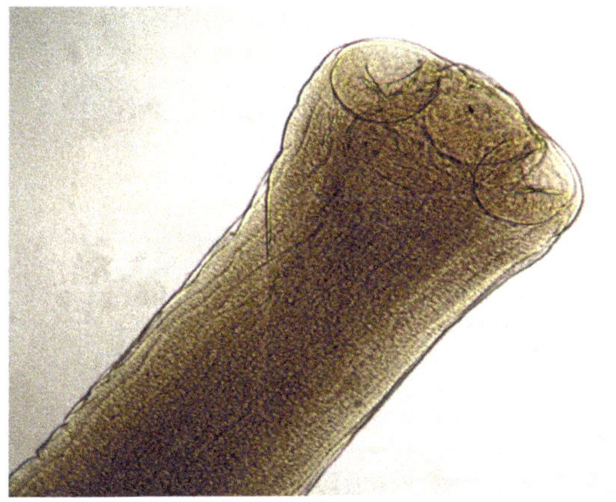

screening an active ID practice used to select a subpopulation for testing based on each individual's recent history of malaria symptoms.

scutellum a term used in relation to arthropods. A small posterior section of the tergum.

scutum a hard protective plate or shield on the dorsum behind the capitulum of hard ticks. The scutum is much more extensive in male than in female ticks.

seam cell specialized epithelial cells that lie along the apical midline of the hypodermis.

seasonality the pattern of changes in malaria transmission over a given calendar year.

second intermediate host the host infected by trematode cercariae, if applicable to a given parasite, and in which the metacercariae develop, often within a cyst.

second mitochondria-derived activator of caspases (SMAC)-mimetic a drug that mimics the activity of natural antagonists of the inhibitor of apoptosis protein (IAP), resulting in apoptosis and inactivation of the nuclear factor-κB (NF-κB) survival signalling pathway.

second set rejection accelerated rejection of an allograft in an already immune recipient.

secondary amplification response primary short interfering RNAs (siRNAs) hybridize with complementary mRNA and can facilitate the generation of secondary siRNAs through the actions of RNA-dependent RNA polymerases thereby amplifying the impact of the original double-stranded RNA and/or primary siRNAs.

secondary antibody (immune) response the qualitatively and quantitatively improved immune response which occurs upon the second encounter of primed lymphocytes with a given antigen.

secondary contact the co-occurrence in one area of two taxa that were previously geographically isolated and had accumulated some genetic divergence.

secondary immunodeficiencies immunodeficiency diseases resulting from a known but non-genetic cause.

secondary lymphoid organs essential immune organs in which antigen-driven proliferation and differentiation of B and T lymphocytes takes place (e.g. lymph nodes and spleen). They serve as the sites of naive T cell activation and induction of adaptive immune responses.

secondary metabolites organic compounds that are not directly involved in the normal growth, development and reproduction of an organism.

secondary reservoir host or minor host a host that can transmit infection but cannot maintain parasite transmission in the absence of the primary host(s).

secondary sex ratio the ratio between the number of males and females at time of birth.

secondary sexual characteristics phenotypic traits that are genetically and/or environmentally based and appear in sexually mature animals. They often differentiate between the sexes, but they rarely have a direct reproductive function and are believed to be the results of sexual selection.

secondary sexual traits sex-specific traits used in mating competition that are not directly involved in reproduction. Some examples include the colourful feathers of many bird species, antlers in several ungulates, and songs of many insects, frogs and birds.

second-site mutations mutations introduced in organisms that are already genetically compromised for a given pathway or process in order to isolate mutations that either suppress or enhance the effect of the first mutation.

secretory component a protein found on mucosal cells that functions as an IgA receptor and, on binding to IgA, protects IgA against proteolytic digestion in the intestine.

secretory IgA dimeric IgA found in seromucus secretions.

sedentary endoparasite commonly, plant parasitic nematodes that, as adults, are stationary and reside in a fixed feeding location within the host.

Seed region the minimal region on the small guide RNA that enables target recognition and binding.

segmental duplications regions of the genome of a given species that are >1 kb and at least 90% identical to each other at the sequence level. They are at a fixed copy number within a species.

segregating variant often termed a polymorphism; a mutation that is not fixed in the population.

segregation the normal biological process whereby the two pieces of a chromosome pair are separated during meiosis and randomly distributed to the germ cells.

selectins a family of cell surface adhesion proteins that bind cells to glycoproteins on vascular endothelium.

selection bias a bias in assignment or a confounding variable that arises from study design rather than by chance. These can occur when the study and control groups are chosen so that they differ from each other by one or more factors that may affect the outcome of the study.

selection coefficient (s) a measure of the relative fitness of a genotype, it ranges $0 \leq s \leq 1$.

selection pressure evolutionary pressure that allows certain genotypes to outcompete others. In the case of malaria, resistance to antimalarials disseminates owing to the selective survival advantage that resistant parasites have in the presence of the drug. In a given population, the greater the proportion of parasites that are exposed to antimalarials at concentrations that allow proliferation only of resistant parasites, the greater the selection pressure.

selective bio-orthogonal labelling a technique that allows the labelling of a specific target (protein) in the cell, but not any other structure.

selective breeding a process in which new or improved strains of plants or animals are developed, mainly through controlled mating or crossing and selection of progeny for desired traits.

selective neutrality a situation in which allelic variation has no effect on fitness, and therefore evolves under no selective pressure (i.e. only affected by mutation and random gene frequency drift within populations).

selective sweep the process by which a selectively advantageous mutation is increased in frequency driven by strong positive natural selection eliminating neutral variation (that does not affect fitness) among neighbouring (linked) sites.

selectivity the ability of ion channels to recognize and transport only certain solutes. Complex structure- and charge-based interactions among the channel, solutes, and associated water molecules may allow remarkable discrimination and selective transport, as is critical for cellular functions.

self-cure in many cases laboratory and farm animals expel nematode infections from their intestine, a phenomenon known as 'self-cure'. To date the mechanisms which drive 'self cure' are still not known but Th2 driven responses, which may involve mast cells and goblet cells, have been shown in some nematode/host infection models. Trichostrongyles (order Strongylida, family Trichostrongylidae) infect sheep when L3 larvae are ingested from contaminated pasture. Infection induces 'self cure' within a few weeks followed by strong protective immunity and studies in sheep have also indicated that Th2 responses are necessary to decrease adult gastrointestinal nematode survival and reduce faecal egg counts.

selfish operon model the view that grouping genes into operons might increase the fitness of this gene set by facilitating horizontal gene transfer (HGT) and/or vertical inheritance. The same argument applies to gene clusters.

self-renewal the ability of stem cells to continue to divide and replenish themselves indefinitely.

self-tolerance the ability of the immune system to prevent or halt responses against self-antigens; in other words, the prevention of autoimmunity.

self-transcribing active regulatory region sequencing (STARR–seq) genome-wide fragmentation of genomic DNA with subsequent cloning into reporter vectors to identify enhancers that act as regulators of gene expression.

semiallogenic trophoblast indicates that the cells that will form the placenta (trophoblasts) share half of the genes with the father and half of the genes with the mother. The genes that come from the father are allogenic (different) from the mother and therefore they have the potential to activate the mother's immune system.

semicarbazone metaflumizone a contact insecticide acting as an antagonist of Na^+ channels. Binding its receptor blocks the entry of Na^+, which results in the inhibition of nerve activity. Insects are paralysed and die. This insecticide is used as a spot-on formulation on cats and dogs. It is combined with amitraz in the dog formulation to provide a broad-spectrum on both fleas and ticks.

semi-field enclosed environments, ideally situated within the natural ecosystem of a target disease vector and exposed to ambient environmental conditions, in which all features necessary for its life cycle completion are present that reflect end-user conditions.

semi-professional antigen-presenting cell (APC) APCs with facultative professional APC status; they are able to induce primary immune responses if properly activated.

senescence the process of ageing.

sensitivity the proportion of actual positives which are correctly identified as such (i.e. the percentage of sick people who are identified as having the condition).

sensitivity analysis (economic studies) the fraction of genes within a biological process that are correctly assigned to that process by a prediction method.

sensitization natural or artificial induction of an immune response by exposure to an antigen. Essentially the same as immunization.

sensory papillae sensory projections from cells or tissues, often containing cilia.

sentinel animal an animal that may be used as an indicator of the presence of a pathogen.

septum a layer of peptidoglycan that is formed at approximately mid-cell (in eubacteria) that generates a physical barrier between nascent daughter cells.

sequela (plural **sequelae**) a diseased condition resulting from a previous disease.

sequence *See* **base sequence**.

sequence assembly a process whereby the order of multiple sequenced DNA fragments is determined.

sequence barcodes short unique DNA sequences, usually inserted in synthetic oligonucleotides, that are used to keep track of sample identity during nucleic acid amplification procedures prior to sequencing. A particular type of sequence barcode is Unique Molecular Identifier used for counting absolute numbers of DNA or RNA molecules that are originally present in a sample before nucleic acid amplification by methods such as PCR.

sequence polymorphism variation between individuals in a species of nucleotide identity at specific positions in the genome.

sequence similarity the extent to which nucleotide (or protein) sequences are related. The extent of similarity between two sequences can be based on per cent sequence identity and/or conservation (http://www.ncbi.nlm.nih.gov/Education/BLASTinfo/glossary2.html.

sequence-tagged site (STS) a segment of DNA of known sequence, which can be amplified and detected by PCR.

sequencing determining the order of bases in a length of DNA. Provides information on where genes start and stop and where mutations or changes have occurred. It also allows you to translate the sequence of bases within a gene into what amino acids it codes for, and therefore what protein is produced.

sequencing coverage the quotient of total bps sequenced by genome size. If sequencing read coverage is unbiased (follows a Poisson distribution), this statistic indicates the average number of reads covering a given position in the genome. For example, 3× and 8× coverage levels are expected to result in at least one read covering 95% and 99.95% of genomic positions, respectively.

sequencing libraries sets of fragmented DNA extracted from one or more individuals that serve as the template for subsequent sequencing.

sequencing-by-ligation (SBL) a next-generation sequencing method based on sequential ligation of fluorescently labelled probes starting from an adaptor primer bound to a template DNA fragment. Multiple rounds of ligation, detection and fluorophore cleavage are performed using a pool of differently labelled oligonucleotides. In each cycle, only the oligonucleotide that is complementary to the query sequence is ligated and its specific fluorescence detected. SBL constitutes the core of the SOLiD and Complete Genomics technology.

sequencing technology the instrumentation and procedures used to determine the order of nucleotides in DNA.

sequential or serial monogamy one individual has no more than one sexual partner at a time but can have multiple partners in a lifetime.

serial sectioning immune-electron tomography sequential thick sections are used for imaging by tomography following immuno-gold labelling of a specific protein before fixation. Tomograms are then merged to obtain a large volume 3D model.

serial sectioning transmission electron microscopy (TEM) a process in which multiple sequential thin sections are collected and imaged by TEM. The images can later be aligned to form a 3D model.

serial surface viewing by dual beam microscope image acquisition technique in which a focused ion beam is used to mill slices as thin as 10 nm from a sample embedded in plastic. Between two millings a scanning electron beam is used to acquire an image of the newly revealed surface. Thus, by repeating this 'slice and view' procedure, an entire 3D volume can be acquired.

serine proteases (proteinases) a group of endoproteases from both eukaryotic and prokaryotic sources that share a common reaction mechanism based on formation of an acyl-enzyme intermediate on a specific active serine residue. Serine proteases are irreversibly inactivated by naturally occurring inhibitors (serpins).

seroconversion acquisition of detectable serum levels of antibodies to a given antigen.

seroconversion rate the rate at which a population develops antibodies to an antigen (or antigens), i.e. *Plasmodium falciparum*, in a given time-frame, either as a result of infection or of immunization.

serodiagnosis diagnosis of an infectious disease based on serum levels of antibodies to the infection.

serology the science of antibody detection.

serotonin (also known as 5-hydroxytryptamine or 5-HT) biogenic amine neurotransmitter synthesized from tryptophan.

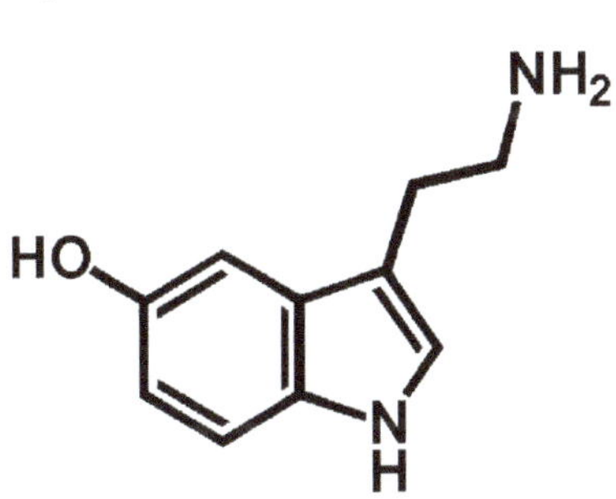

serum the clear, yellow fluid part of vertebrate blood after the fibrin (and clotting factors) has been removed.

serum resistance-associated (SRA) gene gene responsible for *T. brucei rhodesiense* resistance to human serum. The encoded protein promotes trypanosome lytic factor 1 (TLF1) degradation.

serum sickness a type of hypersensitivity response (hypersensitivity type III) to the administration of foreign serum as a result of the development of immune complexes in the bloodstream, which in turn result in an inflammatory reaction. Serum sickness was originally induced following therapy with large doses of antibody from a foreign source, e.g. horse serum.

server a computer that provides data to other computers. It may serve data to systems on a local area network (LAN) or a wide area network (WAN) over the internet. Many types of servers exist, including Web servers, mail servers, and file servers. Each type of server runs software specific to its particular purpose.

severe malaria syndrome defined by the presence of *P. falciparum* (of any density) and of at least one of the following symptoms or signs: severe malarial anaemia, impaired consciousness, multiple convulsions, prostration, respiratory distress, jaundice, circulatory collapse, haemoglobinuria, pulmonary oedema, abnormal bleeding.

severe malarial anaemia symptom complex clinically indistinguishable from severe malaria potentially caused by a variety of different organisms.

sex chromosome a type of chromosome that participates in sex determination. Humans and most other mammals have two sex chromosomes, the X and the Y. Females have two X chromosomes in their cells, while males have both X and a Y chromosomes in their cells. Egg cells all contain an X chromosome, while sperm cells contain an X or Y chromosome. This arrangement means that it is the male that determines the sex of the offspring when fertilization occurs.

sex linked is a trait in which a gene is located on a sex chromosome. In humans, the term generally refers to traits that are influenced by genes on the X chromosome. This is because the X chromosome is large and contains many more genes than the smaller Y chromosome. In a sex-linked disease, it is usually males who are affected because they have a single copy of X chromosome that carries the mutation. In females, the effect of the mutation may be masked by the second healthy copy of the X chromosome.

sex-specific selection difference in magnitude but not direction of selection between the sexes, for example, if a trait experiences stronger selection in one sex, or if a trait is sex-limited and therefore only subject to selection in one sex. Compare with sexually antagonistic selection.

sexual antagonism opposing direction of selection between males and females for a particular heritable trait that has a positive genetic correlation between the sexes. In contrast to intra-locus sexual conflict, sexual antagonism can involve different traits in each sex, and is therefore a more inclusive term.

sexual dimorphism difference between males and females of the same species, such as size, ornamentation, and behaviour.

sexual maturity stage at which an adult parasite is able to produce sperm and/or ova in the definitive host.

sexual reproduction reproduction of an organism that involves the fusion of specialized sex cells or gametes (which are haploid) to form diploid progeny. It is important in bringing new vigour to a species by the mixing of genetic material from the parents to give a genetically different organism.

sexual selection differential reproductive success resulting from competition for fertilization. Competition for fertilization occurs through direct competition between members of the same sex or through the attraction of one sex to the other.

sexually antagonistic selection difference in direction (and possibly magnitude) of selection between the sexes, for example, if a trait experiences positive selection in one sex and negative selection in the other.

sham treatment an intervention given in the control group of a clinical trial, which is preferably identical in appearance and feel to the experimental treatment and believed to lack any disease specific effects (e.g. detuned ultrasound or random biofeedback). Placebo is used for pills, whereas sham treatment is used for devices, psychological, and physical treatments.

Sheather's flotation method a method for examining faeces for the presence of protozoan cysts, worm eggs or larvae by mixing with a saturated solution of sodium chloride or sugar and collecting a sample from the top of a column for microscopic examination.

sheep ked, *Melophagus ovinus* one of the most widely distributed and important external parasites of sheep. They are wingless dipterans. The adult is ≈7 mm long, brown or reddish in colour, and covered with short, bristly hairs. The head is short and broad, and the legs are strong and armed with stout claws. They can cause considerable economic damage by feeding on the blood of sheep which leads to weight loss and unthriftiness.

shoaling behaviour the aggregation behaviour of aquatic organisms, most often used in reference to fish.

short interspersed nuclear elements (SINEs) a subtype of transposable elements reverse-transcribed from RNA molecules. They do not encode a functional reverse transcriptase protein and cover a substantial portion of primate genomes, including all Alu sequences.

short stumpy (ST) non-proliferating forms arising from long slender forms in the mammalian host. Short stumpy forms are pre-adapted for uptake by the tsetse, where they differentiate into PF.

short tandem repeat (STR) a repetitive nucleotide sequence that consists of many copies of a short sequence in tandem, for example, CAGCAGCAGCAG. STRs are frequently known as microsatellites.

short-chain fatty acids (SCFAs) fatty acids with five or fewer carbons, including formic (one carbon), acetic (two carbons), propionic (three carbons), butyric (four carbons), and valeric (five carbons) acid. SCFAs are products of fatty acid beta oxidation as well as bacterial fermentation, for example in the human colon.

shotgun library construction a laboratory technique for determining the DNA sequence of an organism's genome. The method involves breaking the genome into a collection of small DNA fragments that are sequenced individually. A computer program looks for overlaps in the DNA sequences and uses them to place the individual fragments in their correct order to reconstitute the genome. *See* **genomic library**, **library**.

sialic acids (SAs) a generic term for N- or O-substituted derivatives of neuraminic acid. The most common member, N-acetylneuraminic acid (Neu5Ac), is the terminal sugar found on the oligosaccharide attached to IgG at Asn297. They are recognized by sialic acid binding Ig-like lectins (Siglecs); cell surface receptors found on immune cells, e.g. CD22 (Siglec-2) found on B cells that are implicated in the inhibitory mechanism behind intravenous immunoglobulins (IVIg).

sibling recurrence risk the ratio of the probability that a sibling of an individual affected by a disease will also be affected compared to the risk of disease in the general population.

sibling species closely related 'sister' species.

SIDER (short interspersed degenerate retroelement) SIDERs are short non-autonomous heterogeneous elements related to retrotransposons identified in trypanosomes (ingi/RIME, L1Tc/NARTc, and DIRE).

siderophore an iron-binding compound produced by bacteria that aids in iron acquisition, particularly when this resource is limiting.

SIgA secretory immunoglobulin A.

sigma factors transcription factors that target RNA polymerase to specific gene promoters during the initiation of transcription.

sign any objective evidence of disease, e.g. fever, diarrhoea, or skin rash.

signalment that part of the veterinary medical history dealing with the animal's age, sex, and breed.

signature secondary metabolic genes genes encoding enzymes that give rise to the skeletal structures of the different classes of secondary metabolite (e.g. for Non-ribosomal peptide synthases (NRPS), polyketide synthases, terpene cyclases) that are subsequently modified by tailoring enzymes. These 'signature' genes are presumably the founders of secondary metabolic gene clusters.

significance level (or type 1 or α error) the probability of rejecting a true null hypothesis; 5% is normally seen as the acceptable level of significance. α and β are required to be as small as possible (and therefore power as large as possible) but there is a trade-off between the two.

silent substitution mutation in a coding/expressed region of the DNA that produces no change in the amino acid coded for (because of the redundancy of the genetic code). Also known as synonymous substitution.

simple sequence repeat (SSR) a DNA tract consisting of a relatively short bp motif that is repeated in tandem.

simulation combination of multiple models into a meaningful representation of a whole system that can be used to predict how the system will behave under various conditions. Simulations can be used to run *in silico* experiments to gain first insights, form hypotheses, and predict outcomes before conducting more expensive physical experiments.

SINE (short interspersed nuclear element) a non-autonomous retrotransposon that requires the enzymatic machinery of LINE for its mobilization.

single nucleotide polymorphism (SNP) DNA sequence variations that occur when a single nucleotide (A, T, C or G) in the genome sequence is altered. *See* **mutation**, **polymorphism**.

single schizogony cycle (SSC) it corresponds to a full asexual cycle of the parasite.

single-cell FISH allows direct imaging of individual molecules preserving spatial organization and was developed into a quantitative method. It is now possible to quantify up to a dozen transcripts simultaneously by virtual 'colour' barcoding. Improved protocols use a multitude of singly labelled probes against the same target leading to increased signals.

single-exposure chemoprotection (SEC) a target product profile that has been defined to guide the discovery and development of the next-generation antimalarials that will protect vulnerable populations from infection by *Plasmodium* spp.

single-exposure radical cure and prophylaxis (SERCaP) a target product profile designed to guide the discovery and development of new combination therapies that will be administered to patients as a single dose, curing them from all malaria infections as well as providing protection against post-treatment reinfection.

single-molecule RNA fluorescence *in situ* hybridization (smFISH) an *in situ* hybridization method by which individual RNA molecules are specifically visualized in cells or tissues using pools of 20–50 complementary short DNA oligonucleotides, each of which is coupled to the same fluorophore. Single RNA molecules appear as diffraction-limited fluorescence spots, which can be accurately counted using unbiased automated procedures.

single-nucleotide variant (SNV) a variant that consists of a change at a single nucleotide position in the DNA sequence. *See* **single-nucleotide polymorphism**.

single-pulse infections infection trials carried out in a laboratory setting in which fish are exposed only once to parasites before measurements are taken.

sinusoidal intraluminal/inside liver or bone marrow sinusoid.

sinusoids small, fenestrated blood vessels found in the periphery of the lobules of the liver. Sinusoids are lined with endothelial and Kupffer cells.

sitamaquine oral drug which is an 8-aminoquinoline analogue; proposed as possible treatment for visceral leishmaniasis.

site-directed mutagenesis introduction of one or more mutations at specified positions in a gene of interest.

skin test a diagnostic procedure, usually applied to detect allergy, that induces a local inflammatory response following intradermal inoculation of an antigen or allergen.

sliding filament theory a model of muscle contraction postulating that thin filament (mostly actin)-containing I-bands slide past the myosin-containing A-bands to generate force.

slightly deleterious mutations deleterious mutations with such a small selective disadvantage (less than the inverse of twice the effective population size) that their evolutionary fate is influenced more by genetic drift than selection.

slippage replication a mutation process whereby a simple sequence tandem (microsatellite) repeat grows by addition or subtraction of the 'beads' of simple units that make up the 'necklace'. A dinucleotide AC repeat would grow by addition or subtraction of AC units.

slow pace-of-life organisms that mature slowly and produce only a few offspring;

each offspring represents a large portion of the total investment in reproduction.

slow-reacting substance of anaphylaxis (SRS-A) a group of leukotrienes released by mast cells during anaphylaxis that induces a prolonged constriction of smooth muscle. This prolonged constriction is not reversible by treatment with antihistamines.

SM1 transgene a synthetic gene driving the blood-inducible, gut-specific secretion of a 12-amino-acid peptide (named Salivary gland- and Midgut-binding peptide 1) that inhibits the development of *Plasmodium berghei* strongly in *Anopheles stephensi* mosquitoes.

small haemoglobin vacuoles double membrane vesicles that originate from the cytostome and carry haemoglobin into the food vacuole.

small interfering RNAs (siRNAs) class of sncRNAs (≈20–25 nucleotides) notable for their role in the RNA interference pathway. They are derived from endogenous loci or foreign genetic material and are involved in silencing of invading viruses, TE activity, and in some cases protein-coding genes.

small molecule a low molecular weight compound (usually below 500 Da). The term is usually used to distinguish small chemical compounds from larger biomolecules – for example, peptides.

small myristoylated protein (SMP) one of a highly conserved family of monoacylated or dually acylated proteins that are localized to distinct membrane domains; first identified in *L. major*.

small RNAs (sRNAs) non-coding RNA molecules that are typically 50–250 nucleotides in length.

small ruminants lungworms lungworms in small ruminants (sheep and goat) are caused by three nematode species, namely *Dictyocaulus filariae*, *Protostrongylus rufescens* and *Muellerius capillaries*.

smallhold pig farming rearing of pigs is mainly conducted by farmers who each have only one to three pigs.

smear specimen for microscopic study, the material being spread thinly and unevenly across the slide with a swab or loop, or with the edge of another slide, e.g. blood smear for trypanosome sighting, faecal smear USA: faecal) for the presence of worm eggs. Protozoan cysts or protozoan oocysts.

smolt juvenile salmonid fish physiologically adapted for living in seawater. The term is applied to fish immediately prior to their first migration from freshwater to the sea, and can be applied to farm fish within their first year in seawater.

SNP ascertainment (single nucleotide polymorphism) identification of SNPs that should be placed on a genotyping array to ensure representative coverage of the genome.

SNP chip (single nucleotide polymorphism) a tool for efficiently determining the genotype of tens of thousands of SNPs from a single organism.

social behaviour an action that affects other individuals. It is the interaction among individuals of the same species, for example, collaboration within a well-defined group such as a colony of eusocial insects.

social hierarchy an arrangement within a group of animals, such as rabbits, where some individuals are dominant over others. The more dominant an animal, the more likely it is to have preferred access to mates and sources of food.

social monogamous individuals or species in which we can observe pairs formed by only one male and only one female. That does not exclude the potential occurrence of mate changes.

social monogamy the observation of a pair formed of only one male and only one female, but cases of mate changes (or divorces) can be observed.

sociobiology the study of the biology and evolution of social behaviour.

sodium antimony gluconate (SAG) a pentavalent antimonial that was the drug of choice for treatment of kala-azar or visceral leishmaniasis (VL). After being converted into the trivalent form, it enters into macrophages/parasites and shows a dual effect. It causes an apoptosis-like death in *Leishmania* parasites by generation of reactive oxygen species, depletion of thiols, modulation of energy metabolism, and by inhibition of DNA topoisomerases.

Additionally in macrophages, it induces a generation of reactive oxygen and nitrogen species by modulation of mitogen-activated protein kinases. owing to irregularity of use and poor patient compliance, resistance against this drug has emerged. Increased presence of drug efflux pumps and raised levels of non-protein thiols are the major molecular mechanisms responsible for drug resistance in parasites.

soft clipping substrings at either end of reads that were not aligned with a local alignment algorithm and are thereby excluded in the subsequent analysis.

soft selection in a spatially structured environment, the result of frequency- and density-dependent regulation occurring locally before genotypes migrate and mix.

soft ticks soft ticks (family Argasidae) are so-called because they lack a dorsal scutum in the nymphal and adult stages; larvae may present a dorsal plate, which is of taxonomic relevance.

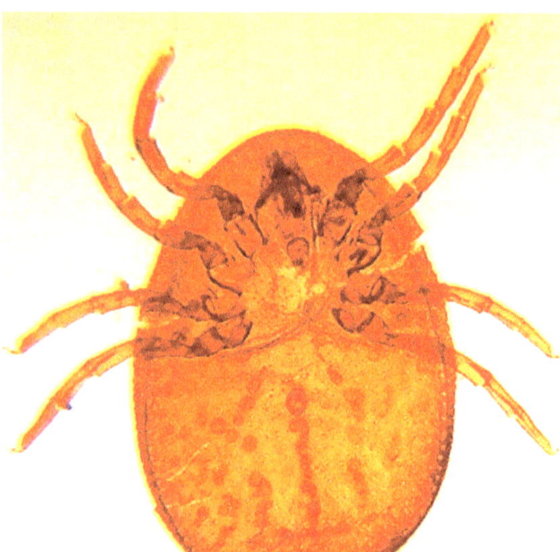

soil-transmitted helminths (STHs) refer to the intestinal worms infecting humans that are transmitted through contaminated soil. Soil-transmitted helminth infection is found mainly in areas with warm and moist climates where sanitation and hygiene are poor, including in temperate zones during warmer months. These STHs are considered neglected tropical diseases (NTDs) because they inflict tremendous disability and suffering yet can be controlled or eliminated. Soil-transmitted helminths live in the intestine and their eggs are passed in the faeces of infected persons. If an infected person defecates outside (near bushes, in a garden, or field) or if the faeces of an infected person are used as fertilizer, eggs are deposited on soil. *Ascaris* and hookworm eggs become infective as they mature in soil.

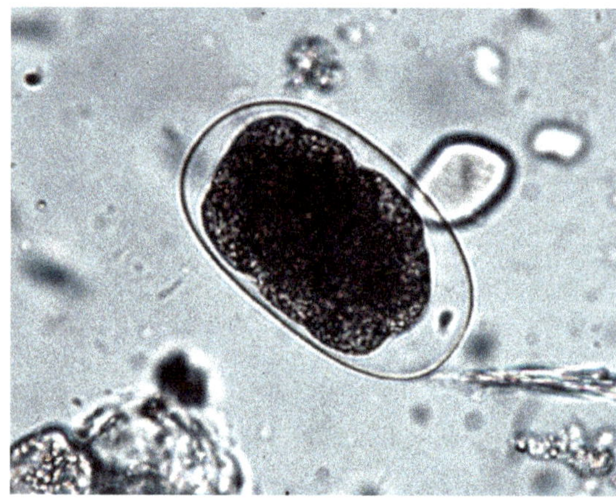

solubility pump system of physical processes [e.g. changes in water temperature, ocean circulation, and gradient of carbon dioxide (CO_2) spanning the ocean depth] that influences the ocean's uptake of CO_2 from the atmosphere. In combination with ocean circulation, the solubility pump results in net CO_2 emissions at the equator and net CO_2 drawdown at high latitudes.

soluble adult worm antigen (SWA) partial purification of *S. mansoni* adult worm antigenic proteins obtained by centrifugation methods.

soluble egg antigen (SEA) partial purified antigenic proteins of *S. mansoni* eggs obtained by centrifugation methods.

solute carrier family 11A1 (SLC11A1) encoded by the *SLC11A1* gene in humans, expressed in monocytes, macrophages, and neutrophils, and an integral phagolysosomal protein composed of 12 hydrophobic transmembrane domains. Functionally, SLC11A1 acts as a divalent transition metal (iron and manganese) transporter on the endosomal membrane and is involved in iron metabolism and host resistance to certain pathogens. It was previously referred to as natural resistance-associated macrophage protein 1 (NRAMP1).

somatic antigens antigens associated with cell (i.e. bacterial) bodies.

somatic cell is any cell of the body except sperm and egg cells. Somatic cells are diploid, meaning that they contain two sets of chromosomes, one inherited from each parent. Mutations in somatic cells can affect the individual, but they are not passed on to offspring.

somatic hypermutation the enhanced occurrence of point mutations in the recombined immunoglobulin variable region V(D)J genes which occurs following antigenic stimulation and acts as a mechanism for increasing antibody diversity and affinity.

somatic migration migration of a parasite through the host's body (often lungs) into the systemic circulation distributed throughout the body and encyst in the tissues.

SOS response a global response to DNA damage in which cell growth is arrested and DNA repair and mutagenesis are induced. The key proteins that are involved are RecA and LexA.

Southern blotting a laboratory technique used to detect a specific DNA sequence in a blood or tissue sample. A restriction enzyme is used to cut a sample of DNA into fragments that are separated using gel electrophoresis. The DNA fragments are transferred out of the gel to the surface of a membrane. The membrane is exposed to a DNA probe labelled with a radioactive or chemical tag. If the probe binds to the membrane, then the probe sequence is present in the sample.

space of disse gap between the liver sinusoidal endothelial cells (LSECs) and hepatocytes in which there is a low-density extracellular matrix.

sparganum the second larval stage of order Pseudophyllidian tapeworms that is characterized by its elongated shape and lack of a cystic cavity. *See* **plerocercoid**.

spatial autocorrelation spatial analysis that tests whether observations are geographically distributed and correlated over some distance in space. Spatial autocorrelation indicates if a variable is aggregated, randomly distributed, or dispersed.

spatial repellent (SR) refers to any molecules (e.g. insecticide or repellent) that prevent an arthropod from entering a space occupied by a potential human host. SRs are expected to reduce encounters between humans and vectors, thereby eliminating or reducing the risk of pathogen transmission.

spatial resolution the geographic scale at which an analysis is carried out. For instance, the presence of a vector may be analysed at a local or regional scale, which is a fine spatial resolution. However, the presence of such a vector may be reported for a large area, such as a country, ignoring the variety of environmental factors that may impact on its distribution.

spatially resolved omics a new technology-driven field in which high-throughput genomic, epigenomic, transcriptomic, proteomic and possibly other omic data are collected from cells or tissues by means that preserve positional information and analysed using high-dimensional data analysis techniques. So far, most of pioneering efforts in this newly developed field have focused on RNA measurements, but measurements at the DNA and protein levels are predicted to ensue soon.

specialist parasite adult parasite, or its developmental stage (s), with a narrow host spectrum.

speciation the process through which populations descending from a single ancestor eventually achieve reproductive isolation.

species smallest group commonly used in biological classification and into which a genus is divided. Species are sometimes further divided into subspecies (races). Generally, no more than one type of organism is present in one species. Members of a species may breed with one another, but cannot generally breed with members of another species. Rarely, very closely related species interbreed to produce a hybrid.

species group an assemblage of closely related species that are usually grouped because of their morphological similarities.

species jump transmission of a parasite from one host species into another, typically novel host species.

species selection the hypothetical process of sorting of species, which might involve selection for survival of those that possess particular irreducible, emergent, species-level characteristics.

species-specific pertaining to individuals of only one species. For example, a pesticide that is species-specific affects only one species.

species-specific hybridization (SSH) hybridization of RNA from one species to a microarray that contains DNA (cDNA or oligomers) from the same species (i.e. the target and the reference species are the same).

specific isotypic response humoral immune response mediated by specific immunoglobulin isotypes and subclasses i.e. anti-*Schistosoma* cercarial, egg and adult worm antigens. Levels of IgM, IgA, IgG1, IgG2, IgG3, IgG4, and IgE are detectable in exposed individuals and during acute and chronic phase of infection.

specific parasites are only adapted to live in a fixed host and within a certain parts of the body, e.g. *Taenia saginata* in small intestine of man while it's larval stage (*Cysticecus bovis*) is found in the muscle fibres of cattle.

specific pathogen-free (SPF) mice mice that are provided by laboratory animal vendors or are generated in a home laboratory, that have a guaranteed health status and are free of particular pathogens.

specificity the proportion of negatives which are correctly identified (i.e. the percentage of well people who are identified as not having the condition). A test with a specificity of 0.98 or 98% will be negative in 98 of every 100 healthy or non-infected individuals.

specimen a small sample or part taken to show the nature of the whole, such as a small quantity of urine for urinalysis, or a small fragment of tissue for microscopic study.

speciose a group comprising many species.

spectral karyotype (SKY) a graphic of all an organism's chromosomes, each labelled with a different colour. Useful for identifying chromosomal abnormalities. The SKY technique makes it easier for scientists to detect chromosomal abnormalities, as compared with a conventional karyotype.

spherical bodies a distinct organelle found in *Babesia* and *Theileria*, considered analogous to dense granules found in other apicomplexan parasites. Up to four spherical bodies are found per *Babesia* merozoite. Proteins localized to these organelles are released after parasite invasion into the host cell but no function has yet been ascribed to any of these.

spicule (=copulatory spicule) an elongate, chitinous structures of male nematodes used in holding open the vulva of the female during copulation and transfer of sperm.

spike controls special control transcripts that are mixed into a biological sample before microarray hybridization. Because the quantity added is known, the control probe signal can be used for accurate normalization.

spinosad part of a family of insecticides indicated for the prevention and treatment of flea infestations. It is for dogs 14 weeks of age or older. It does not kill other internal or external parasites. Spinosad works by activating nicotinic acetylcholine receptors in the flea nervous system, causing an overload of the flea nervous system, leading to involuntary muscle contractions, prostration with tremors, and finally paralysis and death of the fleas. Spinosad also has effects on GABA receptor function that may contribute further to its insecticidal activity. Spinosad is relatively low in toxicity to mammals and birds, and, although moderately toxic to fish, this toxicity represents a reduced risk to fish when compared with many other available insecticides. Side effects are rare but may include vomiting, decreased appetite, lethargy, diarrhoea, cough, excessive thirst, redness of skin, hyperactivity or drooling.

Spirocerca lupi a nematode that causes formation of nodules that may transform into sarcoma in the walls of aorta, oesophagus and stomach of infected canids.

spirorchiids members of a family (Spirorchiidae) of digenetic trematodes that live in the venous system of turtles and crocodiles. They are closely related to Schistosomatidae and Sangunicolidae.

spite carrying out an action that harms a recipient but does not increase an individual's own direct fitness.

splenic dendritic cells (DCs) DCs reside in the spleen. These are isolated from mice for use in *in vitro* experiments.

splice acceptor sites (SAS) mark the 5′ end of the 5′ untranslated region (UTR) of each gene. The 39-nt spliced leader is trans-spliced to the 5′ nucleotide of each gene, which is referred to as the SAS. The SAS predominantly follows the dinucleotides AG.

splice site location in the DNA sequence where RNA removes the non-coding areas to form a continuous gene transcript for translation into a protein.

spliced leader (SL) a 39-nt sequence that is trans-spliced onto the 5′ end of all mRNA transcripts and onto the 5′ of some non-protein-coding RNA.

spliced leader RNA (SL RNA) a class of short transcripts (approximately 100 nt in length) which participate in mRNA maturation in kinetoplastids. An SL RNA molecule includes a 39 nt conserved mini-exon which is transferred onto the 5′ end of each mRNA by trans-splicing. SL RNA genes are arranged as clusters of multiple tandem repeats (0.2–1.0 kb in length). The intergenic regions of the repeats are nearly identical within the same species but are highly variable between species.

spliceosome a macromolecular RNA–protein complex that is responsible for intron removal and that consists of U1, U2, U4, U5 and U6 small nuclear ribonucleoproteins (snRNPs) and many auxiliary protein factors.

splicing regulatory elements (SREs) RNA motifs in precursor RNA that have regulatory roles in splice site selection.

splicing signals essential sequences in the pre-mRNA for recognition by the core splicing machinery.

spore a resistant stage in spore-forming bacteria that is formed internally by the mother cell.

sporocyst in trematodes, the first larval stage in the developmental cycle in the snail, intermediate host. A sac-like structure with a germinal lining that produces secondary larval stages that develop in the snail host.

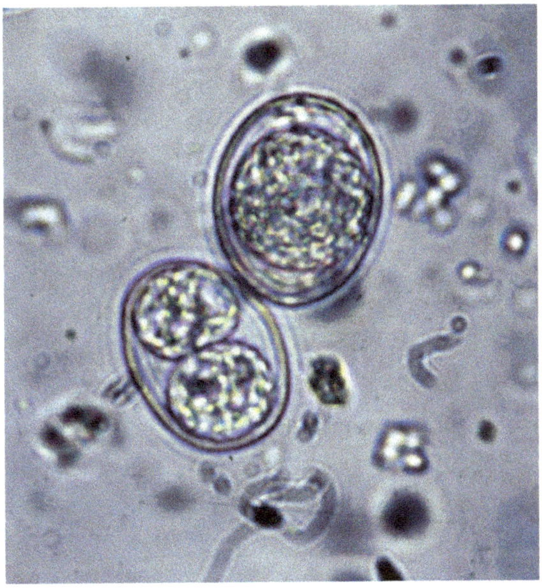

sporogony in apicomplexans, sporogony is the replication by binary fission of a zygote, resulting in production of sporozoites (infective stage).

sporozoite the daughter cells formed by the oocyst. Each oocyst produces between 500 and 9555 sporozoites, which are carried through the haemocoel of the mosquito from where they invade the salivary glands. The female mosquito injects the sporozoites into the vertebrate host when taking a bloodmeal.

sporulation development of sporozoites from the stages of sexual development.

spring rise the increase in egg production from worms in the spring months; leads to increased numbers of infective larvae on the pastures.

spurious infection (false infection) an organism or egg found on faecal examination that is from a source outside of the individual and is not related to an infection of the individual being examined; often derived from eating part of a host infected with a non-human parasite. Just passing through! Eggs of *Dicrocoelium dendriticum* may represent a spurious infection but an actual infection must be ruled out since man may become an accidental host of this liver fluke.

sRNAs (small RNAs; sRNAs) are stable RNAs that are typically less than 250 nucleotides in length and have a defined structure. These RNAs can target mRNAs or proteins to modulate their activity. The best-characterized bacterial sRNA is the 6S RNA of *Escherichia coli*, which controls the activity of RNA polymerase59. Most characterized sRNAs regulate the expression of mRNAs via base-pairing interactions.

SSU (small subunit) the small subunit of the ribosome, which is 16S ribosomal RNA for the Eubacteria and the Archaea and 18S rRNA for the Eukarya.

stable isotope isotope that does not undergo radioactive decay.

stable transfection permanent or long-term introduction of foreign DNA into a cell, either as an episome or integrated into chromosomal DNA. Generally, the initial stability is maintained using a selectable marker.

stage any particular form in the life cycle of a parasite which can be distinguished from all of its other forms.

stage L3 larvae typical the infective larval stage of many nematodes, e.g. larvae that develop in an insect vector of filaria parasites, are deposited on the skin of the host at the time of the bite, and give rise to infection in the human host or the larval stage that is infective for the mammalian host in *Ascaris* or hookworm.

standard error (SE) the standard deviation of the sampling distribution of an estimated statistic (e.g. an arithmetic or geometric mean).

standard membrane feeding assay (SMFA) a method of feeding mosquitoes through an artificial membrane (parafilm) in a device filled with human blood in the absence or presence of compounds. This assay is used to assess the transmission-blocking potential of compounds in the drug discovery and development pipeline.

standard neutral model a common null model, in which all mutations are assumed to be selectively neutral, and the population is at equilibrium.

standardized mean difference (SMD) a measure of effect size used when outcomes are continuous (such as height, weight, or symptom scores) rather than dichotomous (such as death or myocardial infarction). The mean differences in outcome between the groups being studied are standardized to account for differences in scoring methods (such as pain scores). The measure is a ratio; therefore, it has no units.

staple length the length of the individual fibres of cotton. Affects the quality of the fabric that is made from it.

starling equilibrium equilibrium between hydrostatic and oncotic forces (the so-called Starling forces) in the movement of fluid across capillary membranes.

stasis long episodes of no net morphological change in a species lineage.

STAT signal transducers and activators of transcription, transcription factors activated by Janus kinases that regulate several cellular processes, including growth, differentiation and immune activation.

statistical significance the findings of a study are unlikely to have arisen because of chance. Significance at the commonly cited 5% level ($P<0.05$) means that the observed difference or greater difference would occur by chance in only 1/20 similar cases. Where the word 'significant' or 'significance' is used without qualification in clinical evidence, it is being used in this statistical sense.

stealth organism a microorganism that is difficult to detect using routine laboratory diagnostic tests.

steatorrhea presence of fat in the faeces as a result of malabsorption.

stem cell unspecialized cells from which other cell types develop. When a stem cell divides, each new cell has the potential either to remain a stem cell or to become a specialized cell with a distinct function, such as a muscle cell or blood cell. Stem cells serve as a repair system for the body, replacing old and damaged cells. Research using stem cell therapy to treat various diseases is currently under way.

step cline a cline with a large or steep step, where a cline is defined as a gradation in one or more characteristics within a taxon, especially between populations.

stepping-stone model model of dispersal that states that gene flow occurs between adjacent subpopulations, and only the adjacent ones. The model considers that the population is subdivided into an infinite number of panmictic subpopulations (discrete subpopulations). Under a stepping-stone model, the genetic differentiation between two subpopulations is expected to increase with the geographic distance between the subpopulations, such as under the isolation-by-distance model.

stepwise logistic regression analysis a method to test for main effects and interaction effects between loci to assess which variants can explain the association of a region. The most associated variant is put in the model and each one of the others added until there is no improvement in the model.

stepwise mutation microsatellite variation appears to result from slippage in replication, which is most likely to add or delete a single repeat unit (steps of one). As a result, alleles more similar in size will presumably be more closely related. This additional 'phylogenetic' information can be used in assessing genetic differentiation or genetic distance.

sterile **1** incapable of reproduction. **2** free from bacteria or other microorganisms-germ free.

sterile cure (radical cure) the complete (100%) elimination of the pathogen from the host organism. In a parasitic disease, sterile cure never happens.

sterile insect technique (SIT) a method of biological control, whereby overwhelming numbers of sterile male insects are released. The sterile males compete with the wild males for female insects. If a female mates with a sterile male then it will have no offspring, thus reducing the population in the next generation. Repeated release of insects can eventually eliminate a population, leading to eradication.

stochastic relating to a series of random events.

stochastic lineage sorting a process explaining why gene trees are not always identical to species trees. This is a consequence of ancestral polymorphisms (genetic variation at a locus in the ancestor of derived species) sorting randomly during the formation of new species.

stochastic model a description of an event or process within a probabilistic framework that includes 'chance' events in the form of random measurement error. It is contrasted by a deterministic model in which random error is inconsequential or non-existent.

stoichiometry ratio of molecules in a structural complex or chemical reaction.

stop codon a trinucleotide sequence within a mRNA molecule that signals a halt to protein synthesis. The genetic code describes the relationship between the sequence of DNA bases (A, C, G, and T) in a gene and the corresponding protein sequence that it encodes. The cell reads the sequence of the gene in groups of three bases. Of the 64 possible combinations of three bases, 61 specify an amino acid, while the remaining three combinations are stop codons.

stop codon readthrough a translational recoding event in which a proportion of ribosomes read a stop codon as a sense codon, which results in a carboxy-terminally extended protein.

stop–go criteria predefined minimum efficacy threshold on which decisions are based in dual-choice decision patterns that define whether further investment in exploring that tool or strategy is warranted.

STR (short tandem repeats) short DNA sequences that are repeated in a head-to-tail manner. They are useful in DNA profiling.

strain genetically distinct variant of a given species.

strain tracking a process capable of identifying the ultimate origin of parasite isolates, for example, by determining the origin of imported cases or sources of infection. It requires a high resolution molecular-typing method, such as kinetoplastDNA fingerprinting or multilocus microsatellite typing, to distinguish between closely related strains.

strand bias a variant that is significantly more likely to occur within reads that originate from one of the two strands of DNA.

strand-switch regions (SSRs) regions between polycistronic transcription units that are located on opposite strands of the DNA. Strand-switch regions contain RNA polymerase II transcription start or transcription termination sites, depending upon the divergent or convergent orientation of the adjacent transcription units.

stratification division into groups. Stratification may also refer to a process to control for differences in confounding variables, by making separate estimates for groups of individuals who have the same values for the confounding variable.

strength of inference the likelihood that an observed difference between groups within a study represents a real difference rather than mere chance or the influence of confounding factors, based on both p values and confidence intervals. Strength of inference is weakened by various forms of bias and by small sample sizes.

stress any environmental factor, or combination of factors, that has adverse effects on the structure or behaviour of an organism.

stress fibres contractile actomyosin cables that usually span two points of attachment to the substrate.

striated fibre assemblins proteins that form fibres that organize flagellar basal bodies and anchor them to the rest of the cytoskeleton, in other words to microtubules.

strobila a chain of tapeworm proglottids or segments consisting of scolex, neck, and immature, mature, and usually gravid proglottids.

strobilation is an asexual process of forming segments in a tapeworm in which new proglottids are continuously formed in the neck just below the scolex. Along the length of the worm the proglottids increase in size and maturity, developing from immature to mature, carrying fully functional and active sexual organs), to the 'gravid' stage in which essentially the entire proglottid uterus is filled with eggs.

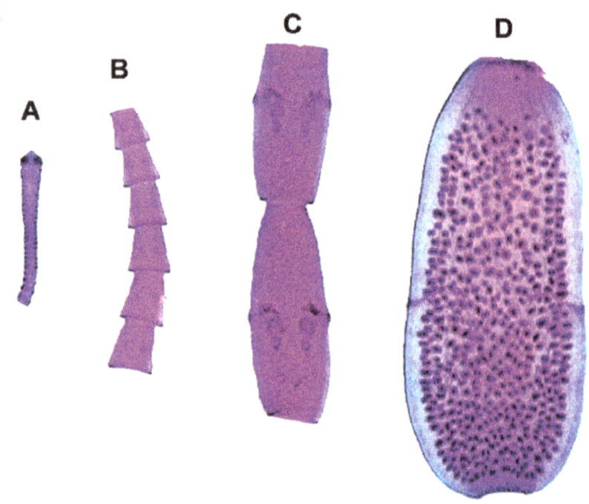

strobilocercus a larva of cyclophyllidian tapeworms (i.e. *Taenia taeniaeformis*).

strongyloidiasis (threadworm) a parasitic disease caused by nematodes in the genus *Strongyloides* that enter the body through exposed skin, such as bare feet. *Strongyloides* is most common in tropical or subtropical climates. Most people who are infected with *Strongyloides* do not know they are infected and have no symptoms. Others may develop a severe form and, if untreated, become critically ill and potentially die. Different species occur in each host. *Strongyloides ransomi* in pigs, *S. westeri* in horses, *S. papillosus* in sheep and cattle, *S. fülleborni*, in chimpanzees and baboons and may produce limited infections in humans and *S. stercoralis* in humans. All are parasites of the small intestine. They are thread-like and less than 1 cm in length. Only female worms are present in the intestine and so eggs are produced by parthenogenesis. The eggs are thin shelled and embryonated. The larvae that hatch out may develop into infective or non-parasitic forms. The latter become free-living males and females which live in decaying organic material and produce fertilized eggs that give rise to infective larvae. Transmission occurs when infective larvae enter the host either by ingestion or by skin penetration.

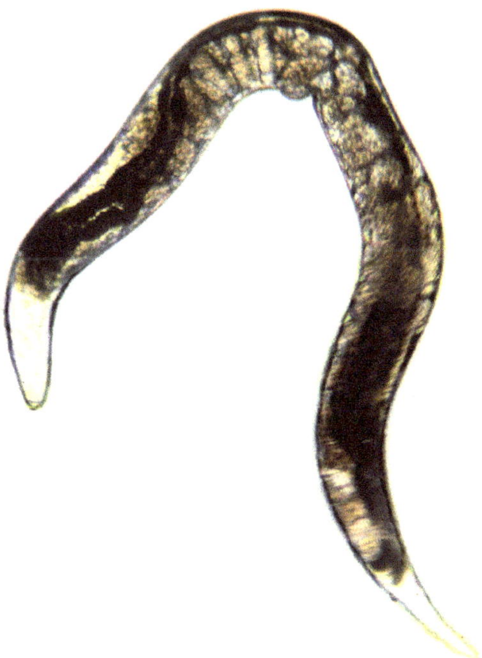

strongylosis (redworm infection) nematodes commonly seen in the large intestine of equids. Two nematode subfamilies: the Strongylinae (large strongyles) and Cyathostominae (small strongyles, small redworms, cyathostomes, or cyathostomins). Strongylinae includes *Strongylus vulgaris*, *S. edentatus*, and *S. equinus*, which can migrate through the body, and *Triodontophorus* spp., and *Oesophagodontus robustus*, which do not. The cyathostomins include more than 40 non-migratory species. Larval stages rather than adults are responsible for the main pathogenic effects. General strongylosis: ill-thrift, weight loss, poor hair coat, and impaired performance. Verminous arteritis (caused by *Strongylus vulgaris*): colic, diarrhoea, pyrexia, inappetence, depression.

structural alert a chemical moiety, or functional group, that is known to impart undesirable properties to a molecule – for example, a group known to be reactive or unstable *in vivo*.

structural annotation the process of identifying gene elements such as coding regions, gene structure, regulatory motifs and ORFs. *See* **genome annotation**.

structural genomics the effort to determine the 3D structures of large numbers of proteins using both experimental techniques and computer simulation.

structural sequence variation a genomic alteration that changes the number of copies or the arrangement of regions of the genome.

structured data data that require the user to select from prespecified categories or limits the user's ability to put data in a narrative free text format. Structured data are easier to analyse (compared with free text) and have the advantage of guiding the user to complete all required elements rather than assuming that they will remember all the requirements in a narrative note.

structured illumination microscopy (3D SIM) a super resolution technique in which illumination with multiple interfering beams of light permits optical tomography of fluorescently labelled samples at a resolution of ≈100 nm in the vertical plane and ≈300 nm in the axial plane.

stylet a hollow mouth spear in a plant parasitic nematode that the nematode pushes into plant cells in order to inject digestive enzymes. The stylet is then used by the nematode to suck and consume the liquefied contents of the plant cell.

subarachnoid space (SAS) the space between the arachnoid mater and the pia mater. It originates from the cisterna magna, is located around the brain, and extends into the spinal cord. The space is kept open by a trabecular system.

subcapsular beneath the capsule.

subclinical a degree of parasitism which interferes with production but is not evident by physical and visual examination of the animal. In practice, it is a state of parasitism usually diagnosed by a positive production response to the administration of an anthelmintic.

subcutaneous pertaining to the area beneath the layers of the skin. Some drugs are given by subcutaneous injection.

subfamily a taxonomic rank below the family and above the genus levels.

subfunctionalization the partitioning of an ancestral function among multiple members of a duplicated gene family. For example, assume the product of an ancestral gene functioned in all plant tissues. If this gene duplicated, one member of the new gene family may now subfunctionalize to only function in root tissues, while another member of the family may subfunctionalize to only function in leaf tissues.

subgroup analysis analysis of a part of the trial/meta-analysis population in which it is thought the effect may differ from the mean effect. Subgroup analysis should always be prespecified. Otherwise, they provide weak evidence as they are underpowered so are more suited for hypothesis generation. If many tests are done on the same data this increases the chance of finding spurious effects and some kind of correction is needed (e.g. Bonferroni). Given independent data, and no underlying effect, one time in 20 a significant result would be expected by chance.

submetacentric a chromosome in which the arms are of unequal length. *See* **centromere**.

submicroscopic gametocyte carriage infections in which sexual gametocytes that are detectable by molecular methods, but not microscopy, are produced.

submicroscopic malaria infections infections that do not necessarily produce gametocytes and are detectable by molecular methods but not microscopy. Submicroscopic infections are almost exclusively asymptomatic.

subpellicular microtubules a cage-like array of microtubules that subtend the plasma membrane (pellicle) and run parallel to the long axis of the cell.

subpleural nodules spotted or coalescing lesions located beneath the pleural surface of the lungs.

subspecies group of organisms within a species that have certain characteristics not possessed by other members of the species. Breeding may occur between members of different subspecies, also termed race.

substitution a type of mutation due to replacement of one nucleotide in a DNA sequence by another nucleotide or replacement of one amino acid in a protein by another amino acid. The term also refers to the replacement of one amino acid in a protein with a different amino acid. *See* **mutation**.

subtelomeric variably sized regions of eukaryotic chromosomes adjacent to, and directly interior of, telomeres. The genomic features that they contain usually define them. They are typically composed of highly polymorphic repetitive DNA sequences that are situated adjacent to gene-rich areas. In *Plasmodium* species, subtelomeric regions often contain genomic repeats and species-specific genes involved in evasion of the immune response of the host.

subtomographic averaging an electron microscopy technique to visualize structures in three dimensions (3D). The specimen stage is tilted at regular increments and a series of transmission electron images is taken, which are then processed to generate a high-resolution (up to 5 nm) 3D image.

subtype (clade) a subgroup of genetically related parasites. For example, *Toxoplasma gondii* can be classified into three major genotypes: Type I, Type II, and Type III lineages. Parasites within each lineage can then be further classified by subtype.

suckers a term used in relation to trematodes. Adhesive organs. There are generally two suckers: oral and ventral (also called an acetabulum).

sucking disk the area that covers the ventral side of *Giardia* trophozoites. It provides a mechanism by which the parasite attaches to the mucosa of the small intestine.

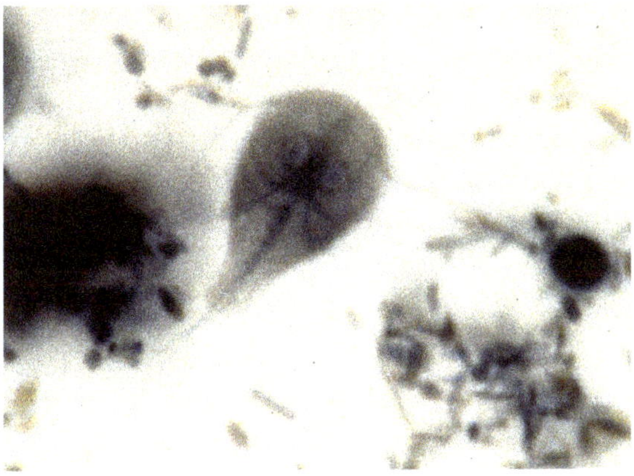

sucrose gradient fractionation separation of molecules (in this case protein) by buoyancy in a sucrose gradient; often used in fractionation of raft components, which are more buoyant than surrounding membrane.

sulfonamides although sulfonamides were the first effective anticoccidial agents developed, acquired drug resistance, especially in the poultry industry, has limited their use as coccidiostats. They are still effective anticoccidial drugs in small animals and ruminants. Sulfonamides work by interfering with protozoal folate synthesis. They competitively inhibit the bacterial enzyme dihydropteroate synthase, which catalyses the formation of dihydropteroate. Inhibition of folate synthesis results in decreased production of the nucleotides needed by coccidia for DNA synthesis. Mammalian cells usually use preformed folate, thereby conserving selective toxicity.

super enhancers specialized large *cis*-regulatory regions identified in embryonic stem cells and cancer cells that regulate the principal genes involved in cell identity and disease.

super resolution light microscopy techniques that use various approaches to break the diffraction limit of light and obtain images in enhanced resolution.

superantigen an antigen that reacts with all the T-cells belonging to a particular T-cell receptor V region family, and therefore stimulates (or deletes) a much larger number of cells than does a conventional antigen.

superinfection infections in which an individual is infected by more than one parasite clone.

superoxide dismutase enzyme that protects cells from oxidative damage by catalysing the transformation of superoxide (a harmful species of oxygen) into oxygen and hydrogen peroxide.

super-resolution microscopy a form of light microscopy that allows the acquisition of images with a resolution not limited by the diffraction limit of light (that is, in the range of dozens of nanometres or less). Super-resolution microscopy methods include stimulated emission depletion (STED), saturated structured illumination microscopy (SSIM), STORM and photoactivated localization microscopy (PALM).

superspreader a highly infectious individual who transmits a parasite to an extraordinarily large number of susceptible hosts (for a precise mathematical definition).

supervised analysis an approach to gene expression analysis in which some aspects of the experimental samples are used to drive the analysis to identify a pattern (signature) of gene expression that characterizes the difference in the samples.

supervised machine learning (SML) a branch of artificial intelligence that involves the design and use of algorithms that can learn from data without being explicitly programmed. For instance, a SML system could be trained on proteins with known immunogenicity to learn to distinguish between vaccine and non-vaccine candidates. After learning, it can then be used to classify uncharacterized proteins.

support vector machine (SVM) supervised learning models that recognize patterns in data sets and that are used for classification and regression analyses.

suppression a mechanism for producing a specific state of immunological unresponsiveness by the induction of suppressor T cells. This type of unresponsiveness is passively transferable by suppressor T cells or their soluble products.

suppressor cells T lymphocytes that are claimed to suppress the response of other cells to antigen, thus being involved in the immune regulation. Their existence is disputed.

suppressor gene a gene that can suppress the action of another gene.

surface antigens markers found on the parasite surface that elicit a host immune response.

surface plasmon resonance a technique used to measure the interaction of macromolecules at a surface via changes in the refractive index. It is based upon changes in the angle of reflected light which occur upon ligand binding to an immobilized target molecule on a biosensor chip. This permits the observation of protein–protein interactions (such as antibody binding to an antigen) in 'real-time', i.e. by continuous monitoring of the association and dissociation of the reversible reaction.

surrogate a person or animal that functions as a substitute for another. In the case of a surrogate mother, a woman or female animal carries an embryo and ultimately gives birth to a baby that was formed from the egg of another female.

surrogate outcome/endpoint intended to capture the treatment effect of an important clinical endpoint but does not directly measure the clinical benefit of the intervention, substitutes something we can measure for something we want to measure. Compare with clinical outcome.

surrogate outcomes outcomes not directly of importance to patients and their carers but predictive of patient-centred outcomes.

surrogate system organism or cellular components used to study a particular mechanism or express a protein naturally produced by a different organism.

surveillance the practice of close medical or other supervision of contacts to permit prompt recognition of infection or illness but without restricting the movements of the individual.

surveillance software computer programs used by facilities to identify infections based on admission, procedure, culture, and clinical data. These are usually offered by third party vendors not associated with facility's electronic medical record and require interfacing to the facility's source systems.

survival analyses illustrated as 'Kaplan–Meier' plots, these are the most commonly used statistical method of assessing differences in the delay to appearance of an outcome measure such as patent infection, for example, related to the treatment of groups with or without a specific outcome modifier, such as vaccination.

survival curve a graph of the number of events occurring over time or the chance of being free of these events over time. The events must be discrete and the time at which they occur must be precisely known. In most clinical situations, the chance of an outcome changes with time. In most survival curves the earlier follow-up periods usually include results from more patients than the later periods and are therefore more precise.

susceptibility a condition of the body that increases the likelihood that the individual will develop a particular disease. It is influenced by a combination of genetic and environmental factors.

swimmer's itch dermatitis and skin hypersensitivity reaction caused when the cercariae of non-human avian schistosomes penetrate skin. It appears as a skin rash caused by an allergic reaction to certain parasites that infect some birds and mammals. These microscopic parasites are released from infected snails into fresh and salt water (such as lakes, ponds, and oceans). While the parasite's preferred host is the specific bird or mammal, if the parasite comes into contact with a swimmer, it burrows into the skin causing an allergic reaction and rash. Swimmer's itch is found throughout the world and is more frequent during summer months. Most cases of swimmer's itch do not require medical attention. *See* **cercarial dermatitis**.

swine lungworm lungworm infection of pigs is associated with *Metastrongylus apri*. Adults are found in bronchi and bronchioles. The parasite has an indirect life cycle, requires an earthworm as an intermediate host. Common in feral and pastured swine. Young animals are more susceptible. Verminous bronchitis, parasitic pneumonia, persistent cough that may become paroxysmal, dyspnoea, lethargy, weight loss.

swinging crossbridge model the first model of the myosin power stroke, which suggests that ATP-dependent changes in the actin–myosin crossbridge angle would cause the thin filaments to slide past the myosin. *See* **sliding filament theory**.

switch sequences highly conserved repetitive sequences which mediate class switching in the immunoglobulin heavy chain gene locus.

sylvatic refers to forest or a wooded area; used as an adjective to describe the location of a disease cycle in the wild. *See* **feral**.

symbiosis (sym = together, bios = life) an ecological relationship between two different organisms or species (e.g. a microbe and a mammalian host) in which both parties benefit (e.g. flagellates in the gut of termites).

sympatric two (or more) species or populations living in the same geographic area and regularly encountering one another. Cf. **allopatric**.

sympatric distribution when two species or populations are considered to exist in the same geographic area and thus regularly encounter one another.

sympatric speciation share the same geographic area but because of reproductive isolation, they do not exchange their genes.

sympatric species (or population) species (or population) occupying fully or partially in overlapping habitats.

sympatry the phenomenon whereby two or more species or populations exist in the same geographic area and thus regularly encounter one another (the opposite term is allopatry).

symptomatic HIV infection stage of HIV infection during which signs or symptoms of the infection begin to appear. The onset of symptoms signals the transition from asymptomatic HIV infection to full blown AIDS.

symptomatic malaria infections infections that are accompanied by fever and/or other symptoms that are indicative of malaria. These infections are, almost without exception, detectable by microscopy or rapid diagnostic test.

symptomatic treatment non-specific therapy of a disease that is designed to reduce the symptoms or the effects.

synapomorphy a derived trait (apomorphy) that is shared by two or more taxa and their most recent common ancestor.

synaptic pruning the process of removing neuronal synapses, usually during development.

synaptic vesicles neuronal vesicles that store and release neurotransmitters or neuropeptides at the synapse.

synchrotron research facility that accelerates charged particles and uses an increasing magnetic field to keep the particles in a circular path. Electromagnetic radiation emitted by the high-energy, accelerated particles can be used in a variety of scientific applications.

syncytiotrophoblast trophoblast cells differentiate into two types of cells: the cytotrophoblast and the syncytiotrophoblast. The syncytiotrophoblast is the layer that promotes the invasion of the uterine wall and will later form part of the placenta.

syndecans a family of transmembrane, chondroitin and heparan sulfate-modified proteoglycans with four members in vertebrates, and related genes in *Drosophila* and in *C. elegans*.

syndinians several genera of parasitic flagellates (e.g. *Amoebophrya*, *Hematodinium* and *Syndinium*) that diverge near the phylogenetic origin of dinoflagellates and perkinsids. The life cycle of syndineans consists of zoospores that penetrate a host (e.g. copepods, polycistines and dinoflagellates) and develop into an undifferentiated sporangium. The growing sporangium eventually destroys the internal compartment of the host and differentiates into many minute zoospores (ranging from about 3 to 20 µm in length, depending on the species) that escape the host in order to infect a new one.

syndrome a collection of recognizable traits or abnormalities that tend to occur together and are associated with a specific disease.

syngamy multiplication (reproduction) by a sexual process There are many gradations of the sexual process among parasitic species.

syngeneic genetically identical members of the same species, e.g. a fully inbred strain of mice. (same as isogeneic)

synonymous mutation a nucleotide difference in a gene that does not alter the encoded protein sequence due to the degeneracy of the genetic code.

synovial fluid fluid present in the cavities of joints to reduce friction during joint movement.

syntenic pertaining to the presence of collinear homologous DNA sequences in related chromosomal regions.

synteny presence of the same loci (usually genes) on corresponding chromosomes in different genomes.

synthetic biology field of biological research and technology that combines science and engineering with the goal of designing and constructing new biological functions and systems not found in nature. Essential synthetic biology tools include DNA sequencing, fabrication of genes, modelling how synthetic genes behave, and precisely measuring gene behaviour.

synthetic interaction screens genetic screens in which two perturbations are combined to assess the possible synergistic and epistatic effects between the two genes perturbed.

systematic immunity certain parasites stimulate systemic immune responses. In these cases antigen sensitized cells, (T and B) may be stimulated locally but hone to other organs or circulate in the bloodstream.

systematic review a review in which specified and appropriate methods have been used to identify, appraise, and summarize studies addressing a defined question.

systemic pertaining to the entire body or affecting the body as a whole.

systemic infection an infection that results in spread to many organs of the body affecting the whole body. Systemic drug therapy involves giving a drug that travels through the bloodstream and affects cells throughout the body.

systemic lupus erythematosus (SLE) systemic autoimmune disease characterized by antinuclear Abs, often including Abs reactive against DNA.

systemic memory response attributable to the immune system mediated by circulating memory T or B cells, whereby a second encounter with an antigen induces a heightened state of immunity.

systems biology an interdisciplinary approach that attempts to develop and test holistic models of living systems. A 'top-down' systems approach uses quantitative modelling to identify and describe the underlying biosynthetic and regulatory networks of a system, whereas a complementary 'bottom-up' approach attempts to model the systems-wide phenotypes that emerge from component interactions.

syzygy association of gamonts (pre-gametes) end to end or in lateral pairing prior to the formation of gametes, found in most gregarine protists and, perhaps, piroplasms.

T cell immunoglobulin and mucin domain 3 (Tim3) expressed by several lymphocytes subsets, including IFN-γ secreting T helper 1 cells, negatively regulates Th1 response.

T cells lymphocytes with several functions in the immune system, such as cytotoxic (expressing CD8 on their surface), helpers (expressing CD4 on the surface), regulatory and memory. Major role is cell-mediated immunity.

T follicular helper (Tfh) cells are a newly described CD4 T cell type with a distinct differentiation lineage. Tfh cells are distinguished by their expression of CXCR5 (a chemokine receptor), IL-21, and ICOS (inducible T cell costimulator). Tfh cells are critical for B cell help through IL-21 expression.

T helper 1/T helper 2 cell (TH1/TH2) a classification of $CD4^+$ T cells on the basis of the patterns of cytokines that they secrete. TH1 cells secrete large amounts of interferon-γ and associated pro-inflammatory cytokines. TH2 cells secrete large amounts of interleukin-4 and associated cytokines that promote antibody production by B cells. TH1/TH2 cytokines can cross-regulate each other's responses. An imbalance of TH1/TH2-cell responses is thought to contribute to the pathogenesis of various infections, allergic responses and autoimmune diseases.

T helper cell subsets (Th1/Th2/Th17/Treg) subsets of a CD4 T cell lineage, which promote various types of immune response. Th1: cell mediated cytotoxic response (via IFN-γ). Th2: B cell mediated antibody response (via IL-4, IL-5 and IL-13). Th17: antifungal response (via IL-17A). Treg: anti-inflammatory response (via IL-10).

T lymphocyte (T Cell) a type of lymphocyte. There are two major types of T lymphocytes: CD8 cells (cytotoxic T lymphocytes) and CD4 cells (helper T lymphocytes); both T cell types are essential for a healthy immune system. HIV infects and destroys CD4 cells, gradually destroying the immune system.

T regulatory cells (Tregs) a proportion of T cells usually identified by the expression of the transcription factor FoxP3 and a high level of expression of CD25. They can be either $CD4^+$ or $CD8^+$ and are important in controlling heightened immune activation or inflammation. They act either through cell–cell interaction such as cytotoxic T lymphocyte antigen 4 (CTLA-4) or by the release of immunoregulatory cytokines such as IL-10 and TGF-β.

tachycardia abnormal rapid heartbeat. In adult humans, a rate over 100 beats per minute is usually considered tachycardia.

tachypnoea abnormal increased rate of breathing.

tachyzoite asexual stage showing rapid growth by endodyogeny in the life cycle of certain protozoan parasites, such as *Toxoplasma gondii*. Tachyzoites are motile forms of coccidian parasites that can actively invade and infect different types of cell throughout the host, including neural, macrophage, fibroblast, vascular endothelial, muscle, and liver cells. Tachyzoites can burst host cells by active multiplication, leading to cell death.

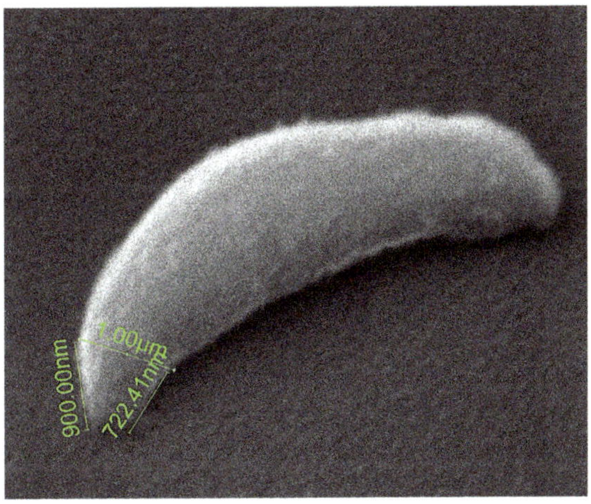

taeniasis (or taeniosis) a parasitic infection caused by the tapeworm genus *Taenia*. Humans can become infected with these tapeworms by eating raw or undercooked beef (*T. saginata*) or pork (*T. solium* and *T. asiatica*). *Taenia solium* tapeworm infections or ingestion of *T. solium* eggs can lead to cysticercosis, which is a disease that can affect various tissues and neurocysticercosis can cause seizures.

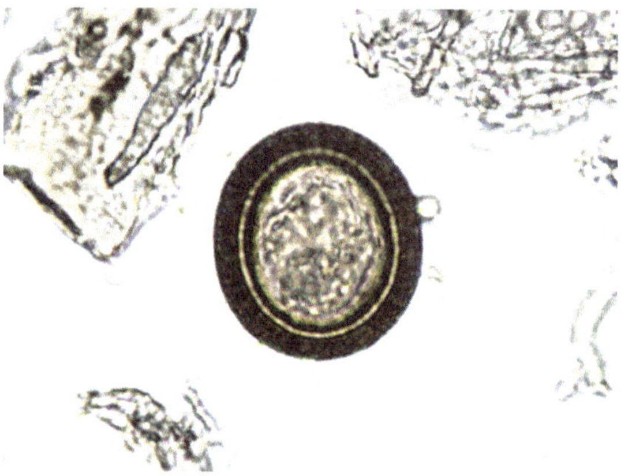

tafenoquine a new generation 8-aminoquinoline that may replace primaquine with a better efficacy and safety profile. Currently in Phase II clinical trial the drug is expected to be a single-dose treatment that will eliminate *P. vivax* hypnozoites from the liver of the host.

tailoring enzymes enzymes responsible for individual modifications to a secondary metabolite skeleton (such as a polyketide carbon chain, non-ribosomal peptide or a terpene ring structure), typically oxidoreductases, methyltransferases, acyltransferases and glycosyltransferases.

Tajima's D test summarizes the spectrum of substitutions found in an alignment of DNA sequences. It can be used to make preliminary inferences about deviations from the neutral model of molecular evolution, such as population growth and natural selection.

tandem mass spectrometry (MS–MS) a technique for obtaining detailed information about the ions present in a complex mixture by coupling two stages of mass analysis: the first selects a component ion to fragment and the second determines the masses of the fragment ions. It involves multiple steps of MS selection. Between each step, a fragmentation and selection of molecular ions occurs. Tandem MS can be carried out 'in space' when the two MS steps are physically separated in specific instruments (triple quadrupoles or QQQ, or a quadrupole and a time-of-flight analyser in QTOF instruments). Tandem MS can be carried out 'in time' in ion trap analysers. Different modes of analyses are classically achieved with tandem MS, including the precursor ion scan and neutral loss scan modes.

tandem repeat sequences a sequence of two or more DNA bps that is repeated in such a way that the repeats lie adjacent to each other on the chromosome. Tandem repeats are generally associated with non-coding DNA. In some instances, the number of times the DNA sequence is repeated is variable. Such variable tandem repeats are used in DNA fingerprinting procedures.

tannins the term 'tannin' refers originally to the capacity of some mineral or plant compounds, to tan animal hides (to transform them into leather). Restricted to plant products, an operational definition refers to water-soluble phenolic natural products that can precipitate proteins from aqueous solution. Their molecular weight ranges from 500 to 20,000. Depending on the structure, they can be categorized into two major groups: the hydrolyzable tannins and the condensed tannins.

TAP (transporters associated with antigen processing; TAP-1 and TAP-2) are molecules which carry antigenic peptides from the cytoplasm into the lumen of the endoplasmic reticulum for incorporation into MHC class I molecules.

tapeworm a cestode composed of a scolex (holdfast), a neck, where growth of new proglottids occurs, and a chain of proglottids known as a stroblia. Proglottids may be immature, mature, and gravid (full of eggs).

target candidate profile (TCP) pharmacodynamics, pharmacokinetics, and safety parameters that a drug candidate should exhibit to fulfil the cure or prophylaxis criteria.

target host the host of concern. With respect to zoonotic diseases, it is obviously the human, but for all other diseases, it is the alternative to reservoir hosts, and is the host selected for concern by investigators.

target product profile (TPP) all the desired attributes of the final drug or drug combination, including pharmacodynamics, pharmacokinetic, safety and non-biological properties, including cost, stability, dosage, package, etc. Usually, the TPP briefly summarizes the specific studies (both planned and completed) that will supply the evidence for each conclusion about that product.

target regions the region of interest defined for enrichment. The genomic co-ordinates of a target region are used to design the capture baits, probes, or primers for enrichment and vary by exome sequencing kit.

target species the species under study.

targeted chemo-elimination (TCE) the administration of antimalarial treatment to entire populations in a selected area with the goal of eliminating malaria.

targeted gene replacement substitution of one allele for another in a genome, at the same genomic location; allows for comparisons between alleles in a homogeneous genetic background without position effects.

targeted mutagenesis deliberate change in the genetic structure directed at a specific site on the chromosome. Used in research to determine the targeted region's function. *See* **mutation**, **polymorphism**.

targeted selective treatment (TST) individually determined treatment to animals within a flock or herd.

targeted treatment (TT) some flocks are left untreated while other flocks are drenched.

TATA-box consensus DNA sequence 5′-TATAAA-3′ found within the proximal and/or core promoter of approximately a quarter of the genes in eukaryotes and archaea. This sequence is typically bound by the TATA-box binding protein to initiate transcription by the RNA polymerase.

taxon (plural taxa) group of organisms linked by common ancestry. Taxa can range in scale from populations to kingdoms. The categories (e.g. phylum, order, family, genus, or species) used to classify animals and plants.

taxonomic pertaining to the classification of organisms, not necessarily based on their evolutionary relationships.

taxonomically restricted genes (TRGs) a more careful definition of 'orphan' or 'novel' genes that takes into consideration not only the absence of sequence similarity to genes or proteins in other organisms, but also the narrow phylogenetic distribution of these genes. This definition is synonymous to 'lineage-specific genes'.

taxonomy hierarchical classification system for naming and grouping organisms based on evolutionary relationships.

TAZ1 telomere-associated in *Schizosaccharomyces pombe*. It regulates telomere length and function; required for the repression of telomere-adjacent gene expression and for normal meiosis or sporulation.

T-bet transcription factor functionally linked to the differentiation and activity of Th1 cells and inhibition of Th2 and Th17 cell differentiation.

TCA cycle (also tricarboxylic acid cycle, Krebs cycle or citric acid cycle) a series of enzyme-catalysed chemical reactions occurring in mitochondria that generate energy (ATP) through the oxidation of acetate derived from carbohydrates, fats and proteins.

T-cell exhaustion (T-cell depletion) the gradual decrease in T-cell function that can occur with chronic infections and cancers. T-cell exhaustion weakens the immune system, making it difficult for the body to fight off infections or kill cancer-causing cells.

T-cell receptor (TCR) the heterodimeric antigen receptor of the T lymphocyte exists in two alternative forms, consisting of α and β chains, or γ and δ chains. The $\alpha\beta$ TCR recognizes peptide fragments of protein antigens presented by MHC molecules on cell surfaces. The function of the $\gamma\delta$ TCR is less clearly defined but it can recognize native proteins on the cell surface.

T-dependent antigen an antigen that is able to induce antibody synthesis only in the presence of lymphokines released by helper T cells.

technology transfer the process of transferring scientific findings from research laboratories to the commercial sector.

tegument complex and dynamic membrane structure of trematodes and cestodes that covers the entire surface of adult worms. Its apical surface consists of two tightly opposed bilayer membranes, whereas the inner surface of the tegument is bounded by a single lipid bilayer basal membrane. The tegument is a syncytium without lateral membranes and is connected by cytoplasmic tubular bridges with cell bodies (cytons), which are located beneath the muscle layers.

***Teladrosagia circumcincta* (brown stomach worm)** *T. circumcincta (formerly called Ostertagia ostertagi)* are Strongylida (bursate) nematode. Like *H. contortus*, *T. circumcincta* is a parasite of the abomasum. However, unlike *H. contortus* because it dose not feed on blood nor are the females as prolific. Adult measures 0.6–1 cm in length, slender, and brown-coloured worms.

teleost a bony fish belonging to the infraclass Teleostei, the most species-rich of the three major groups of ray-finned fishes.

telomerase the enzyme that directs the replication of telomeres. Telomerase makes the DNA ends of linear chromosomes. Telomerase is made up of an RNA portion and a protein portion, and probably dates back to the RNA world.

telomere a region of repetitive DNA located at the end of chromosomes. Telomeres are usually maintained by telomerase, a ribonuclear protein particle (RNP) that includes an RT enzyme and an RNA molecule that is reverse transcribed. This specialized structure is involved in the replication and stability of linear DNA molecules.

temporary parasite a parasite that lives free of its host during part of its life cycle.

ten-eleven translocation (TET) DNA-binding enzymes that are methylcytosine dioxygenases in mammals and include TET1, TET2 and TET3. They were named for the genetic variant found in the TET1 gene sequence in acute myeloid and lymphocytic leukaemia.

teneral immature, young adult fly that has not yet taken a blood-meal.

tension zone a hybrid zone maintained primarily by selection against hybrids or rare genotypes, rather than by adaptation of alternative alleles to alternative environments.

TER the telomerase RNA component that contains a short sequence encoding the cognate telomere repeat that serves as the template for reverse transcription by TERT.

teratogenic pertaining to birth defects and developmental malformations. Some antiretroviral (ARV) drugs should be avoided during pregnancy because they can potentially have teratogenic effects on the developing fetus.

teratogenicity some drugs, such as albendazole, are teratogenic and should never be given to pregnant animals.

terpenes a class of volatile, non-polar secondary metabolites that are structurally related to isoprene and that are produced by plants and some insects primarily as chemical signals.

TERT the protein component of telomerase, the enzyme that elongates telomeres by copying a template sequence within its RNA component (TER).

tertian malaria malaria characterized by a 48-hour periodicity, associated with infections caused by *Plasmodium falciparum*, *Plasmodium vivax* and *Plasmodium ovale*.

tertiary lymphoid tissues organized clusters of naive lymphocytes that form during chronic inflammation and can prime adaptive immune responses.

tertiary sex ratio the ratio between the number of male and female mature organisms.

test of neutrality a test that compares observed patterns of diversity with expectations generated under the assumption of overall selective neutrality.

test prevalence the proportion of individuals in a population that test positive to a diagnostic test at a single time point. Test prevalence is only equal to true prevalence if the test used is a perfect gold standard.

tetrahydrofolate (THF) a cofactor that is involved in transfer reactions of single carbon groups. It is important not only in acetogenesis but also in the metabolism of amino acids or nucleic acids.

tetrahydropyrimidines pyrantel and oxantel are active against larval and adult forms of nematodes, but are not active against migrating or arrested larvae, and have no action on cestodes. The mechanism of action is similar to that of the imidazothiazoles. Pyrantel preparations are usually formulated as either the tartrate or embonate salts (in the USA the embonate salt is known as the pamoate salt). Pyrantel tartrate is more water soluble and better absorbed from the gastrointestinal tract. The embonate (pamoate) salt is less water soluble than the tartrate salt and more poorly absorbed, and is thus less toxic to the host, and is considered safe to administer to young, sick or pregnant animals.

tetraploid having four complete sets of homologous chromosomes in a cell. In the case of diplomonads, these are divided into two diploid nuclei.

tetrasomy the presence of four copies of a given chromosome.

tetrathyridium worm like larva with an invaginated scolex found only in Mesocestoididae.

Th1/Th2 cell response $CD4^+$ T helper cell responses based on cytokine secretion profiles. Th1 cells produce interferon-γ, IL-2, and TNF-α; whereas, Th2 cells produce IL-4, IL-5, IL-10, and IL-13. The different cytokines affect different classes of blood cells, resulting in different immune system responses.

Th1/Th2 paradigm Th1 and Th2 are lineages of $CD4^+$ T cells. The ability of each to regulate the growth of the other can lead to dominance of one lineage over the other. Th1 cells are associated with the development of cell-mediated immunity, whereas Th2 cells are associated with allergy and immune responses to parasitic infection.

Th17 $CD4^+$ T cells that produce IL-17; combinations of IL-1β, IL-6, and TGFβ or IL-23 induces their development; IL-17 promotes neutrophil homeostasis.

Th2 biasing when the immune response is pushed to favour a Th2-type response, favouring the production of Th2-associated cytokines and antibodies.

thapsigargin potent inhibitor of sarcoplasmic–endoplasmic reticulum (SERCA) calcium ATPase.

the 'big gulp' described as an actin independent mechanism by which ring stage parasites fold on themselves and take the first 'gulp' of host cell cytoplasm.

the 3Rs (reduction, refinement, and replacement) an ethical framework for the use and care of animals in scientific experiments. This includes the replacement of animals wherever possible, the refinement of husbandry and procedures during experiments, and the reduction in terms of improving approaches and increasing efficiency of experiments.

***Thelazia callipaeda* (Spirurida, Thelaziidae)** a whitish nematode of 0.5–1.8 cm length infecting the conjunctival pouches and the lachrymal ducts of dogs, cats, rabbits, wild carnivores and humans. Adult and pre-adult stages cause mechanical damage to the conjunctival and corneal epithelium, provoking lacrimation, conjunctivitis or even keratitis, uveitis and corneal ulcers. Lacrimal secretions are of significance for the transmission of first stage larvae (L1) to the intermediate and vector host, the Phortica flies (Diptera, Drosophilidae). Attracted by ocular discharge, the flies ingest L1 while feeding on conjunctival secretions of an infested animal and, after approximately 3 weeks, they also transmit the infective third stage larvae (L3) to a definitive host. Without treatment, adult worms can persist up to one year behind the third eyelid, rudimentary in humans.

thelaziosis (eyeworm) *Thelazia rhodesii* a disease mainly seen in summer and autumn when the flies are active. It is usually more common in ruminants than horses and worms may be more abundant in beef than in dairy cattle. The flies deposit larvae on the conjunctiva when feeding on fluid around the eye. Infection is often inapparent but they may cause excessive lacrimation, and photophobia. Eyeworm in cattle is differentiated from infectious keratitis by observing the adult worm in the conjunctival sac or demonstrating first-stage larvae in eye washings. Therapy involves the manual removal of the worms after application of topical anaesthetics or organophosphate. In cattle, ivermectin, doramectin, or levamisole are effective. The face flies and other *Musca* species are intermediate host for the parasite, and therefore fly control is essential for reducing infections.

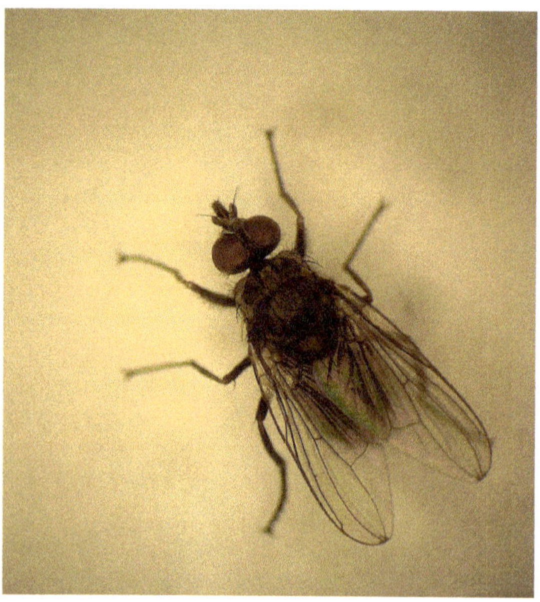

thelytokous the parthenogenetic production of female offspring from unfertilized eggs in Hymenoptera, some fish salamanders, and reptiles.

therapeutic cloning generally referred to as somatic cell nuclear transfer technology. It involves replacing the nucleus of an egg cell with the nucleus from a cell from a patient's body and allowing it to develop to form a blastocyst. The embryonic stem cells from the inner cell mass are then harvested and used to establish a cell line that has the same genetic makeup of the patient. These cells can then be directed to develop into the tissue needed for transplant.

therapeutic index (TI) a ratio that compares the blood concentration at which a drug becomes toxic and the concentration at which the drug is effective. The larger the therapeutic index (TI), the safer the drug is. If the TI is small (the difference between the two concentrations is very small), the drug must be dosed carefully and the person receiving the drug should be monitored closely for any signs of drug toxicity.

therapy any medical intervention to cure a disease. This includes the use of medicinal products, to eliminate an existing parasite infection/infestation.

therapy-related secondary malignancies malignancies triggered by cancer treatment, including therapy-related myelodysplastic syndrome and acute myeloid leukaemia. These diseases have been linked to treatments with topoisomerase inhibitors. Potential causes include mutations such as chromosomal translocations that result from the inappropriate repair of treatment-induced DNA breaks.

theratypes a recently invented term used to classify disease-associated DNA variants according to the molecular-based treatment to which they respond.

therians a subclass of mammals that comprises placental mammals (eutherians) and marsupials (metatherians). Mammals that give birth to live young without using a shelled egg. The remaining major mammalian lineage is that of the egg-laying monotremes (prototherians).

thermal cycle the 'engine' or PCR machine, in which the PCR is performed.

thermal response curve phenotypic response over a range of temperatures; the outcome can be a process (e.g. enzyme activity), trait (e.g. size), or measure of fitness. The curve is characterized by its cold and warm extremes (CT_{min} and CT_{max}), the difference between these extremes (operative range), and the peak (T_{opt}). The position on the curve corresponding to the mean habitat temperature experienced by an organism is defined as T_{hab}. If the curve is based on the response of a single genotype over a range of environments then this is equivalent to a reaction norm.

thermoneutral zone (TNZ) this represents the optimal range of environmental temperatures for an endotherm, where the basal rate of heat production is in equilibrium with the rate of heat loss to the external environment and the organism does not have to use large amounts of energy to control its body temperature.

thiacetarsamide an organic arsenical used as a heartworm adulticide in dogs and cats. The drug's exact mechanism of action is unknown but may be related to reaction of the arsenic with sulfhydryl groups in essential parasite energy systems. The drug is only effective against adult heartworms and has no activity against microfilariae. It is given intravenously; extravascular administration must be avoided since the drug is a severe vesicant. It is metabolized in the liver and eliminated equally in faeces and urine. The drug concentrates in the liver and kidneys and can be associated with nephrotoxicity and hepatotoxicity. Other side effects include gastrointestinal disturbances. The drug has a very narrow therapeutic index and seems to be more toxic when administered to cats.

thin layer chromatography (TLC) method of separating molecules for analysis.

thioester-containing protein 1 (TEP1) protein found in the haemolymph of mosquitoes, which has structural similarity to complement factor C3. Upon cleavage, it binds to the surface of *Plasmodium* and tags it for elimination.

third complement receptor (CR3) an integrin on the surface of phagocytes, including macrophages and neutrophils that primarily detect iC3b during pathogen recognition. The role of CR3 is varied, and encompasses particle clearance, leucocyte adherence, and actin reassembly.

third-generation sequencing a new progression of sequencing technology that aims to improve throughput and reduce sequencing cost and time. The main goals of third-generation sequencing are to eliminate the DNA amplification step before sequencing and to enable real-time signal monitoring.

THP1 macrophage a monocyte cell line derived from a patient diagnosed with acute monocytic leukaemia. This cell line is grown in suspension and is a non-adhesive type. Treating this cell line with PMA helps the cells to adhere and differentiate into macrophages.

threshold density the minimum population density of hosts at which transmission of an infectious agent can be sustained.

threshold of polygyny this is reached when females' reproductive success is greater by mating with an already mated male than with an unmated one.

thrombocytopenia a decrease in the number of platelets in circulating blood.

thrombosis formation of a blood clot inside a blood vessel. It occurs as a consequence of blood vessel injury and can obstruct the flow of blood through the circulatory system.

thymine (T) one of four chemical bases in DNA, the other three being adenine (A), cytosine (C) and guanine (G). Within the DNA molecule, thymine bases located on one strand form chemical bonds with adenine bases on the opposite strand. The sequence of four DNA bases encodes the cell's genetic instructions. *See* **bp**, **nucleotide**, **purine**, **pyrimidine**.

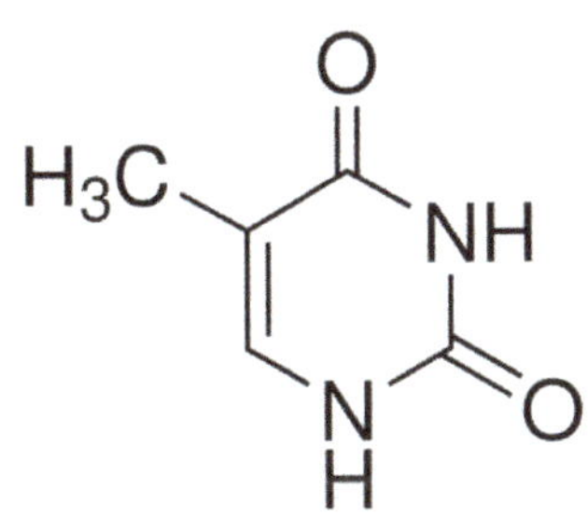

thymocyte an immature (developing) T cell in the thymus.

thymus an organ of the lymph system where T lymphocytes (T cells) develop and mature. The thymus is important for normal immune system development early in life and is at its largest size at puberty. The thymus declines in size and function during adult life, eventually being replaced by fat.

tick haematophagus acari whose bites promotes activation of an inflammatory process.

tick infestation ticks may act as causative agent of disease or may be carriers and vectors of pathogens such as viruses, bacteria, protozoa and even helminthes.

tick paralysis a neurological disease affecting a variety of animals, including domestic animals, companion animal pets, and humans.

tick pyaemia of sheep this disease affects lambs born on or introduced to a tick-infested area, usually at the time tick activity is at its greatest (usually in the spring). The disease is caused by the bacterium *Staphylococcus aureus*, normal skin flora, which enters the bloodstream by the biting action of the blood-sucking nymphal stage of the tick *Ixodes ricinus*. Diagnosis is based on the history and necropsy findings, bacterial culture of infected joint. Parenteral administration of penicillin or tetracycline can be effective, provided the lesions are not too advanced. Local treatment of affected joints is often disappointing, because by the time lesions are noticed abscess formation may be advanced and irreversible damage to the joint may have occurred. Long-acting antibiotics, to which particular strains of *S. aureus* agents are sensitive, can be used prophylactically, though care should be taken not to under-dose, which could promote bacterial resistance.

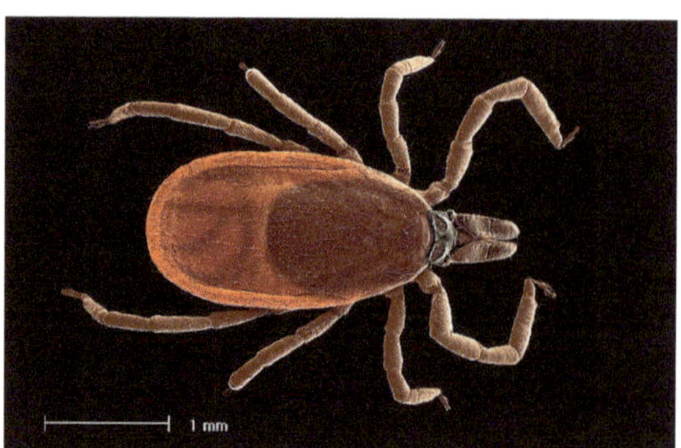

tidal volume the lung volume representing the normal volume of air displaced between normal inspiration and expiration when extra effort is not applied.

tight junction (TJ) specific structures present between cells such as brain microvascular cells that 'seal' the monolayer and prevent passage of macromolecules between the luminal and abluminal sides of the cells. It has a crucial role in maintaining the integrity of the BBB. It is composed of molecules such as vascular endothelial (VE)-cadherin, zona occludens 1, junction adhesion molecules 1 to 3, or occludin.

time since the MRCA if the times back to a common ancestor differ between two individuals, then the average is used.

time-location testing a proactive ID tactic aimed at testing (and subsequently treating) a hotpop using prior knowledge of their occupation-specific location. This might include visits to mines, fishing docks, and forest camps, as well as mobile military units.

time-of-flight (TOF) and quadrupole time-of-flight (QTOF) mass analysers a TOFanalyser is part of a mass spectrometer, in which the mass-to-charge ratio (m/z) of ions is determined after their acceleration by an electric field of known strength and record of the time it takes the ion to reach a detector. TOF MS has a very high mass accuracy. A QTOF is a tandem MS instrument in which an initial quadrupole filters sample ions based on their mass-to-charge ratio (m/z). After fragmentation, the filtered ions are then separated in a TOF analyser.

T-independent antigen an antigen which induces antibody synthesis in the absence of lymphokines released by T cells; the antibodies are generally of the IgM isotype and typically lack memory.

tissue culture the separation of cells from each other and their growth in a container of liquid nutrients.

tissue cyst the vehicle in coccidian parasites such as *Toxoplasma*, which holds the bradyzoites, usually found in brain, muscle, or eye tissue. Tissue cysts vary in size depending on the number of bradyzoites they hold. Raw or undercooked meat with tissue cysts is an important source of infection for definitive and intermediate hosts.

tissue-specific immunity the immune response specific to the tissue site. This response is specific towards a specific antigen and is regulated by dendritic cells, macrophages, T cells, and B cells.

titration the measurement of the level of specific antibodies in a serum by testing increasing dilutions of the serum for antibody activity.

titre the reciprocal of the last dilution of a titration giving a measurable effect, e.g. if the last dilution giving significant positive reaction is 1:256, the titre is 256.

t-loops telomeric loops, duplex loops generated by invasion of the G-rich strand in double-stranded telomeric DNA. The loops are maintained by proteins, such as human TRF2, and mask telomeres from the cellular machinery that detects DSB.

TMIC aggregate time for which a drug exceeds the minimum inhibitory concentration (MIC).

Tn5 transposase a member of the RNase superfamily of proteins that harbours retroviral integrases to catalyse random insertion of transposon DNA into target DNA.

tolerance 1 diminished or absent capacity to make a specific response to an antigen; usually produced as a result of contact with that antigen under non-immunizing conditions. **2** the ability to tolerate a drug when given as prescribed.

tolerant host could be easily parasitized with certain species of parasites.

tolerogen an antigen used to induce tolerance. Often depends more on the circumstances of administration (e.g. route and concentration) than on any inherent property of the molecule.

toll-like receptor (TLR) pathways TLR signalling plays an essential role in the innate immune response. Activation of TLR signalling through recognition of pathogen-associated molecular patterns leads to the transcriptional activation of genes encoding proinflammatory cytokines, chemokines and co-stimulatory molecules, which subsequently control the activation of the antigen-specific adaptive immune response.

toll-like receptor 9 (TLR9) transmembrane receptor which recognizes pathogen associated molecular patterns (PAMPs), such as DNA. TLR9 is expressed in several immune cells such as monocytes, dendritic cells, and B lymphocytes.

toltrazuril a triazinetrione derivative that has activity against all intracellular coccidian stages. The mode of action is unknown, but could be similar to diclazuril. It is used for the treatment of coccidiosis in broilers and broiler breeder chickens.

tomographic reconstruction a computational process in which a series of projected images acquired from multiple directions is used to compute a 3D volume data set.

tomography image acquisition technique involving gathering projection data from multiple directions and reconstructing it to obtain a 3D volume.

topoisomerase 2 (TOP2) an enzyme that regulates DNA topology by cutting a pair of strands from a DNA helix to allow another unbroken helix to pass through, followed by resealing of the broken ends. TOP2 activity is essential for many DNA transactions and represents the target of several chemotherapeutic drugs.

topoisomerase IV a topoisomerase enzyme that can introduce double-strand breaks to relax supercoils but also is responsible for unlinking circular DNA molecules after replication.

TOR complex contains the target of rapamycin (TOR), a serine/threonine kinase together with raptor (a substrate binding subdomain) and polypeptides and plays a major regulatory role in autophagy by regulating the activity of autophagy related proteins.

total internal reflection fluorescence microscopy (TIRF) technique that allows investigating fluorescent objects that are placed within about 100 nm above a substrate surface by reflecting a laser beam from the bottom of that substrate (usually a glass slide), thus generating a limited so-called evanescent field above the surface.

totipotent cells capable of forming a completely new embryo that can develop into a new organism. For example, a fertilized egg is totipotent.

tough decoy RNA a hairpin-shaped decoy RNA containing two miRNA recognition sites that compete with a miRNA target, leading to long-term miRNA loss-of-function.

toxic poisonous.

toxisome a vesicle that contains toxic factors and that is expelled from the cell surface (for example, shed microvilli).

toxocariasis a common zoonotic infection mostly caused by *Toxocara canis* and less frequently by *Toxocara cati*. The two main clinical presentations of toxocariasis are visceral larva migrans (VLM) and ocular larva migrans (OLM).

toxoid a non-toxic derivative of a toxin (often formalin treated) used as an immunogen for the induction of antibodies capable of cross-reacting with the toxin.

Toxoplasma gondii an obligate intracellular cyst forming coccidian parasite of the phylum Apicomplexa capable of infecting essentially any warm blooded vertebrate species. This protozoan pathogen is the causative agent of toxoplasmosis, an infectious zoonotic disease with worldwide prevalence. The capacity of *T. gondii* to undergo sexual and asexual replication implies that both recombinant and clonal isolates occur. The majority of *T. gondii* isolates from human and domestic animals in North America, Europe, and Africa are grouped into three main clonal lineages, designated as types I, II, and III. A fourth clonal lineage, type 12 (formerly type A and type X), has been identified in North America where it is most commonly detected in wildlife. In contrast, a diversity of genetically distinct 'atypical' *T. gondii* isolates differing from the archetypal clonal lineages appear to predominate in South America and geographically isolated regions.

toxoplasmosis an opportunistic infection caused by the parasite *Toxoplasma gondii*. It is most commonly transmitted by contact with infected cat faeces, eating contaminated meat, or drinking contaminated water. Toxoplasmosis can also be transmitted from an infected mother to her child during pregnancy. Symptoms, if any, include swollen lymph nodes and muscles aches that last for a month or more. In fetuses and people with weakened immune systems, toxoplasmosis can cause severe damage to the brain (*T. gondii* encephalitis), eyes, and other organs. In people with HIV, toxoplasmosis of the brain is an AIDS-defining condition.

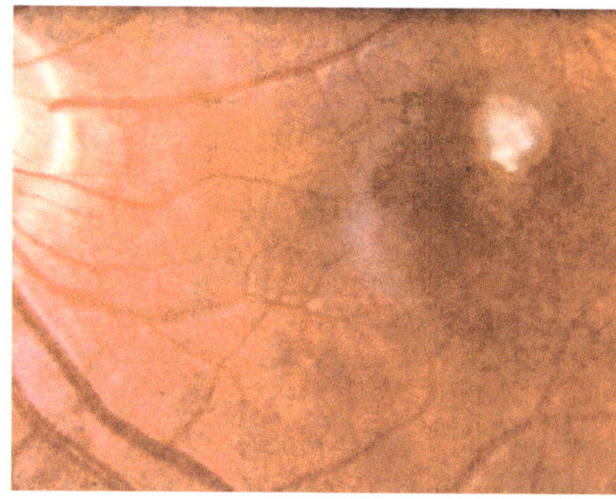

Tpase (transposase) the enzyme responsible for the catalysis of transposition.

Tr1, peripherally induced T regulatory cells regulatory population of CD4 T cell that develops in the periphery after infection or immune insult, exhibits specificity for non-self-antigens and exerts suppressive function through secretion of IL-10 and/or TGF-β. Unlike iTregs, Tr1 cells do not express Foxp3.

tracer flux a method that estimates membrane permeability of a radiolabelled solute by quantifying the rate movement across a membrane.

tracheal migration migration in the host by a parasite into alveoli, up the respiratory tree, swallowed and then mature in the intestine.

tracheal rings incomplete rings of hyaline cartilage found in the anterior two-thirds of the tracheal wall.

traction force microscopy (TFM) a technique that allows calculating the forces exerted by objects/cells on the substrate. To this end, cells are plated on a flexible gel (e.g. polyacrylamide) that contains fluorescent beads. When the cell pulls on the substrate, the beads are displaced and indicate the direction and strength of the applied force, which can be determined computationally.

tradeoff a situation where the gain of one advantage involves the corresponding loss of another.

TRAF (TNF receptor-associated factors) a family of proteins primarily involved in the regulation of inflammation, antiviral responses and apoptosis.

trait a specific characteristic of an organism. Traits can be determined by genes or the environment, or more commonly by interactions between them. The genetic contribution to a trait is called the genotype. The outward expression of the genotype is called the phenotype.

***trans* conformation** the conformation of a glycosidic linkage of receptor analogues, in which the two carbon atoms at either end of the glycosidic bond are oriented in opposing directions.

***trans*-acting factors** molecules, such as proteins or RNAs that can elicit an effect in an interacting partner molecule.

transcellular as opposed to the term paracellular, transcellular transport occurs through the epithelial cells. This can be in the form of transcytosis, as in M cells, or in the specific case of PP LysoDCs, through cellular processes extended through the M cell apical membrane.

transcellular diapedesis migration of cells through endothelial or epithelial cells forming a pore through the cell body.

transcript RNA molecule (mRNA) generated from a gene's DNA sequence during transcription.

transcription **1** the conversion of a DNA nucleotide sequence into a RNA nucleotide sequence by complementary base pairing. **2** the process of making an RNA copy of a gene sequence. This copy, called a mRNA molecule, leaves the cell nucleus and enters the cytoplasm, where it directs the synthesis of the protein, which it encodes. *See* **translation**.

transcription activator-like effector nucleases (TALENs) genome-editing tool using nucleases that are fused to a DNA-binding module, which can be customized for sequence-specific targets.

transcription factor protein that binds to regulatory regions in the genome and helps control gene expression.

transcriptional attenuators RNA structures that cause transcriptional termination only in the presence of a metabolic end-product, such as an amino acid.

transcriptional burst bursts of mRNA production resulting from the promoter of a gene switching between periods of prolonged inactive (or 'off') states and short-lived active (or 'on') states.

transcriptional regulatory network a network representation of transcriptional events where nodes represent TFs and TGs and links denote regulatory interactions between TFs and TGs, mediated by the binding of a TF to the promoter region of its TG.

transcriptome global analysis of expression levels of all RNA transcripts present in a cell at a given time.

transduction the conversion of a signal from one form to another.

transepithelial the term is used to characterize transport across the epithelium, regardless of whether transcellular or paracellular. This can be in the basolateral-to-apical or in the opposite direction.

transfection the introduction of foreign DNA into a host cell. *See* **cloning vector**.

transfer RNA (tRNA) a small RNA molecule that participates in protein synthesis. Each tRNA molecule has two important areas: a trinucleotide region called the anticodon and a region for attaching a specific amino acid. During translation, each time an amino acid is added to the growing chain, a tRNA molecule forms bps with its complementary sequence on the mRNA molecule, ensuring that the appropriate amino acid is inserted into the protein.

transferrin iron binding glycoproteins in blood plasma that reversibly bind iron to regulate levels of free iron. In trypanosomatids such as *T. brucei*, endocytosed via transferrin receptors in the flagellar pocket.

transforming growth factor beta (TGFβ) pleiotropic anti-inflammatory cytokine that suppresses T helper cell activity and promotes the differentiation and maintenance of Tregs.

transgenesis introduction of an exogenous gene (transgene) into a living organism in order that it will exhibit a desired phenotype, transmittable to its descendants.

transgenic an organism with one or more genes that have been transferred to it from another organism using recombinant DNA techniques. Animals usually are made transgenic by having a small sequence of foreign DNA injected into a fertilized egg or developing embryo. Transgenic plants can be made by introducing foreign DNA into a variety of different tissues. *See* **genetically modified organism**.

transient transfection short-term introduction of foreign DNA (e.g. a reporter gene) into a cell. Segregation of this DNA is not selected for and thus does not get carried through mitosis/meiosis, and the number of cells containing that DNA diminishes over time.

transition a point mutation in the DNA in which replacement is by a similar nucleotide i.e. a purine (A and G) by a purine or a pyrimidine (C or T) by a pyrimidine. Transitions happen more often than transversions. The dissimilar rates of mutation can be incorporated in phylogenetic inference by various weighting schemes.

translation elongation factors proteins that facilitate the steps involved in the protein translation elongation. EF-Tu allows the correct aminoacyl-tRNA to enter the ribosome, EF-Ts is the nucleotide exchange factor for EF-Tu, and EF-G is the translocase that allows the mRNA and tRNA to move one codon along the ribosome after each round of elongation.

translation initiation factors a large group of proteins responsible for effective initiation of protein synthesis. In eukaryotes, eIF-4A (eukaryotic initiation factor) is a helicase that prepares the mRNA for ribosome binding eIF-4B is responsible for the recognition of the mRNA cap, eIF-4E and eIF-4G direct ribosomes to the mRNA cap, eIF-5A stimulates formation of the first peptide bond in translation, and eiF1 prevents premature association of the large and small ribosomal subunits to increase specificity in initiation codon recognition.

translocation a type of chromosomal abnormality in which a chromosome breaks and a portion of it reattaches to a different chromosome. Chromosomal translocations can be detected by analysing karyotypes of the affected cells.

transmammary transmission transmission via the milk/mammary gland. Puppies become infected with the larvae of *Ancylostoma caninum* while nursing. Some percentages of larvae that enter a dog persist in the dog's muscles. These larvae are reactivated (probably hormonally) towards the end of gestation and make their way to the mammary glands where they are transferred to the suckling puppies.

transmigration pathway a cellular process in which circulating lymphocytes bind to and then migrate across blood vessel walls to enter body tissues.

transmission any mechanism by which an infectious agent is spread from one host or reservoir to another host. This may be direct or indirect (that is, vehicle-borne, vector-borne, or airborne).

transmission disequilibrium test family-based test of allelic association that looks for a bias in transmission of particular alleles from heterozygous parents to affected offspring.

transmission hot spots areas in which transmission intensity is higher than that in the surrounding area.

transmission potential average number of transmission stage parasites produced during the life time of infection; a feature of infection duration and productivity.

transmission risk the qualitative description of an individual's or population's ability to spread parasite.

transmission stage the life cycle stage of a pathogen or parasite that is transmitted between hosts.

transmission threshold a threshold variable in vector-borne infections that defines the vector biting density below which infection cannot persist.

transmission time time frame occurring from the initiation of feeding by an arthropod vector and the inoculation of an infectious agent in the host. It may vary widely according to vectors and pathogens.

transmission-blocking intervention (TBI) this can be in the form of a drug, vaccine, or other measure that reduces the probability that the malaria parasite is successfully transmitted through the mosquito vectors.

transmission-blocking vaccines unlike the canonical vaccines, which protect the vaccinee, vaccines in this group do not protect the vaccinated individual from infection. These 'alternative' vaccines aim at preventing the spread of the pathogen from an infected individual to a non-infected individual, eliciting herd immunity if a large proportion of the target population is immunized.

transovarial passage the passage or transmission of a microorganism from parent arthropod to offspring arthropod (e.g. tick and *Babesia*).

transport host an organism that merely carries the non-developing parasite for sometimes without further development to the next host (e.g. earth worm may ingest eggs or larvae and disseminate them as they passing through it's gut). *See* **paratenic host**.

transporter a protein that permits or promotes movement within the cell or across membranes. Can act as a permease or as a mediator of transport vesicle fusion events.

transporter protein proteins that bind fragments of endogenous antigen and carry them to newly assembled MHC class I molecules in the endoplasmic reticulum.

transporters associated with antigen processing (TAP1 and TAP2) ATP-binding cassette proteins involved in transporting short peptides from the cytosol into the lumen of the ER, where they associate with MHC class I molecules.

transposable element (TEs) mobile genetic elements (DNA sequences) that can move within and between genomes. TEs are grouped into two major categories: Class I elements (retrotransposons) transpose through an RNA intermediate, whereas Class II elements (DNA transposons) transpose directly as DNA. Class I elements are further categorized into long terminal repeat (LTR) and non-LTR retrotransposons.

transposase-based library construction a procedure to generate a sequencing library using the transposase Tn5 to insert common transposon sequences to DNA. DNA segments are then amplified by annealing of primers to the transposon sequences.

transposons segments of DNA that can move around to different positions in the genome of a single cell. These DNA elements that are variable in structure 100 and 101, but the typical composite transposon is a region of DNA flanked by insertion sequence (IS) elements. The IS transposase (a type of recombinase or integrase) catalyses transposon excision and integration into other DNA elements (e.g. chromosome or plasmid). They vary in size depending on cargo genes (the variable component of the transposon such as antibiotic resistance or detoxification genes or virulence determinants) that can be encoded within the transposon.

trans-sialidase (TS) enzyme expressed by all African trypanosomes and *Trypanosoma cruzi*, and used to transfer host sialic acids (Sas) onto trypanosome surface molecules.

trans-stadial across or between stages of a process or disease.

trans-stadial passage passage or transmission of a microorganism from one life stage to the successive one. In the case of ticks, trans-stadial transmission occurs from larva to nymph and from nymph to adult.

trans-stadial vector transmission occurs when an infection is picked up by one stage in the vector's life cycle and transmitted to succeeding stages in its metamorphosis.

trans-stimulation phenomenon in which the presence of a substance at the trans face of the membrane (i.e. the face towards which transport is occurring) can stimulate the flow of substrate from the cis to the trans direction.

transverse arcs curved, antiparallel actomyosin bundles that interact with dorsal stress fibres and flow inward towards the cell centre.

transversion a point mutation in the DNA in which replacement is by a dissimilar nucleotide,i .e. a purine (A or G) is replaced by a pyrimidine (C or T) or vice versa. *See* **transition**.

treatment application of medicinal products (medication) as deemed necessary based on a given diagnosis.

tree incongruence when phylogenetic trees derived from different loci or different genetic markers are very different for the same set of organisms.

tree-based methods model-free or non-parametric machine-learning approaches for regression and classification analyses by recursive partitioning of variables into tree structures. Popular applications in epistasis studies include random forest, random jungle, classification and regression trees.

T_{reg} (regulatory T cells) subsets of $CD4^+$ T cells that regulate the immune response. They can either be generated during maturation in the thymus, or can be induced in the periphery from naive T cells. Several subsets exist, producing inhibitory cytokines such as IL-10, transforming growth factor β (TGFβ), and IL-35, or inducing co-inhibition via, for example, cytotoxic T-lymphocyte antigen 4 (CTLA-4).

trematoda class Trematoda of the phylum Platyhelminthes which are parasitic, leaf-shaped flatworms (flukes) mainly in the digestive tract of all classes of vertebrates. Most are hermaphroditic, with male and female sex organs in a single body except the schistosomes. Their digestive systems are incomplete and have oral opening and ceca (caecum, a sac-like tube), but no anus. Their life cycle is complex; snails serve as first intermediate hosts and other aquatic animals or plants as second intermediate hosts. Most adult parasitic trematodes are endoparasites. Exceptions include the monogenetic trematodes, which are ectoparasites of fish, amphibians, and reptiles.

TRF1 and TRF2 TTAGGG repeat-binding factors 1 and 2, which bind to telomeric double-stranded DNA as homodimers and to other telomeric proteins. TRF1 is involved in the cis-inhibition of TERT, whereas TRF2 protects chromosomes against end-to-end fusion and has a role in cell cycle progression and in t-loop formation.

trisomy the presence of three copies of a given chromosome.

triatomid bug a vector of Chagas disease.

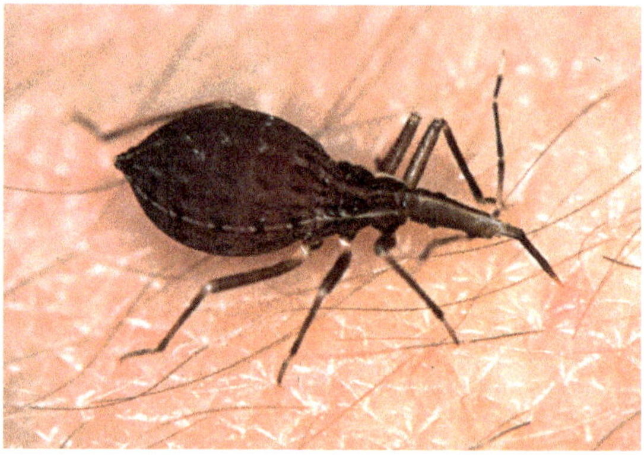

tribe a group of closely related genera ranking below the subfamily level.

tricarboxylic acid (TCA) cycle aerobic metabolic cycle used by procyclic but not bloodstream trypanosomes to release energy from amino acids.

Trichinella spiralis* and *Trichinella britovi are the two most common species of the genus *Trichinella* persisting in the European wildlife.

trichinellosis also called trichinosis, a food-borne zoonotic disease transmitted primarily by the consumption of undercooked or raw pork meat.

Trichodina a group of dorsal-ventrally flattened oval ciliated protozoan parasites of marine and freshwater species of fish. A distinguishable feature of these parasites is the presence of a prominent denticular or 'tooth-like' internal cytoskeleton ring. Heavy infestation can cause serious pathology and even death especially in small fish and fry.

trichomonad a member of the Trichomonadida subgroup of parabasalid protists.

trichomoniasis is a very common STD that is easy to cure. Most men and women who are infected with trichomoniasis do not exhibit symptoms. Without treatment, trichomoniasis can increase a person's chances of getting or spreading other STDs. Pregnant women with trichomoniasis can deliver premature (early) or low birth weight babies.

Trichostrongyles bursated nematodes that inhabit alimentary tract of animals and birds. They are small, hair-like worms that measure up to 0.7 mm long.

Trichostrongylus tenius a nematode infecting chicken, guinea fowl, pheasant, quail, and many others.

trichurid a nematode belonging to the Trichuridae family.

***Trichuris trichiura* (whipworm)** adult worms are whip-like with thin anterior ends that wind through the mucosa. Adults (30–45 mm in length) are found in the caecum and proximal colon. Three species of whipworms are found in ruminants: *Trichuris ovis*, *T. discolor* and *T. globulosa*, while *T. suis* occurs in pigs. Whipworms in farm livestock are usually considered relatively harmless. In fact, induced *T. suis* infections are evaluated in human medicine for amelioration of chronic inflammatory bowel disease, since whipworm induced Th2 immune responses dampen harmful Th1 activity in some patients.

TRIF-dependent signalling LR-mediated signalling pathway dependent on the adapter protein TRIF (TIR-domain-containing adapter-inducing interferon-β) and independent of MyD88.

triglycerides fat molecules composed of a sugar alcohol (glycerol) linked to three molecules of fatty acids through ester bonds between the hydroxide groups of the glycerol and the carboxyl group of the fatty acid.

triplet repeat diseases a class of hereditary disorders (including fragile X, Huntington's disease, spinocerebellar ataxia and cleidocranial dysplasia) originally characterized by 'genetic anticipation', a peculiar pattern of inheritance in which symptoms become more severe and appear at an earlier age as the disease is passed from one generation to the next. The cause is now understood to be extreme pathological expansion of DNA triplets that encode homopolymeric amino acid stretches, commonly glutamine or alanine.

trisomy having three copies of a particular chromosome in each somatic (body) cell instead of the normal two copies. This leads to certain conditions, for example Down syndrome (trisomy 21) or Edwards syndrome (trisomy 18).

Tritrichomonas a genus of trichomonad flagellates that are commensals or parasites of mammals and amphibia. Examples include *Tritrichomonas foetus*, *Tritrichomonas augusta* and *Tritrichomonas muris*.

tRNA synthetase one of a family of enzymes that catalyse the ligation of an amino acid to its cognate tRNA to form an aminoacyl-tRNA (also referred to as charged tRNA) ready for use in protein translation. In most cases a distinct tRNA synthetase is required for each amino acid.

tRNAs these non-coding adaptor-like RNA molecules carry amino acids to the ribosome and facilitate protein synthesis by complementary base pairing with the codons in the mRNA template.

troglostrongylosis the disease caused by nematodes of the genus *Troglostrongylus* in wild felids. In the domestic cat this occasional condition is often characterized by severe coughing, dyspnoea, polypnoea, respiratory failure, and death.

Troglostrongylus brevior a neglected feline lungworm species, which has been increasingly reported in the Mediterranean area, although scant data are available on the respiratory alterations it causes in cats.

trogocytosis a process whereby lymphocytes (B, T and NK cells) conjugated to antigen-presenting cells extract surface molecules from these cells and re-express them on their own cell surface.

trombiculiasis an infestation caused by larval members of the family Trombiculidae, common called chigger mites.

trophic niche is the space occupied by an organism within the environment in relation to its food and the consumers that prey upon it.

trophic transmission transmission by direct ingestion, usually by ingestion of one host by the next.

trophozoite **1** also known as the 'vegetative stage', this term is used to denote the feeding and dividing form of many protozoan parasites. Trophozoites are usually non-infectious. **2** intracellular stage of the parasite asexual life cycle, before nuclear and cellular division begins (e.g. *Eimeria*). **3** motile, active vegetative stages in the host within the life cycle of many protozoan (e.g. *Giardia*).

tropism preferential colonization of specific tissues by the parasite.

troposphere region of the atmosphere closest to the Earth's surface.

true prevalence the proportion of individuals in a population which have the disease or infection at a single time point.

trypanolytic factor a component of mammalian serum that causes trypanosome lysis. In humans the trypanolytic factor is a subcomponent of the densest subfraction of HDL, HDL3, which contains ApoL1 and Hpr. More specifically it can refer to the particular protein responsible for lysis.

***Trypanosoma brucei (T. brucei)* chromosomes** these are organized as 11 pairs of 1–5 Mb chromosomes containing most of the VSG and other essential genes, in addition to a variable number of intermediate chromosomes ranging in size from 200–500 kb and 100 minichromosomes varying from 50–150 kb, which contain mainly non-transcribed VSG genes and the 177-bp minisatellite.

Trypanosoma brucei brucei a protozoan parasite and the causative agent of the wasting disease nagana in cattle in sub-Saharan Africa. This organism is not infective to humans. Most laboratory strains, including the TREU 927 and Lister 427, whose genomes have been sequenced, belong to the subspecies *T. brucei brucei*.

Trypanosoma brucei gambiense a human-infective subspecies of *T. brucei*, most commonly found in central and western Africa, which causes slow-onset chronic trypanosomiasis. This subspecies accounts for >95% of sleeping sickness cases.

Trypanosoma caninum (T. caninum) a newly described species is found in various endemic regions in Brazil, but data regarding its prevalence rates are limited.

Trypanosoma cruzi the causative agent of Chagas disease, a zoonotic infection that can be transmitted to humans by blood-sucking triatomine bugs.

Trypanosoma equiperdum (related to T. brucei) the aetiological agent of dourinein in equines and related animals spread primarily by sexual transmission. Vertical transmission has been described.

Trypanosoma evansi the etiological agent of the disease known as trypanosomiasis (trypanosomosis), 'Mal das Cadeiras' or 'Surra', which affects horses. Cattle, sheep, goats, donkeys, cats and pigs are also susceptible hosts. Canines, capybaras (*Hydrochaeris hydrochaeris*), coatis (*Nasua nasua*) and vampire bats (*Desmodus rotundus*) are reservoirs and occasionally may manifest clinical signs. Additionally, bats are considered vectors of this disease. The transmission of the parasite is essentially mechanical, since trypomastigotes are transferred from one host to another through blood-sucking insects (flies of the families Tabanidae and Muscidae) or artificially by needles contaminated with infected blood. Clinical signs include weight loss, pale mucous membranes, intermittent fever, cough, swelling in the lower parts of the body, enlarged superficial lymph nodes, muscle atrophy, incoordination, hindquart's paresis, difficulty to get up and proprioceptive.

Trypanosoma brucei rhodesiense a virulent human-infective subspecies of *T. brucei*, most commonly found in southern and eastern Africa, which causes acute human African trypanosomiasis. Adding one specific gene (SRA) to *T. b. brucei* is sufficient to convert it to *T. b. rhodesiense*.

trypanosomatids a group of parasitic protozoa that infect mammals (*Trypanosoma* spp. and *Leishmania* spp.), plants (*Phytomonas* spp.) and insects (*Crithidia* spp.).

trypanosome lytic factors (TLF) human serum high-density lipoproteins that cause lysis and killing of *T. brucei* parasites that infect animals. The toxic components of TLF1 are apolipoprotein L-1 and haptoglobin-related protein. Human-pathogenic *T. brucei* subspecies have evolved mechanisms to resist killing by these factors. *See* **trypanolytic factor**.

trypanotolerance the ability to survive, reproduce and remain productive under trypanosomiasis risk without the need for chemicals to control the vector or drugs to control the parasite. Trypanotolerance can be either innate resistance or resilience to trypanosome infections.

trypomastigote a morphotype characterized by a prolonged cell shape with apposed flagellum emerging from a lateral opening of the posterior flagellar pocket and the post-nuclear kDNA. It is a form of trypanosome, including bloodstream, metacyclic, and procyclic stages, in which the kinetoplast is posterior to the nucleus.

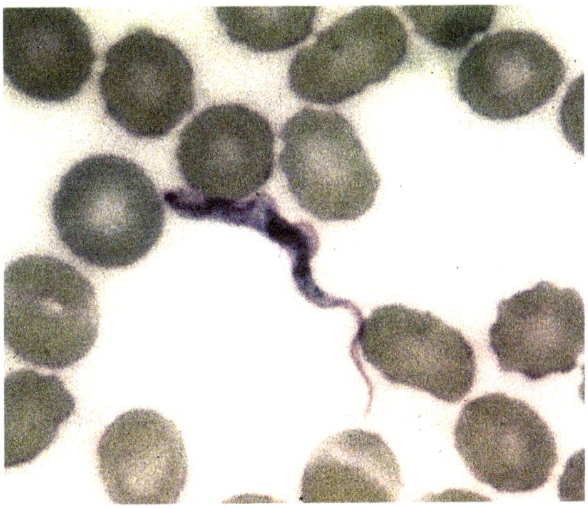

trypomastigote small surface antigen (TSSA) a surface glycoprotein expressed in the trypomastigote stage of the *T. cruzi* life cycle and which is involved in host cell invasion.

TSD (target site duplication) LINE flanking sequences corresponding to the sequence between the two single-strand nicks performed by a LINE-encoded endonuclease at the target site of the element insertion.

tsetse fly the blood feeding insect of the genus *Glossina* in the superfamily Hippoboscidae. It acts as a host (vector) of the African trypanosomes *T. b. brucei* and *T. congolense* in sub-Saharan Africa.

tuberculin a protein fraction of the *Mycobacterium tuberculosis* that is used in a skin test for tuberculosis.

tubulovesicular network (TVN) one of two main features of the exomembrane system. It is formed by extensions of the parasitophorous vacuole membrane into the host cell cytoplasm. The TVN has a role in protein trafficking from the parasite to the host cell.

tumour necrosis factor (TNF)-α cytokine produced by macrophages, natural killer cells, and T cells which promotes inflammatory pathways. Through TNF receptors, it promotes cytotoxicity, induces angiogenesis, bone resorption, and thrombotic processes.

tumour necrosis family receptor 1 (TNFR1) a member of the TNFR superfamily that contains a cytoplasmic death domain. It serves as the main receptor for TNF-α and plays roles in apoptosis and NF-κB activation.

tumour suppressor gene directs the production of a protein that is part of the system that regulates cell division. The tumour suppressor protein plays a role in keeping cell division in check. When mutated, a tumour suppressor gene is unable to do its job, and as a result uncontrolled cell growth may occur. This may contribute to the development of a cancer.

TUNEL (terminal deoxynucleotidyl transferase biotin-dUTP nick end labelling) an enzymatic method for labelling the ends of broken DNA strands. Detects the DNA cleavage characteristic of apoptosis.

two-allele mate choice different alleles cause mating in each of two environments. This can only lead to assortative mating when mating alleles are in linkage disequilibrium with alleles under disruptive selection.

two-dimensional (2D) cell culture systems classical culture systems in which cells settle on a 2D plastic or glass surface.

type 1 diabetes mellitus an autoimmune disease in which there is inflammation of the endocrine pancreas leading to the destruction of β cells (by antibodies) of the islets of Langerhans and insulin insufficiency.

type 1 T helper cell (Th1) cells functionally distinct subset of CD4 T helper cells associated with IFN-γ and TNF-α expression and resistance to intracellular pathogens.

type 2 T helper (Th2) cells functionally distinct subset of CD4 T helper cells associated with IL-4, IL-5 and IL-13 expression and the induction of allergic responses. Th2 cytokines mainly defend a host against extracellular parasitic infection (e.g. multicellular helminths with IgE, and mucosal with IgA).

type 17 T helper (Th17) cells functionally distinct subset of CD4 T helper cells associated with IL-17 expression and resistance to bacterial and fungal pathogens.

type I error the errors that result when there are false positives or when falsely rejecting the null hypothesis.

type I nitroreductases FMN-dependent NTRs that catalyse the reduction of nitroaromatic compounds by a two-electron mechanism. Type I NTRs are oxygen-insensitive, being able to carry out reactions under both aerobic and anaerobic conditions.

type I ostertagiasis clinical parasitic disease occurs in calves when they ingest enough *Ostertagia ostertagi* L3 (third-stage larvae).

type II error the errors that result when there are false negatives or incorrect failure to reject the null hypothesis.

type II functional response a functional response is the consumption rate of food by a consumer, as a function of prey density. It is a key component of predator–prey theory that models the dynamical relation between predators and their prey. A type II response results from circumstances where food intake rate per consumer saturates with increasing prey density, for example, wolves preying on moose.

type II nitroreductases NTRs that catalyse the reduction of nitroaromatic compounds by a one-electron mechanism. In the presence of oxygen, this reduction leads to the production of superoxide anions.

type II ostertagiasis also known as PPR or Spring rise. Occurs in animals that are old enough to have arrested larvae from previous grazing season.

type-2 cytokines cytokines that stimulate the humoral immune system (e.g. IL-4, IL-5 and IL-10).

typical host one in which the parasite is commonly found and in which the parasite can continue its development (or the appropriate phase of its development) necessary for subsequent completion of its life cycle.

U937 cells an immortalized monocyte-like cell line of human origin.

ubiquitin a small protein found in almost all cellular tissues that helps to regulate the processing of other proteins in the body.

ubiquitination a dynamic post-translational modification (the addition of ubiquitin) critical to many subcellular processes including protein turn over and cell signalling.

ulcer an open sore on the skin or mucous membrane characterized by the disintegration of the tissue and the discharge of pus.

ultra-conserved elements regions of the genome with ≥200 consecutive nucleotides conserved with 100% identity across multiple species. Some of these regions have been argued to have functional relevance as enhancers and to have evolved under purifying selection.

unbalanced gametes gametes that carry chromosomal duplications or deficiencies.

uncomplicated malaria an illness that is caused by the malaria parasite and is characterized by fever and other general unspecific symptoms, such as headache and malaise, but lacking systemic or organ complications, which threatens the life of the patient.

underdominant mutation a mutation that causes reduced fitness when heterozygous.

underdominant selection selection against heterozygotes, as might occur with chromosome rearrangements.

ungulates mammals with hooves.

unilocular cyst containing only one cavity. In echinococcal tapeworms, an intermediate host larval stage (cyst) having only a single cavity.

uninuclear having only one nucleus.

uniseriate in a single row.

unmappable reads the reads that are not aligned to the organism genome.

unnatural host a synonym for abnormal host.

unresponsiveness the inability to respond to antigenic stimulus. Unresponsiveness may be specific for a particular antigen, or broadly non-specific as a result of damage to the immune system, for example after whole body irradiation. *See* **tolerance**.

unspecialised having no specific function.

unstable malaria transmission malaria transmission is considered unstable when the force of infection is insufficient to produce moderate or high levels of infection each year. Areas of unstable malaria show substantial variations in the incidence of malaria from year to year and are prone to epidemics.

unsupervised analysis a form of gene expression analysis that involves discovery of empirical structure (patterns) in a given data set without regard to prior knowledge of the underlying biology.

uPA (urokinase-type plasminogen activator) serine protease that activates plasminogen into plasmin, resulting in fibrinolysis.

uPA receptor or CD87 glycosylphosphatidylinositol-anchored receptor for urokinase-type plasminogen activator.

upper respiratory tract the nasopharynx, oral cavity, and throat.

upstream open reading frames (uORF) short open reading frames located upstream of coding sequences with a role in translational regulation of transcripts.

upwards incorporation an extension of the life cycle by incorporating a new host higher up the food chain, which then becomes the definitive host, with the original definitive host becoming an intermediate host.

uracil (U) one of four nitrogenous pyrimidine bases that are part of RNA. The other three bases are adenine (A), cytosine (C), and guanine (G). In DNA, the base thymine (T) is used in place of uracil. Uracil is capable of forming a bp with adenine. *See* **bp**, **nucleotide**, **pyrimidine**.

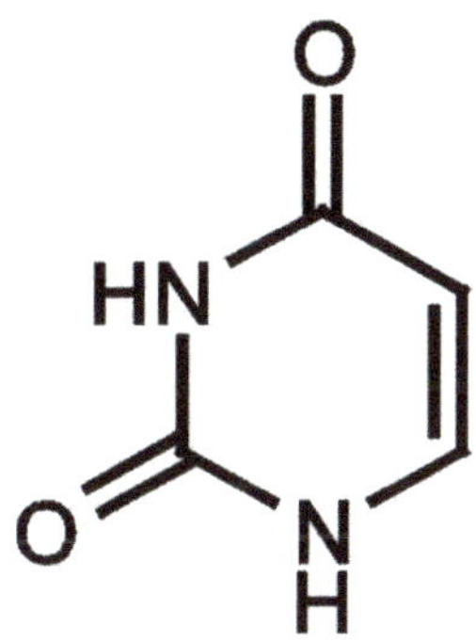

urethritis inflammation of urethra, the canal through which urine is discharged.

urinalysis physical, chemical, and microscopic examination of urine. Urinalysis can detect a specific substance, such as glucose or blood, in the urine which can be helpful in diagnosing a disease or monitoring the body's response to treatment.

urinary schistosomiasis *S. haematobium* is the causative agent of urinary schistosomiasis that is characterized by the presence of eggs and blood in urine.

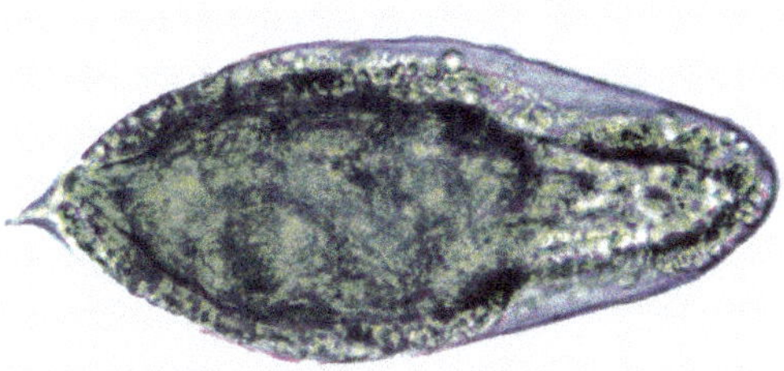

urocanic acid an intermediate of histidine catabolism with two isoforms of which the *trans* form is the naturally occurring form. *Trans*-urocanic acid is isomerized into the *cis* form by UVB light. *Trans*-urocanic acid acts as a natural sunscreen, providing a low-level protection against UVB-induced DNA damage. *Cis*-urocanic acid is proposed to have immunosuppressive properties because it activates T regulatory cells, decreases the density of Langerhans cells, and alters their morphology. Also, it can initiate production of reactive oxygen species and cause oxidative DNA damage.

urticaria an allergic skin reaction with changing, raised, itchy rashes and burning sensations. When deeper skin layers are affected, the inflammatory reaction produces a swelling of the tissue (= angioedema). Acute urticaria/angioedema can be a manifestation of food allergy, when exposed patients display specific IgE for a specific food agent. The onset and disappearance are often sudden.

user interface a system that controls the way information is shown to the user and how the user interacts with the system.

use-scenario the outcome-oriented categorization of the interaction between a user, the setting and a diagnostic tool.

utilization measure of the percentage of delivered intervention being used.

UTRs (untranslated regions) regions of mature cellular mRNAs or viral RNA genomes that do not code for proteins, including the 5′- and 3′-UTRs. UTRs contain information for the regulation of translation and mRNA stability. These regions, which are most commonly located at 5′ or 3′ ends, often contain cis-acting RNA elements that regulate RNA activities.

uveitis inflammation of the uvea, the pigmented, vascular layer of the eye.

vaccination or immunization artificial introduction of a killed or attenuated pathogen or its antigenic products to stimulate a protective immune response in the host against later infection by the corresponding pathogen. Vaccination can be intended either to prevent a disease (a preventative vaccine) or to treat a disease (a therapeutic vaccine).

vaccine a preparation that contains either whole disease-causing organisms such as parasites, bacteria, or viruses, which have been killed or weakened, or parts of such organisms, used to induce specific protective immunity against the disease that the infectious organism causes. Vaccine preparations can be natural, synthetic or derived by recombinant DNA technology.

vaccine adjuvants substances added to the vaccine which can increase the systemic and local immune response to the particular vaccine.

vaccine effectiveness a measure of the ability of the vaccine to prevent disease in the 'real world'. Vaccine effectiveness is proportional to the vaccine efficacy, but is influenced primarily by how well the target groups are immunized. Many factors can influence vaccine effectiveness, including the population immunized (age, prior exposure, genetic make-up, etc.) and vaccine characteristics (mode of delivery, accessibility, stability, etc.).

vaccine efficacy the percentage reduction of disease incidence in a vaccinated group compared to an unvaccinated group. The vaccine efficacy is measured by double-blinded clinical trial, thereby measuring the 'best-case scenario' under strict and controlled conditions.

vacuole a membrane bounded space or cavity in the cytoplasm of a cell usually functioning in collecting and digesting food taken into the organism and eliminating wastes. Vacuole contents and/or morphology may be helpful in identification of some of the intestinal protozoa (e.g. *Entamoeba*, *Iodamoeba*). *See* **contractile vacuole**, **endocytosis**.

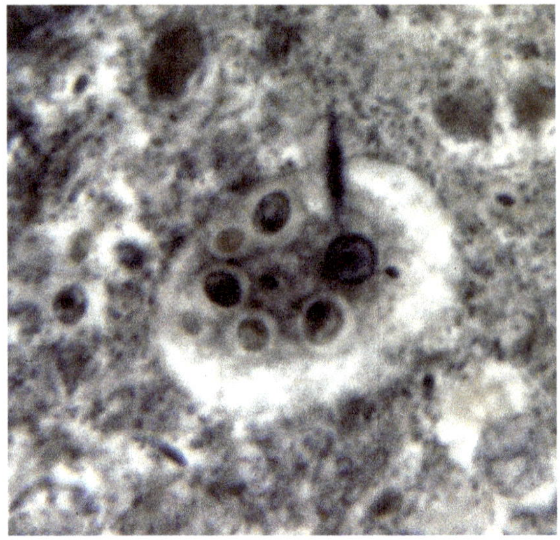

vaginitis inflammation of the vagina.

validation testing any new system or upgrade against data known to be correct. Validation is crucial to implementing any new system, maintaining systems after upgrades or changes to any of the data sources that feed into a system.

validity the degree to which the results of a study are likely to be true, believable and free of bias. This is entirely independent of the precision of the results and does not predict the results. The internal validity of a study refers to the integrity of the experimental design.

Van der Waals bonds weak intermolecular bonds resulting from the attraction between electron-rich regions of one molecule and electron-poor regions of another.

var genes *P. falciparum* genes encoding PfEMP1, which shows antigenic variation.

variable (V) gene segments genes that rearrange together with D (diversity) and J (joining) gene segments in order to encode the variable region amino acid sequences of immunoglobulins and T-cell receptors.

variable region (V region) part of the immunoglobulin or TCR peptide chains where the amino acid sequence shows significant variation between molecules.

variable surface antigens (VSAs) proteins, encoded for by the parasitic genome that are expressed on the surface of parasitized erythrocytes. VSAs promote the adhesion of parasitized erythrocytes to the endothelium. VSAs are also antigenic for the host immune system and, therefore, the parasite has evolved a variety of VSA genes to mitigate host immune defences.

variable surface glycoprotein coat a highly variable physical barrier made up of approximately 10^7 identical GPI anchored molecules, which shield the plasma membrane and other surface proteins from interaction with antibodies and toxins.

variable surface glycoprotein expression sites a group of approximately 20 telomeric expression sites (ESs) used for VSG expression for antigenic variation in *Trypanosoma brucei*.

variable-number tandem repeat (VNTR) short tandem repeats of a repetitive nucleotide sequence that are dispersed throughout the genome. The repeated sequence varies in length and copy number.

variance heterogeneity difference in variance of a quantitative trait among the three possible genotypes of a biallelic SNPs in the presence of genetic interactions. It can be used to screen for potential interacting SNPs.

variant surface glycoprotein (VSG) a major surface coat protein of the trypomastigote that mediates antigenic variation.

Varroa an external parasitic mite that attacks the honey bees *Apis cerana* and *Apis mellifera*. The adult female mites are reddish-brown to dark brown and oval in shape, measuring 1.00 to 1.77 mm long and 1.50 to 1.99 mm wide.

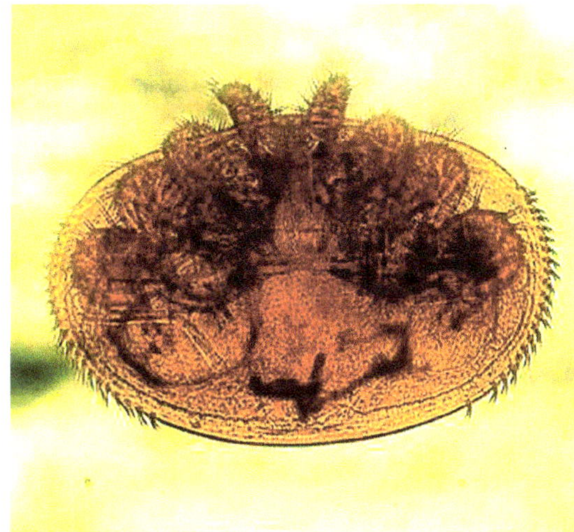

vas deferens a tubular structure that conveys sperm from the epididymis to the urethra of the penis.

vascular addressins cell adhesion molecules present on the luminal surface of blood and lymph vessel endothelium recognized by homing molecules, which direct leucocytes to tissues with the appropriate 'address'.

vascular endothelial growth factor (VEGF) growth factor that stimulates angiogenesis through binding to its receptor, (VEGFR2). Previously also named vascular permeability factor (VPF), for its ability to activate endothelial cells and induce oedema formation.

vasculitis inflammation of blood vessels leading to lesions on the skin and mucous membranes.

vasoactive amines substances including histamine and 5-hydroxytryptamine, which increase vascular permeability and smooth muscle contraction.

vector 1 a plasmid or virus used in genetic engineering or molecular cloning to insert foreign DNA into a host cell. **2** an agent, usually an arthropod or other animal, able to carry pathogens from one infected host to another. When the final or intermediate host is an arthropod it is termed a vector, e.g. ticks act as vector host for *Babesia* and mosquitoes act as vector for malaria parasites. *See* **biological vector**, **cloning vector**, **mechanical vector**.

vector competence the intrinsic ability of a vector to acquire and transmit pathogens from infected hosts to uninfected vectors and from infected vectors to uninfected hosts.

vector control tool (VCT) intervention that reduces the ability of an insect vector to transmit disease.

vector dependent the mode of transmission in which a parasite is dependent on a vector, such as a mosquito or leech, for transmission from one host to the next. Development of the parasite might or might not occur within the vector.

vector independent the mode of transmission of a parasite between hosts that does not depend on a vector for any stage of spread or development. This also includes the direct or indirect transmission of a parasite conventionally thought of as being vector dependent.

vector-borne disease (VBD) a disease dispersed by blood-feeding arthropods and is caused by their transfer of infectious microbes.

vectorial capacity the ability of a vector to transmit a pathogen.

vectorial competence the susceptibility of a vector host (e.g. a mosquito) to be infected by a pathogen.

vectorial system refers to all genetically competent vector species and populations that play an active role in the transmission of an infectious agent/pathogen in an ecosystem (defined in time and space).

vehicle a non-living source of pathogens which infect large numbers of individuals. Common vehicles are food and water.

ventral stress fibres antiparallel actomyosin bundles anchored to focal adhesions at both ends.

venus flytrap module (VFT) a bilobate, ligand-binding extracellular domain found in many G protein-coupled receptors (seven-transmembrane receptor molecules). Closure of the VFT module upon ligand binding resembles the closure of the terminal lobes of the carnivorous plant venus flytrap (*Dionae muscipula*) as they trap insects.

verapamil (VPL) a calcium channel blocker used to treat hypertension and cardiac arrhythmia that is also well known for 'resistance reversal' effects in some strains of chloroquine resistant *P. falciparum*.

verminous mesenteric arteritis a condition caused by migrating *Strongylus vulgaris* larvae, resulting in a defective blood supply to the intestine which causes intermittent colic and sometimes terminal infarction.

verminous pneumonia (lungworm infection or husk) a group of parasitic nematodes that colonize the lower respiratory tract of livestock. They cause high morbidity, high mortality, and high economic costs throughout the world especially in temperate regions. The disease caused by lungworms has many local names including verminous bronchitis, verminous pneumonia, parasitic bronchitis, hoose, husk. Several factors known to affect the development and clinical manifestations of 'husk' include animal age, previous immunization or infection, presence of passively acquired antibody, climatic condition, and anthelminthic treatment. Signs of lungworm infection range from moderate coughing with slightly increased respiratory rates to severe persistent coughing and respiratory distress and even failure. The major lungworms of livestock belong to one of two superfamilies, Trichostrongyloidea or Metastrongyloidea.

vertebrate complement part of the innate immune system of vertebrates, composed of serum proteins including 'Factor C3' that upon activation lead to tagging of a pathogen for destruction.

vertical programmes (also known as stand-alone programmes or the vertical approach) refer to disease control programmes focusing on a single given health problem or a specific demographic population, usually implemented in parallel and in addition to other normal health activities.

vertical transmission 1 the transmission of material from parent to offspring. Maternal transmission is a subtype, involving the specific passage from mother to child. 2 transmission of a parasite from one generation to the next through the egg or *in utero*. See **horizontal transmission**.

vesical plexus a network of veins that are distributed to the bladder and adjacent reproductive organs.

vesicle 1 a small bulla or blister containing clear fluid. 2 a small membrane-delimited sac that buds off of one membrane, carrying cargo, and fuses with another. 3 small circumscribed fluid-filled elevations on the skin.

VH the variable region of the Ig heavy chain, resulting from the recombination of a V, D and J segment.

viable alive and able to reproduce.

viable count measurement of the concentration of live cells in a microbial population.

villi minute, elongated projections from the surface of intestinal mucosa that are important in absorption as they increase intestinal surface. The cells that cover the villi are enterocytes. The smaller villi on enterocytes are called microvilli.

vinca alkaloids class of anticancer drugs originally derived from *Catharanthus roseus*; well known for antimitotic effects via inhibition of tubulin polymerization.

vinegar eel free-living nematode, *Turbatrix aceti*, occasionally occurring as a contaminant in laboratory solutions and home-made vinegar.

virulence 1 the degree of pathogenicity or disease-producing ability of a microorganism. 2 the negative effects caused by the parasite on the host animal in terms of survival or fecundity. 3 the reduction in host health or fitness owing to infection. 4 the relative capacity of a parasite to produce pathogenic effects or to invade. 5 the ability of a pathogen to cause disease in a given host.

virus a genetic element containing either DNA or RNA covered by protein that is able to alternate between intracellular (virus) and extracellular (virion) states, the latter being the infectious state. Inside the infected cell, the virus uses the synthetic capability of the host to produce progeny virus. See **cloning vector**.

visceral pertaining to the internal organs of the body, especially those within the abdominal cavity.

visceral larva migrans a condition caused by migration of nematode larvae in human visceral tissues, commonly caused by accidental ingestion of the eggs of the roundworm *Toxocara canis* or *Toxocara cati*. It is often characterized by high peripheral eosinophilia. The life cycle cannot be completed in the humans because humans are not the natural hosts (i.e. unsuitable for development to the adult stage).

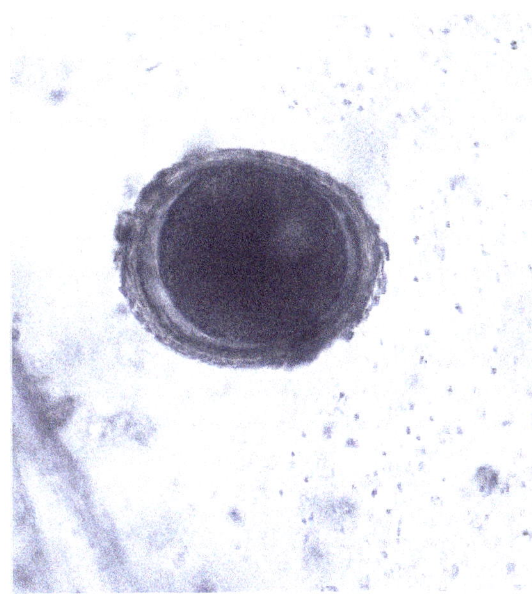

visceral leishmaniasis or leishmaniosis (VL) a sand fly vector-borne disease of dogs and humans caused by infection with parasites of genus *Leishmania*. The parasite infects reticuloendothelial system. In dogs, *Leishmania infantum* (believed to be closely related to *L. donovani*) infections may result in fatal disease and infected domestic dogs are the most important reservoir host for human infections. Leishmaniosis caused by *L. infantum* is endemic in many parts of the world. In humans, affected individuals present with fever, malaise, and weight loss associated with anaemia, hepatosplenomegaly, hypergammaglobulinaemia, and concomitant hypoalbuminaemia, and eventually death if left untreated. *See* **kala-azar**.

viscerotropic the affinity or tendency of an organism to reside within the abdominal or thoracic viscera.

viscus any of the organs within one of the four major body cavities (cranium, thorax, abdomen, and pelvis).

visualization a technique for assessing variation among DNA segments (genetic markers). Methods include radiolabeling (exposure of gels to X-ray film) and various stains (ethidium bromide, silver stains, etc.).

vitamin any organic compound which is necessary for the normal metabolism and growth of living organisms. It functions as a coenzyme or a component of a coenzyme and is required by organisms in small quantities. All organisms require some vitamins of different varieties; they can either synthesize vitamins by themselves or they have to obtain them from the growth medium. Examples of some important vitamins are: nicotinic acid, riboflavin, pantothenic acid, biotin, thiamine, B12, folic acid, etc.

vitamin B12 or cobalamin a water-soluble B vitamin that is important for blood formation and normal functioning of nerves.

vitamin B9 or folic acid a water-soluble B vitamin that is important for cell division and growth. Folic acid is activated when converted to dihydrofolate, which is transformed into tetrahydrofolate, the active form in one carbon pool by folate (also known as the folate cycle).

vitamin D receptor (VDR) a nuclear receptor, which upon activation by dihydroxylated vitamin D_3, forms a heterodimer with retinoid X receptor (RXR) and causes upregulation of arginase 1 and cathelicidin (LL-37), along with subsequent down-regulation of proinflammatory genes.

vitamin D response elements (VDREs) a DNA sequence found in the promoter region of vitamin D regulated genes. It is present in many cell types including monocytes, macrophages, and dendritic cells.

vitellaria or vitelline glands a group of glands in flatworms (Phylum: Platyhelminthes) that secrete yolk materials around the fertilized egg. Their arrangement and position is often a diagnostic feature for digenetic trematodes.

vitelline membrane the innermost layer in the shell of fertilized eggs of helminths.

viviparous a condition in which the embryo develops within the female and is nourished by the mother (no egg shell exists at any time).

VL the variable region of the Ig light chain that results from the recombination of a V kappa or V lambda segment, and a J kappa or J lambda segment.

volutin granules bodies in the peripheral cytoplasm of *Blastocystis* that stain darkly.

Wahlund effect a reduction in homozygosity (increase in heterozygosity) when distinct taxa are analysed jointly or when they hybridize. Whenever subpopulations vary in gene frequency, the population as a whole will show a Wahlund effect. The opposite effect, known as isolate breaking, occurs when divergent populations intermix. In that case, the interbreeds will show an increase in heterozygosity over the Hardy–Weinberg expectation. If two or more subpopulations have different allele frequencies, then the overall heterozygosity is lowered, even if each subpopulation is in Hardy–Weinberg equilibrium.

Walker A ATP-binding motif or P-loop one of the most common and highly conserved sequence and structural motifs found in genomes and is considered to be a fundamental and ancient functional motif in biological systems.

warble flies (or bot flies) the larvae of the bot flies *Hypoderma bovis* and *H. lineatum* which are the causative agents of a skin parasitic disease called hypodermosis.

warbles swellings under the skin produced by insect larvae of *Dermatobia* and *Hypoderma* flies. *See* **cuterebriasis**.

well-supported clade a clade is a group of closely related taxa consisting of a single common ancestor and all the descendants of that ancestor. Methods that resample a particular dataset, for example with replacement (the bootstrap), can provide confidence that particular groupings of clades are robust given the data available. Various methods of phylogenetic inference yield evolutionary trees for visualizing the phylogenetic relationships among taxa.

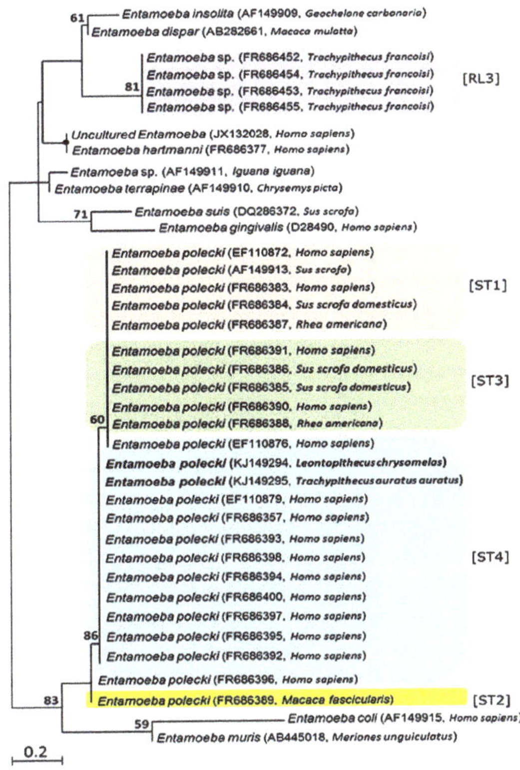

western blot a technique used to identify and detect proteins (based on their ability to bind to specific antibodies) within a complex mixture of proteins after separation by gel electrophoresis. The method involves using gel electrophoresis to separate the sample's proteins. The separated proteins are transferred out of the gel to the surface of a membrane. The membrane is exposed to an antibody specific to the target protein. Binding of the antibody is detected using a radioactive or chemical tag. A western blot is sometimes used to diagnose disease. *See* **northern blot**, **protein**, **Southern blotting**.

whipworms the common name for nematode species of the *Trichuris* genus. Whipworms live in the large intestine and eggs are passed in the faeces of infected host. If the infected host defecates outdoor or if human faeces as used as fertilizer, eggs are deposited on soil. They can then mature into a form that is infective. Whipworm infection is caused by ingesting infective eggs in contaminated food. People with heavy infection can experience painful passage of stool that contains a mixture of mucus, water and blood.

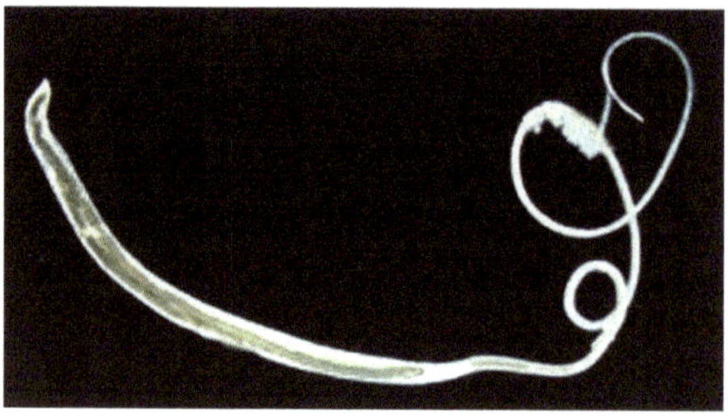

white blood cells (WBCs) immunological cells found in the blood and lymph and are key components of the immune system. WBCs are responsible for co-ordinating an adaptive immune response (TH-cells) and producing antibodies (B-cells), thus help fight infections and diseases. Examples of white blood cells include lymphocytes, neutrophils, eosinophils, basophils, macrophages, and mast cells. *See* **leucocyte**.

white pulp lymphoid-rich area of the spleen.

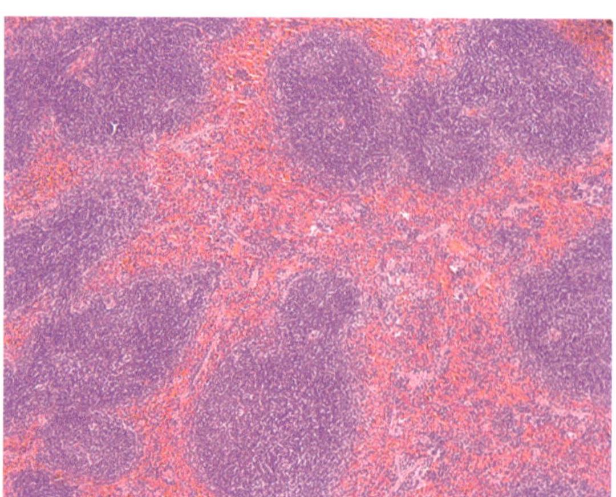

whole exome sequencing (WES) technology of next-generation sequencing that is used to determine all variants of known coding regions (exons) in the genome. Available genome sequences provide reference genomes for comparative sequencing or resequencing technologies, which are used to identify mutations and polymorphisms between organisms.

whole-genome sequencing (WGS) modern WGS relies on one or a combination of second- or third-generation sequencing technologies. Although the entire genome is sequenced, it cannot be assembled completely if a genome contains repeated stretches of DNA that are longer than the length of DNA that can be sequenced in a single read by the sequencing technology employed. Additional parts of the genome may be excluded for some analyses.

widefield epifluorescence microscopy a fluorescence microscopy where the excitatory light is passed through the objective lens and then onto the specimen, giving rise to emitted light which is focused on the detector by the same objective that is used for the excitation.

widespread clustering when a given species is subdivided through its entire range into discrete genetic clusters that are stable in space and time.

wild type or 'average' phenotype often assumed to be the 'normal' phenotype of an organism that is found in natural populations and/or any subpopulation or inbred lines derived from such a population. The genotypes producing such a phenotype are often considered wild type genotypes.

wildlife or sylvatic cycle part of a life cycle of a pathogen involving wild animals.

wild-type parasite the naturally occurring, non-mutated strain of a parasite. When exposed to antiparasitic drugs, wild-type parasite can develop mutations that make the parasite drug-resistant.

winged helix (WH) domain a variant of the highly conserved helix-turn-helix domain, found in prokaryotic and eukaryotic proteins, acting in both replication and transcription. WH is thought to mediate protein–DNA and protein–protein interactions.

Winterbottom's sign an enlargement of posterior cervical lymph nodes in patients with African trypanosomiasis.

within-patch selection the fitness outcome of growth in a given environment. In the context of infection, this is roughly the number of transmission stages a pathogen can produce during an infection cycle.

Wolbachia pipiens an intracellular endosymbiont bacteria of filarial nematodes, which has been implicated in the pathogenesis of filarial diseases, in particular in *Dirofilaria* infections.

Wood–Ljungdahl pathway (WLP) the reductive acetyl-CoA pathway in which two molecules of CO_2 are reduced and joined to form acetyl-CoA and then acetate.

worker a non-reproductive caste in eusocial insects. Workers co-operatively care for the brood of the colony, forage for food, clean up the nest and defend it against invaders. Workers constitute the body of any eusocial insect colony.

World Health Organization (WHO) the agency of the United Nations that provides global leadership on health-related issues. Responsibilities of WHO include shaping the global health research agenda, setting health standards, promoting evidence-based policy options, providing technical support to countries, and monitoring and assessing health trends.

worm count a numerical computation or indication of the total number of worms. A total worm count requires a freshly slaughtered cadaver, collection of intestinal or other fluid in an aliquot sample, counting actual worms and, by multiplication, measuring the total worm burden. In the case of lungs it is necessary to digest the tissue.

worm resistance a significant increase in the ability of worms to tolerate doses of individual drugs which have previously been fatal to the worms.

writer enzyme a protein that places post-translational modifications at specific residues of a histone tail.

X chromosome one of two sex chromosomes. Humans and most mammals have two sex chromosomes, the X and Y. Females have two X chromosomes in their cells, while males have X and Y chromosomes in their cells. Egg cells all contain an X chromosome, while sperm cells contain an X or a Y chromosome. This arrangement means that during fertilization, it is the male that determines the sex of the offspring.

X chromosome dosage compensation a process that compensates for the loss of most Y-chromosome genes after sex chromosome differentiation from ancestral autosomes.

xanthurenic acid (XA) a byproduct of eye pigment synthesis in insects.

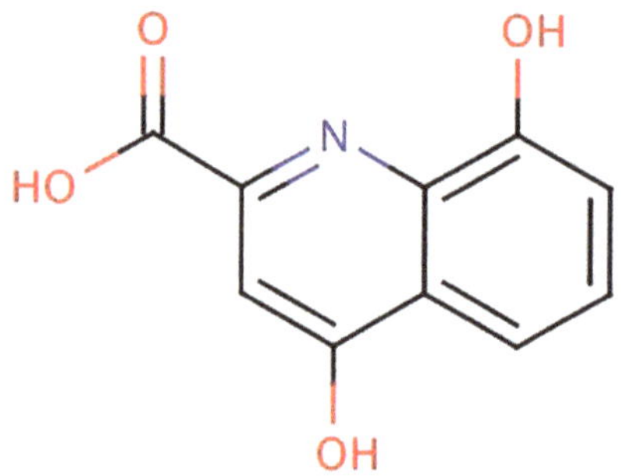

xenobiotic a foreign chemical or substance found in an organism or biological system that is not produced by that organism, but originated from another source. It is often toxic and must be destroyed and/or excreted by the organism.

xenodiagnosis a diagnostic method in which laboratory-reared (uninfected) arthropod hosts are fed on a suspected patient and subsequently examined to recover the parasite. For example, the feeding of laboratory-reared triatomid bugs on patients suspected of having Chagas disease and after several weeks, the faeces of the bugs are checked for intermediate stages of *Trypanosoma cruzi*. Also, it is used as a diagnostic method for the detection of trichinellosis or trichinosis by feeding laboratory-bred rats or mice on meat suspected of being infected with *Trichinella*, and then examining the animals for the parasite.

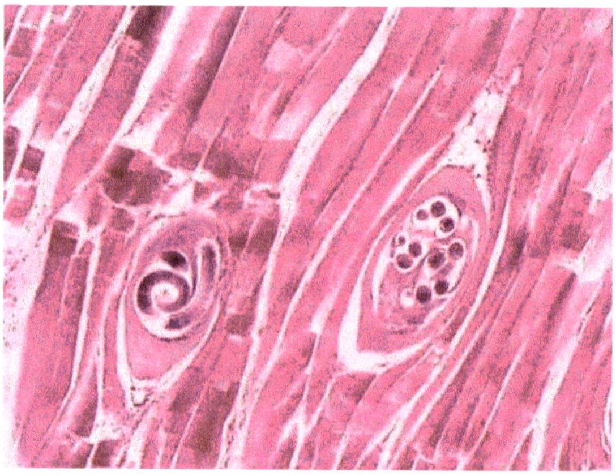

xenogeneic genetic differences between species.

xenograft a tissue or organ transplantation/graft between individuals of different species.

xenologues homologous genes transferred between strains through horizontal gene transfer. See **xenotransplantation**.

xenomonitoring testing of field-sampled vector or intermediate hosts for evidence of ongoing infectious disease transmission within human communities.

xenoparasite an organism not usually parasitic on a particular species, but becomes so because of a weakened condition of the host.

xenotransplantation any procedure that involves the transplantation of live cells, tissues or organs from one species to another, including animal-to-human transplantation.

X-linked a trait where a gene is located on the X chromosome. Humans and other mammalian species have two sex chromosomes, the X and the Y. In an X-linked or sex-linked disease, it is usually the males that are affected because they have a single copy of the X chromosome that carries the mutation. In females, the effect of the mutation can be masked by the second healthy copy of the X chromosome.

Y chromosome one of the two sex chromosomes. Humans and other mammalian species have two sex chromosomes, the X and the Y. Females have two X chromosomes in their cells, while males have X and Y chromosomes in their cells. Egg cells contain an X chromosome, while sperm cells contain an X or a Y chromosome.

yeast artificial chromosomes (YACs) are shuttle-vectors that can be amplified in bacteria and employed for the cloning and manipulation of large deoxyribonucleic acid (DNA) inserts in the yeast *Saccharomyces cerevisiae*. YACs are often used in connection with the mapping and sequencing of genomes. Segments of an organism's DNA, up to one million bps in length, can be inserted into YACs. The YACs, with their inserted DNA, are then taken up by yeast cells. As the yeast cells grow and divide, they amplify the YAC DNA, which can be isolated and used for DNA mapping and sequencing. *See* **cloning vector**.

zebrafish a member of the minnow family of fish; a model organism used to study the development of vertebrates. Zebrafish are useful because the embryo is transparent, it develops outside of its mother, and its development from eggs to larvae happens in just three days.

zero-trans experiment in which no substrate is present at one face (the trans face) of the membrane.

zinc-fingers are structurally diverse, small protein domains in which zinc plays a structural role contributing to the stability of the domain. They are found in proteins that perform a broad range of functions in many cellular processes, such as replication, repair, transcription, translation, metabolism, signalling, cell proliferation and apoptosis.

zoogeography the discipline that is concerned with the geographic distribution of animal species.

zoonosis (plural **zoonoses;** adjective **zoonotic)** diseases that are transmitted from non-human vertebrate animals to humans. For example, people can acquire cryptosporidiosis if they accidentally swallow food or water that is contaminated by oocyst-containing faeces from *Cryptosporidium*-infected animals. People can acquire trichinellosis by ingesting undercooked or raw meat from bear, boar, or domestic pigs that are infected with *Trichinella* parasite. Some dog and cat parasites can infect people (e.g. toxocariosis). Further, people can be infected by the raccoon parasite *Baylisascaris* if they accidentally swallow soil that is contaminated with infected raccoon faeces. Canine leishmaniosis is another zoonotic disease caused by *Leishmania* parasites transmitted by the bite of an infected phlebotomine sand fly.

zoonotic agent an organism that causes a zoonosis. Zoonotic agents include viruses, bacteria, parasites (e.g. *Cryptosporidium*), and fungi.

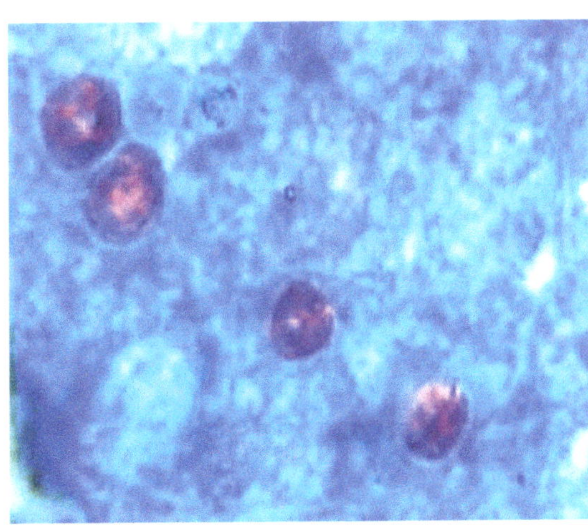

zoophilic preferring animals to humans as a blood source. Mosquitoes exhibit a preference for blood feeding on non-human animals.

zooplankton free-floating, microscopic organisms that drift with water currents.

zygote fertilized diploid cell (egg) that results from the fusion of two haploid, male and female, gametes during sexual reproduction (e.g. *Plasmodium* spp. in the stomach of the mosquito vector).

Appendix 1 Abbreviations and Acronyms Used in Parasitology

A	adenine
AAH	adenine aminohydrolase
ABC	an ATP-binding cassette
Abs	antibodies
ADCC	antibody-dependent cell-mediated cytotoxicity
ADP	adenosine diphosphate
ADSS	adenylosuccinate synthetase
AFC	antibody-forming cell
Ag	antigen
AID	activation-induced cytidine deaminase
AIDS	acquired immunodeficiency syndrome
AK	adenosine kinase
AMP	adenosine monophosphate
AMPDA	AMP deaminase
AMPs	antimicrobial peptides
ANA	antinuclear antibodies
APC	antigen-presenting cell
ApoL1	apolipoprotein L1
APP	acute phase protein
APRT	adenine phosphoribosyltransferase
ARC	AIDS-related complex
ASL	adenylosuccinate lyase
ASPA	Animals (Scientific Procedures) Act 1986
AS-PCR	allele-specific PCR
asRNAs	antisense RNAs
AST	asparate aminotransferase
ATCC	American Type Culture Collection
ATP	adenosine triphosphate
BAL	bronchoalveolar lavage
BBB	blood–brain barrier
BMR	basal metabolic rate
bp	base pair
BSA	bovine serum albumin
BSC	biological safety cabinet
BUN	blood urea nitrogen
BZ	benzimidazole
C	cytosine
CAT	catalase
CBC	complete blood count

CBT	cognitive behaviour therapy
CDC	Centers for Disease Control and Prevention
cDCs	conventional dendritic cells
CE	cystic echinococcosis
CFT	complement fixation test
C_{max}	maximum concentration
CMI	cell-mediated immunity
C_{min}	minimum concentration
CNS	central nervous system
CNV	copy number variation
COPD	chronic obstructive pulmonary disease
CRP	C-reactive protein
CSF	cerebrospinal fluid
CSFs	colony stimulating factors
CT	computed tomography
CTL	cytotoxic T lymphocyte
CVL	canine visceral leishmaniasis
DA	dopamine
DALY	disability-adjusted life year
DBS	dried blood spot
DCF	direct centrifugal flotation
DCs	dendritic cells
DIC	disseminated intravascular coagulation
DNA	deoxyribonucleic acid
EDTA	ethylene-diaminetetraacetic acid
EF	eosinophilic folliculitis
EHA	egg hatching assay
EL3	early third larval stage
ELISA	enzyme-linked immunosorbent assay
EMEM	Eagle's minimum essential medium
e.p.g.	eggs per gram
EPNs	entomopathogenic nematodes
EWAS	environment-wide association studies
FACS	fluorescence-activated cell sorter
FDA	Food and Drug Administration
FECRT	faecal egg count reduction test
FECs	faecal egg counts
fMRI	functional magnetic resonance imaging
FRET	fluorescence resonance energy transfer
FTA	fluorescent treponemal antibody
FZT	fish-borne zoonotic trematodes
G × E	gene–environment interactions

G6PD	glucose-6-phosphate dehydrogenase
GABA	gamma-aminobutyric acid
GALT	gut-associated lymphoid tissue
G-CSF	granulocyte colony-stimulating factor
GDA	guanine deaminase
gDNA	genomic DNA
GEWIS	gene–environment-wide interaction analysis
GFP	green fluorescent protein
GGT	gamma-glutamyl transpeptidase
GI	gastrointestinal
GIN	gastrointestinal nematodes
GIS	geographic information system
GLM	general linear model
Glu	glutamate
GM-CSF	granulocyte–macrophage colony-stimulating factor
GMPR	GMP reductase
GMPS	GMP synthase
GN	gene network
GPI	glycosylphosphatidylinositol
GSD	ground state depletion
GSEA	gene-set enrichment analysis
GSH-Px	glutathione peroxidase
GWAS	genome-wide association studies
HAT	human African trypanosomiasis
HDL	high-density lipoprotein
HGPRT	hypoxanthine-guanine phosphoribosyltransferase
HLA	human leucocyte antigen
Hp	haptoglobin
HPLC	high-performance liquid chromatography
Hpr	haptoglobin-related protein
Hpr–Hb complex	Hpr–haemoglobin complex
HTS	high-throughput screening
HWD	heartworm disease
I.M.	intramuscular
I.P.	intraperitoneal
I.V.	intravenous
IACUC	Institutional Animal Care and Use Committee
IAGNH	inosine-adenosine-guanosine nucleoside hydrolase
ICAM	intracellular adhesion molecule
ICT	immunochromatographic tests
IEL	intraepithelial lymphocytes
IFA	immunofluorescent antibody

IFAT	indirect immunofluorescent antibody test
IFN	interferon
IFN-β	interferon beta
IgA	immunoglobulin A
IgG	immunoglobulin G
IgM	immunoglobulin M
IGNH	inosine-guanosine nucleoside hydrolase
IGRs	insect growth regulators
IL-2	interleukin-2
IMP	inosine monophosphate
IMPDH	inosine monophosphate dehydrogenase
IPM	integrated parasite management
IRB	institutional review board
ITS	internal transcriber spacer region of the ribosomal DNA
IUNH	inosine-uridine (or non-specific) nucleoside hydrolase
kb	kilobase
kbp	kilobase pair
kDa	kilodalton
L3s	third-stage larvae
LAMP	loop-mediated isothermal amplification
LBP	lipopolysaccharide binding protein
LC	liquid chromatography
LCI	live cell imaging
LD	linkage disequilibrium
LDA	larval development assay
LMIA	larval migration inhibition assay
LRS	likelihood ratio statistics
LTR	long terminal repeat
LTS	low-throughput screening
MCF	median corpuscular fragility
MDA	malondialdehyde
MDR1	multidrug resistance protein 1
MHC	major histocompatibility complex
ML	macrocyclic lactone
ml	millilitre
mol. wt.	molecular weight
mRNA	messenger RNA
MRS	magnetic resonance spectroscopy
MS	mass spectrometry
MW	molecular weight
NA	neuraminidase
NADA	new animal drug application

ncRNA	non-coding RNA
NFDS	negative frequency-dependent selection
NHPs	non-human primates
NK	natural killer
NMR	nuclear magnetic resonance
NSAID	non-steroidal anti-inflammatory drug
nt	nucleotide
NTDs	neglected tropical diseases
NTP	nucleoside triphosphate
NTR	non-translated region
ORF	open reading frame
OTC	over-the-counter
PAM	protozoal amoebic meningoencephalitis
PBMCs	peripheral blood mononuclear cells
PCR	polymerase chain reaction
PCV	packed cell volume
pDCs	plasmacytoid DCs
PET	positron emission tomography
PfEMP1	*P. falciparum* erythrocyte membrane protein 1
P-gp	P-glycoprotein
PM	postmortem
PNS	peripheral nervous system
PP	Peyer's patch
PPD	purified protein derivative
PPR	periparturient rise in faecal egg counts
PRPP	phosphoribosylpyrophosphate
qRT-PCR	quantitative real-time PCR
QTLs	quantitative trait loci
RAPD	random amplified polymorphic DNA
rDNA	ribosomal DNA of ribosomal gene
RNA	ribonucleic acid
RNAi	RNA-mediated interference
RNA-seq	RNA sequencing
rRNA	ribosomal ribonucleic acid
SAA	serum amyloid A
SAC	South American camelids
SC	subcutaneous
SELEX	selection of ligands by exponential amplification
SEM	scanning electron micrograph
SGOT	serum glutamic oxaloacetic transaminase
SIgA	secretory IgA
SIM	structured illumination microscopy

SIT	sterile insect technique
SLAMS	stepwise linkage analysis of microarray signatures
SNPs	single nucleotide polymorphisms
SNVs	single-nucleotide variants
SOD	superoxide dismutase
sp.	a species of
SPECT	single photon computerized tomography
SSR	simple sequence repeat
STORM	stochastical optical reconstruction microscopy
TAC	total antioxidant capacity
***Tb*HpHbR**	*Trypanosoma brucei* haptoglobin-haemoglobin receptor
TEM	transmission electron microscopy
TERT	telomerase reverse transcriptase
TLFs	trypanosome lytic factors
TLRs	toll-like receptors
TNF-α	tumour necrosis factor-alpha
Treg	regulatory T cells
TST	targeted selective treatment
URI	upper respiratory tract infection
UTI	urinary tract infection
UTRs	untranslated regions of human genes
VSG	variant surface glycoprotein
WT	wild-type
XMP	xanthine monophosphate
XPRT	xanthine phosphoribosyltransferase

Appendix 2 Common Names of Parasites

Common name	Scientific name
African eyeworm	*Loa loa*
amoebic dysentery	*Entamoeba histolytica*
American dog tick	*Dermacentor* spp.
assassin bugs*	*Triatoma* spp. of the family Reduviidae.
barber's pole worm	*Haemonchus contortus*
bed bug	*Cimex lectularius*
beef tapeworm	*Taenia saginata*
black fly (buffalo gnats)	*Simulium* spp.
black scour worm	*Trichostrongylus colubriformis* and *T. vitrinus*
bot fly (warble fly)	*Hypoderma* spp.
brisket worm	*Stephanofilaria stilesi*
brown dog tick	*Rhipicephalus* spp.
brown stomach worm	*Ostertagia* spp.
cat, rodent tapeworm	*Taenia taeniaformis*
cat biting louse	*Felicola subrostratus*
cat flea	*Ctenocephalides felis*
cattle biting louse	*Damalinia bovis*
cattle bot fly	*Hypoderma bovis*
cattle hookworm	*Bunostomum* spp.
Chagas disease	*Trypanosoma cruzi*
deer fly	*Chrysops* spp.
deer tick	*Ixodes* spp.
dog, rabbit tapeworm	*Taenia pisiformis*
dog-biting louse	*Trichodectes canis*
dog flea	*Ctenocephalides canis*
dwarf tapeworm	*Hymenolepis nana*
ear mite	*Otodectes* spp.
eyeworm	*Thelazia* spp.
equine pinworm	*Oxyuris equi*
equine stomach worm	*Habronema* spp.
equine tapeworm	*Anoplocephala* spp.
feline cytauxzoonosis	*Cytauxzoon felis*
fish/broad tapeworm	*Diphyllobothrium latum*
flea tapeworm	*Dipylidium caninum*
follicular mange	*Demodex* spp.

Common name	Scientific name
fox lungworm	*Eucoleus aerophilus*
gapeworm	*Syngamus trachea*
giant kidney worm	*Dioctophyma renale*
heartworm	*Dirofilaria immitis*
hookworm	*Ancylostoma* spp.
horn fly	*Haematobia irritans*
horsefly	*Tabanus* spp.
human crab louse	*Phthirus pubis*
human flea	*Pulex simulans*
human head louse	*Pediculus humanus capitis*
human pinworm	*Enterobius vermicularis*
hydatid tapeworm	*Echinococcus granulosus* and *E. multilocularis*
large-mouthed bowel worm	*Chabertia ovina*
liver fluke	*Fasciola hepatica*
lone star tick	*Amblyomma* spp.
long-nosed cattle louse	*Linognathus vituli*
lung fluke	*Paragonimus kellicotti*
lungworm	*Metastrongylus* spp. (pig); *Aelurostrongylus* spp. (cat); *Filaroides* spp. (dog); *Angiostrongylus vasorum* (dog)
mosquito	*Culex*; *Anopheles*; *Aedes*
neck threadworm	*Onchocerca cervicalis*
nodular worm	*Oesophagostomum columbianum*
North American Chigger	*Trombicula* spp.
northern hookworm	*Uncinaria* spp.
oesophageal worm	*Spirocerca lupi*
pork tapeworm	*Taenia solium*
rat tapeworm	*Hymenolepis diminuta*
red stomach worm	*Hyostrongylus rubidus*
rodent bot fly	*Cuterebra* spp.
roundworm	*Ascaris suum* (pig); *Toxocara canis* (dog); *Toxocara cati* (cat); *Toxascaris leonine* (dog and cat); *Toxocara vitulorum* (cattle); *Parascaris equorum* (equine); *Bayliascaris procyonis* (raccoon); *Ascaridia galli* (bird)
ruminant tapeworm	*Moniezia* spp.
salmon poisoning fluke	*Nanophyetus salmincola*
sand flies	phlebotomine flies belonging to the genus *Phlebotomus* (Old World) or *Lutzomyia* (New World)
scabies	*Sarcoptes* spp.
scaly face–scaly leg	*Knemidocoptes* spp.

Common name	Scientific name
screw worm fly	*Cochlioma* spp.
seatworm	Pinworm (*Enterobius vermicularis*)
short-nosed cattle louse	*Haematopinus* spp.
spinous ear tick	*Otobius* spp.
stable fly	*Stomoxys* spp.
stick-tight flea	*Echidnophaga gallinacea*
stomach bot fly	*Gastrophilus* spp.
stomach hair worm	*Trichostrongylus axei*
stomach worm	*Physaloptera* spp. (cats and dogs)
subcutaneous worm	*Acanthocheilonema reconditum*
swine kidney worm	*Stephanurus dentatus*
Texas cattle fever tick	*Boophilus* spp.
thin-necked intestinal worm	*Nematodirus* spp.
thorny-headed worm	*Macracanthorhynchus hirudinaceus*
threadworm	*Strongyloides westeri* (equine); *S. ransomi* (pig); *S. papillsous* (ruminant); *S. stercoralis* (human, dog or cat)
trichina worm	*Trichinella spiralis*
walking dandruff mite	*Cheyletiella* spp.
whipworm	*Trichuris suis* (pig); *T. vulpis* (dog); *T. felis*, *T. campanula*, and *T. serrata* (cat); *T. trichiura* (human); *T. bovis* (cattle); *T. ovis* (sheep)
wing louse	*Lipeurus* spp.

*Also known as the conenose bug, kissing bug and Mexican bed bug.

Appendix 3 Word Roots in Parasitology and Biology

Speak the language of parasitology

The language of parasitologists offers intriguing challenges both to parasitologists and to colleagues in other fields. But, there is no such a discipline called 'parasitological linguistics'. Indeed, scientific names and words describing parasite taxa, various morphological structures, biological processes, etc. are often composed of word roots derived from the Greek or Latin language. Knowing some of these word roots makes it easy to learn the terms in which they are used. One particularly common example is the name of taxa. The formal taxonomic names for many parasite groups (from phylum down to species) are often formed from roots that refer to distinctive features of the parasite. For example, *Arthropoda*, a phylum name that means 'joint-' (*arthro-*) 'foot' (*pod*), or the 'jointed-foot'. We compiled, from various sources, a list of Greek or Latin word roots, each root is followed by the meaning in English.

A

a(n)- without
ab- away from
abduc- lead from
acanth- prickle
acer- without horns
acetabul- vinegar cup
aceto- acid
acin- cluster of grapes
acri- acrid, sharp
acro- summit, top
actin- ray
acust- of hearing
ad- towards or near
adipo- fat
adventit- coming to
aegopod- goat-foot
aestiv- summer
affer- carrying to
agrest- rural, wild
agro(st)- field
al(i)- wing
alb- white
altissim- tall
alve- pit, socket
ambly- blunt
ambul- walk
ammo- sand
amnio- fetal envelope
amoeb- change
amphi- both
amplex- clasping
ampulla- flask
amygdala- almond-shaped organ
amyl- starch
an- without
ante- before
aur- ear, gold
austral- southern
auto- self
aux- grow
avi- bird
axi- axis
azyg- unpaired

B

baca- small berry
baccatus- berried
baena- walking
balan- acorn
barb- beard
basi- at the bottom
bat(is)- skate, ray
bdella- sucker
bell- pretty
bi- two
bi(ola)- life
blast- germ
bore- north
botry- small bunch of grapes
bov- ox
brachi- arm
brachy- short
brady- slow
branch- gill
brevi- short

brizo- to nod
bromo- oats, broom, shadow
bronch- windpipe
bucca- mouth cavity
bulbos- bulbous
bulla- bubble, flask
bullat- wrinkled
buno- hill
bursa- pouch, purse

C

caec- blind
calc- stone
calli- beautiful
calor- heat
calyx- cup, chalice
camp- field
cani- dog
cap(it)- head
capill- hair
card- heart
carin- keel
cam- flesh
carotid- producing stupor
cary- nut
cata- down
cauda- tail
cav- hollow
centi- hundred
ceph(al)- head
cera(s)(t)- horn
cerc- horns, lobes, short tail
cerebro- brain
cerv- neck
chaero- to please
chaet- bristle
chamae- on ground
chel- turtle
cheno- goose
chiasm- crossing
chil- lip
corium- leather
corn- horn
coron- crown
corp- body
cort- bark
cost- rib
cotyl- cup
crani- skull
crass- thick
crep- shoe
cribi- sieve
cric- ring
crispus- curled
crist- crest
crosso- fringe, tassel
cruci- cross-like
crypt- hidden
cten- comb
cune- wedge
cusp- lance, point
cuti- skin
cyan- blue
cycl- circular
cyno- dog
cyst- capsule
cyt- cell

D

dactyl- finger
de- remove
deci- ten
decid- falling off
decuss- cross-banding
deka- ten
delph- womb
delphis- dolphin
demi- half
dendr- tree
dent- tooth
derm- skin
dero- persisting
desmo- band
deut- second
dextro- right
di(a)- two
dia- through
didym- twin
digit- finger
dino- terrible
diplo- double
dipso- thirst
dissect- deeply cut
dodeca- twelve
dont- tooth
dors- back
drom- quick running
dubi- doubtful
dulci- sweet
duo- two/double
dur- hard
dys- difficult

E

echino- spiny
echis- viper
eco- house
ecto- outside
effer- carrying away
effusus- loosely spreading
ejacul- throw out
elasmo- plate, flat
elat- tall
-*ell*- diminutive
embio- living
embol- thrown in
endo- inside
engy- narrow
ensi- sword
eo- dawn, east
epi- upon
equi- horse, equal
erect- upright
erythr- red
esculent- edible
ethm- sieve
eu- well, very
eury- wide
ex- out of
exo- outside
extens- draw out
extra- beyond
extrins- coming from

F

fab- bean
falc- sickle-shaped
fasc(ic)- bundle
fatuus- foolish, useless
feli- cat
fer- carry
ficar- fig-like
fil- thread
fistul- hollow, tube
flacc- flabby
flav- yellow
flexu- wavy
fluitans- floating
foen- hay
foetid- foul smelling
foli- leaf
fontan- fountain
foramen- opening
fornix- vault
fund- to pour
furc(ul)- fork

G

gal- milk
gale- weasel
gall- of France
gangli- knot
gano- shining
gast(e)r- stomach, pouch
gemin- twins
gemm- bud
genu- knee, joint
geranium- crane
gerro- shield
gladi- sword
glen- socket
glia- glue
glo(ss/tt)- tongue
glut- sticky
glypt- carved, engraved
gnaph- wool
gnath- jaw
goni- angle
gono- seed
gracil- slender
gram- of grass
gymn- naked
gyn- woman
gyr- turning

H

haem- blood
hamat- hook
haplo- single, simple
hect- hundred
heira- hawk
helios- sun
helminth- worm
helo- wart
hemi- half
hepta- seven
hetero- different
hex- six
hippo- horse
hirsut- hairy
hispid- bristly
histri- actor
holo- complete, whole
homo- man, same
homal- flat
hormon- that which excites
hyal- glassy

hydro- water
hyemal- winter
hyo- u-shaped
hyper- above/over/high
hypo- beneath/under/low

I

ichthy- fish
-ida- like
-idae- ending of zoological family names
ileum- twisted
impar- unpaired
in- in, not
-in- diminutive
-inae- ending of zoological subfamily names
incarnat- flesh coloured
incis- cutting in
inermis- unarmed
inflexis- rigid
infra- under
inter- between
intra- within
iris- rainbow
iso- equal
iter- passage

J

jug- yoked together
jungoljunc- to bind

K

ket- ketone
kilo- thousand

L

labi- lips
labyrinth- tortuous passage
lacer- torn
lacrim- tears
lact- milk
lacuna- space
laevo- left
lagena- flask
lamella- leaf, layer
lanceol(at)- lance-shaped
lati- broad
lecith- egg yolk
lemm- skin, husk
lens- lentil, bean
leo(n)- lion
lepid- scale
lepto- slender
leuco- white
levator- lifter
libo- fat
lign- wood
ligul- strap-shaped
lin- line
ling- tongue
lipo- fat
lith- stone
loph- crest
lumen- cavity
luna- moon
lupus- wolf
lute- yellow
lymph- clear water
lysis- loosen

M

macr- large
macula- spot
magni- large
maj- greater
mal- bad
mala- cheek
malleus- hammer, mallet
mamma- breast
mandib- lower jaw
manu- hand
maritim- of the sea
marsupium- pouch
masseter- chewer
mast- nipple
maxi- large
maxill- jaw
maxim- greatest
meatus- passage
medi- medium
medull- marrow
mega- large
melan- black
meno- moon
mer- part
meso- middle
meta- after
micro- small
milli- thousand
min- smaller
minim- smallest
mirabile- wonderful
mixi- mingle
moll- soft
monil- string of beads

mono- single
montan- of mountains
morph- form
motor- mover
multi- many
mural- walls
muri- mouse
myo- muscle, mouse
myri- countless
myx- mucous
myz- sucker

N

narc- sleep
necro- dead
necto- swimming
nema- hair
nemoral- in woods
neo- new
nephr- kidney
neur- nerve
nictitat- winking
nigr- black
non- not
nona- nine
noth- spurious
notho- southern
noto- back, south
nuch- back of neck
nucle- little nut
nud- naked
nulli- none
nutans- nodding

O

obtect- concealed
obturat- closed
obtusi- blunt
occiden- Western
occip- back of head
oct- eight
odont- tooth
odor- fragrant
oecious- house of
oed- swollen
oen- wine
oesoph- gullet
-oid- like
olecran- skull of elbow
oleo- oily
olfact- smelling
oligo- few
oliva- olive
omaso- paunch
oment- fat skin
omo- shoulder
omphalo- navel
onto- existing
oo- egg
operculum- little cover
ophi- snake
ophthalmo- eye
opistho- behind
opoter- either
or- mouth
orbi- circle
orch- testicle
oriental- Eastern
ornith- bird
ortho- straight
os(ti)- mouth
os(s/t)- bone
ostrac- shell
ot- ear
ovi- sheep
ovo- egg
oxys-, *oxus*- sharp, pointed

P

pan- all
phanero- visible
pher- carry
phil- love
phloe- tree bark
phor- carry
phragm- fence
phren- diaphragm, mind
phyl- tribe
phyll- leaf
physa- bladder
physio- nature
phyto- plant
pil- hair
pinea- pine cone
pinnat- feather-shaped
pisum- pea
pisc- fish
plac- plate, tablet
plagio- oblique
plan- flat
platy- flat, broad
plesio- near
pleth- full
pleur- side

plex- interwoven
plica- fold
plum- feather
pod- foot
pogo- beard
polio- grey
polit- polished
poly- many
pons- bridge
porc- pig
porphyr- purple
porta- gate
post- after
potam- river
potero- drinking cup
praeco- early
pratens- in meadows
pre- before
prim- first
pro- in front of
proct- anus
procumbens- lying flat
profund- deep
proliger- bear offspring
prom- first
proso- front
prostratus- lying flat
prot- first
pseud- false
psycho- mind
pter- wing, fern
pubescens- downy
pubi- sexually mature
pulmo- lung
pygo- rump
pyl- gate
pyo- pus
pyro- heat, fire

Q

quadr- four
quarter- four
quin(t/que)- five

R

rach- spine
radi- root
ram- branching
ran- frog
rapa- turnip
raph- seam
rect- straight
reflex- bent back
ren- kidney
repens- crawling
reptans- crawling
retic- network
retina- little net
retro- behind, backward
rhabd- rod
rhach- spine
rhage- tear, rent
rhin- nose, snout
rhiz- root
rhodo- red
rhynch- snout
-rrh- flow
rode- gnaw
rogos- wrinkled
rostr- beak, prow
rota- tum, wheel
rub(e)r- red

S

sacchar- sugar
saccul- little sac
sacr- sacred
sagitta- arrow
sal- salt
sanguini- bloody
sapon- soapy
sarc- flesh
sativus- cultivated
saur- lizard
scala- ladder
scalene- uneven
schizo- split
scler- hard
scut- shield
scyph- cup
seba- tallow, wax
sect- cut
segetum- in cornfields
selen- moon
sella- saddle, seat
semi- half
sept- seven, wall
serot- late
serrat- saw-toothed
set- hair
sex- six
simi- monkey
sinus- hollow, bay
sipho- tube

sol- sun
soma- body
somni- sleep
specios- showy
sperm- seed
sphen- wedge
sphinct- closing
spinos- spiny
spondyl- vertebra
squalid- squalid
squam- scale
steg- covering
stell- star
steno- narrow
stereo- solid
steril- sterile
stom- mouth
strat- layer
strept- twisted
strictus- upright, stiff
strigos- having stiff bristles
stroph- turning styl- column
sub- below, under
sucr- sugar
sulc- furrow
super- beyond
supin- lying back
supra- above
sutur- seam
sym- with
syn- with
sys- with

T

tachy- fast
tact- touch
taenia- ribbon
talus- ankle
tapetum- carpet
tard- late
tarsus- ankle
tect- covered
tegmen- covering
teleo- complete
temno- cut
tenacul- holding
tentor- spread like a tent
tenuis- slender
ter(ti)- three
teres- round
tetr- four
textilis- of textiles
thalam- chamber, bed
thalass- sea
theca- case
thel- nipple, female
therm- heat
thero- breast, mammal
thyreo- large shield
tinctori- of dyes
torn- cut
tomentos- densely woolly
torpe- numb
toxo- arrow, dart
trab- beam
trachy- rough
trago- goat
trans- across
trema- hole
tri- three
trich- hair
troch- wheel
trop- turning
troph- feed
tussi- cough

U

ultra- above
umbilic- navel
un- one
unc- hook
ungui- nail, claw
ungul- hoof, claw
urens- bum
uro- tail
urs- of bears
utricul- little skin bag
uv(ela)- grapes

V

vagina- sheath
vagus- wandering
vas- vessel
ven- vein
ventr- belly
verd- green
verm- worm
vern- spring
versi- various
vesic- blister
vill- shaggy hair, velvet
virens- green
visc- organs of body cavity
vita- life

vitell- yolk
vitr- glass
vulgaris- common
vulp- fox

X

xanth- yellow
xen- stranger
xer- dry
xyl- wood

Z

zo- animal
zyg- yoke

Appendix 4 The Greek Alphabet

Greek letter	Symbol	English equivalent
Α	α, alpha	a
Β	β, beta	b
Γ	γ, gamma	g
Δ	δ, delta	d
Ε	ε, epsilon	e
Ζ	ζ, zeta	z
Η	η, eta	*ē or e*
Θ	θ, theta	th
Ι	ι, iota	i
Κ	κ, kappa	k
Λ	λ, lambda	l
Μ	μ, mu	m
Ν	ν, nu	n
Ξ	ξ, xi	x
Ο	ο, omicron	o
Π	π, pi	p
Ρ	ρ, rho	p
Σ	σ, sigma	s
Τ	τ, tau	t
Υ	υ, upsilon	u
Φ	φ, phi	ph
Χ	χ, chi	kh
Ψ	ψ, psi	ps
Ω	ω, omega	*ō or o*

Appendix 5 Parasitology and Related Journals

- *Acta Parasitologica*
- *Acta Protozoologica*
- *Acta Tropica*
- *Advances in Parasitology*
- *American Journal of Tropical Medicine and Hygiene*
- *American Journal of Veterinary Research*
- *Clinical Microbiology Reviews*
- *Food and Waterborne Parasitology*
- *Insect Biochemistry and Molecular Biology*
- *Insect Molecular Biology*
- *Infection and Immunity*
- *International Journal for Parasitology*
- *International Journal for Parasitology: Drugs and Drug Resistance*
- *International Journal for Parasitology: Parasites and Wildlife*
- *Journal of Eukaryotic Microbiology*
- *Journal of Helminthology*
- *Journal of Medical Entomology*
- *Journal of Parasitic Diseases*
- *Journal of Parasitology*
- *Journal of Parasitology Research*
- *Journal of Vector Borne Diseases*
- *Korean Journal of Parasitology*
- *Malaria Journal*
- *Medical and Veterinary Entomology*
- *Molecular and Biochemical Parasitology*
- *Nature Reviews Microbiology*
- *Parasite*
- *Parasite Immunology*
- *Parasites and Vectors*
- *Parasitology*
- *Parasitology International*
- *Parasitology Research*
- *PeerJ*
- *PLOS Biology*
- *PLOS Computational Biology*
- *PLOS Neglected Tropical Diseases*
- *PLOS ONE*
- *PLOS Pathogens*
- *Protist*
- *Systematic Parasitology*
- *Ticks and Tick-Borne Diseases*
- *Transactions of The Royal Society of Tropical Medicine and Hygiene*
- *Trends in Parasitology*
- *Veterinary Parasitolology*

Appendix 6 Important Organizations and Databases

A list of organizations and databases that have made significant contributions to parasitology and life science research.

AAVP	American Association of Veterinary Parasitologists (www.aavp.org)
AHS	American Heartworm Society (www.heartwormsociety.org)
ASM	American Society for Microbiology (www.asm.org)
ASP	American Society for Parasitologists (http://amsocparasit.org)
BBSRC	Biotechnology and Biological Sciences Research Council (www.bbsrc.ac.uk)
BSP	British Society for Parasitology (www.bsp.uk.net/home)
CAPC	Companion Animal Parasite Council (www.capcvet.org)
DDBJ	DNA Data Bank of Japan (www.ddbj.nig.ac.jp/)
EBI	European Bioinformatics Institute (http://www.ebi.ac.uk/)
ESCCAP	European Scientific Counsel Companion Animal Parasites (www.esccap.org)
EuPathDB	Eukaryotic Pathogen Database Resources (http://eupathdb.org/eupathdb/)
EVPC	European Veterinary Parasitology College (www.eurovetpar.org)
GenBank	Genetic Sequence Database of the National Institutes of Health (www.ncbi.nlm.nih.gov/genbank/)
GeneDB	Sanger Institute Pathogen Genomics Data Resource (www.genedb.org/Homepage)
MRC	Medical Research Council (www.mrc.ac.uk)
NHGRI	National Human Genome Research Institute
NIAID	National Institute of Allergy and Infectious Diseases (www.niaid.nih.gov/Pages/default.aspx)
NIEHS	National Institute of Environmental Health Sciences (www.niehs.nih.gov)
NIH	National Institutes of Health (www.nih.gov)
NLM	National Library of Medicine (www.nlm.nih.gov)
TDRTargets	Tropical Disease Research Chemogenomics Resource (http://tdrtargets.org/)
UniProt	Universal Protein Resource (www.uniprot.org)
WAAVP	World Association for the Advancement of Veterinary Parasitology (www.waavp.org)
WT	Wellcome Trust (https://wellcome.ac.uk)

Appendix 7 Additional Readings

Boothroyd, J.C. (2009). *Toxoplasma gondii*: 25 years and 25 major advances for the field. Int. J. Parasitol. 39, 935–946. http://dx.doi.org/10.1016/j.ijpara.2009.02.003

Campbell, S.J., Nery, S.V., McCarthy, J.S., Gray, D.J., Soares Magalhães, R.J., and Clements, A.C. (2016). A critical appraisal of control strategies for soil-transmitted helminths. Trends Parasitol. 32, 97–107. http://dx.doi.org/10.1016/j.pt.2015.10.006

Conboy, G. (2009). Cestodes of dogs and cats in North America. Vet. Clin. North Am. Small Anim. Pract. 39, 1075–1090. vi. http://dx.doi.org/10.1016/j.cvsm.2009.06.005

Jacobs, D., Fox, M., Gibbons, L., and Hermosilla, C. (2015). Principles of Veterinary Parasitology. (Chichester, UK: Wiley-Blackwell).

Despommier, D.D., Griffin, D., Gwadz, R.W., Hotez, P.J., and Knirsch, C. (2017). Parasitic Diseases. (New York, USA: Parasites Without Borders, Inc).

Bowman, D.D. (2013). Georgis' Parasitology for Veterinarians. (Missouri, USA: Saunders).

Elsheikha, H.M., and Khan, N.A. (2010). Protozoa traversal of the blood-brain barrier to invade the central nervous system. FEMS Microbiol. Rev. 34, 532–553. http://dx.doi.org/10.1111/j.1574-6976.2010.00215.x

Elsheikha, H.M., and Khan, N.A. (2011). Essentials of Veterinary Parasitology. (Norfolk, UK: Caister Academic Press).

Elsheikha, H.M., and Patterson, J. (2013). Self-Assessment Colour Review: Veterinary Parasitology. (London, UK: Manson Publishing).

Loker, E.S., and Hofkin, B. (2015). Parasitology: a conceptual approach. (New York, USA: Garland Science, taylor & Francis Group, LLC).

Foster, N., and Elsheikha, H.M. (2012). The immune response to parasitic helminths of veterinary importance and its potential manipulation for future vaccine control strategies. Parasitol. Res. 110, 1587–1599. http://dx.doi.org/10.1007/s00436-012-2832-y

Gosling, P.J. (2005). Dictionary of Parasitology. (FL, USA: CRC Press,Taylor & Francis Group).

Mehlhorn, H. (2016). Encyclopedia of Parasitology. (Berlin, Germany: Spriner).

Hunter, C.A., and Sibley, L.D. (2012). Modulation of innate immunity by *Toxoplasma gondii* virulence effectors. Nat. Rev. Microbiol. 10, 766–778. http://dx.doi.org/10.1038/nrmicro2858

Walochnik, J., and Duchêne, M. (2016). Molecular parasitology: protozoan parasites and their molecules. (Berlin, Germany: Springer).

Keiser, J., and Utzinger, J. (2010). The drugs we have and the drugs we need against major helminth infections. Adv. Parasitol. 73, 197–230. http://dx.doi.org/10.1016/S0065-308X(10)73008-6

Korhonen, P.K., Young, N.D., and Gasser, R.B. (2016). Making sense of genomes of parasitic worms: tackling bioinformatic challenges. Biotechnol. Adv. 34, 663–686. http://dx.doi.org/10.1016/j.biotechadv.2016.03.001

Taylor, M., Coop, B., and Wall, R. (2015). Veterinary Parasitology. (Oxford, UK: Wiley-Blackwell).

Sibley, L.D. (2004). Intracellular parasite invasion strategies. Science 304, 248–253. http://dx.doi.org/10.1126/science.1094717

Szempruch, A.J., Dennison, L., Kieft, R., Harrington, J.M., and Hajduk, S.L. (2016). Sending a message: extracellular vesicles of pathogenic protozoan parasites. Nat. Rev. Microbiol. 14, 669–675. http://dx.doi.org/10.1038/nrmicro.2016.110

www.ingramcontent.com/pod-product-compliance
Ingram Content Group UK Ltd.
Pitfield, Milton Keynes, MK11 3LW, UK
UKHW062000041026
15591UKWH00016B/1100

9 781910 190678